Optical Remote Sensing of Ocean Hydrodynamics

Optical Remote Sensing of Ocean Hydrodynamics

Victor Raizer

CRC Press
Taylor & Francis Group
Boca Raton London New York

CRC Press is an imprint of the
Taylor & Francis Group, an **informa** business

MATLAB® is a trademark of The MathWorks, Inc. and is used with permission. The MathWorks does not warrant the accuracy of the text or exercises in this book. This book's use or discussion of MATLAB® software or related products does not constitute endorsement or sponsorship by The MathWorks of a particular pedagogical approach or particular use of the MATLAB® software.

CRC Press
Taylor & Francis Group
6000 Broken Sound Parkway NW, Suite 300
Boca Raton, FL 33487-2742

First issued in paperback 2020

© 2019 by Taylor & Francis Group, LLC
CRC Press is an imprint of Taylor & Francis Group, an Informa business

No claim to original U.S. Government works

ISBN 13: 978-0-367-65646-1 (pbk)
ISBN 13: 978-0-8153-6014-8 (hbk)

This book contains information obtained from authentic and highly regarded sources. Reasonable efforts have been made to publish reliable data and information, but the author and publisher cannot assume responsibility for the validity of all materials or the consequences of their use. The authors and publishers have attempted to trace the copyright holders of all material reproduced in this publication and apologize to copyright holders if permission to publish in this form has not been obtained. If any copyright material has not been acknowledged, please write and let us know so we may rectify in any future reprint.

Except as permitted under U.S. Copyright Law, no part of this book may be reprinted, reproduced, transmitted, or utilized in any form by any electronic, mechanical, or other means, now known or hereafter invented, including photocopying, microfilming, and recording, or in any information storage or retrieval system, without written permission from the publishers.

For permission to photocopy or use material electronically from this work, please access www.copyright.com (http://www.copyright.com/) or contact the Copyright Clearance Center, Inc. (CCC), 222 Rosewood Drive, Danvers, MA 01923, 978-750-8400. CCC is a not-for-profit organization that provides licenses and registration for a variety of users. For organizations that have been granted a photocopy license by the CCC, a separate system of payment has been arranged.

Trademark Notice: Product or corporate names may be trademarks or registered trademarks, and are used only for identification and explanation without intent to infringe.

A Cataloguing in Publication record is available for this title from The Library of Congress

Visit the Taylor & Francis Web site at
http://www.taylorandfrancis.com

and the CRC Press Web site at
http://www.crcpress.com

Contents

Preface

This book is about ocean optics, hydrodynamics, and remote sensing. These three disciplines are combined into one topic titled "Optical Remote Sensing of Ocean Hydrodynamics." This text is intended to be introductory to the subject. The goal is to investigate and demonstrate capabilities of optical remote sensing technology for enhanced observations and detection of ocean environments.

At the moment, many different remote sensors are used for exploring ocean and atmosphere. They are known as microwave radar (active sensor), radiometer (passive sensor), altimeter, infrared scanner, lidar (laser-based sensor), multispectral (hyperspectral) imager and some others. Most operated instruments have a relatively low level of *detectability* of ocean hydrodynamic features due to limitations in payload and data acquisition capabilities. For example, radar images give us incredible information about mid- and large-scale processes in the ocean, such as internal waves, fronts, currents, eddies, and ship wake signatures. Space-based optical sensors provide efficient monitoring of mesoscale (10–1,000 km) meteorological phenomena associated with the ocean and atmosphere interaction, weather conditions, or hurricane situations. Altimeters can measure sea surface heights.

As mentioned in my previous book (*Advances in Passive Microwave Remote Sensing of Oceans, CRC Press, 2017*), "Today, only one remote sensing method remains for the quantitative direct observation of ocean surface features: fine-resolution (few meters and better) airspace optical imagery. Unlike radar or radiometer data, the high visual image quality of satellite optical systems provides detailed information about surface wave processes and their spatial characteristics. However, such optical data are not always readily available for the public."

The new book emphasizes this statement extensively and in a very clear manner, considering the science and practice of optical remote sensing as a powerful tool for global explorations of ocean environments. The book explains consistently how and what type of optical sensors and data analysis should be employed to provide comprehensive ocean studies and advanced applications.

The total scope represents cross-disciplinary scientific topics including elements of hydrodynamics, oceanography, remote sensing, data processing, interpretation and application of optical data. In particular, the book explores observation capabilities of electro-optical technology of high spatial resolution for detection purposes.

We know that the monitoring of ocean features from space is a priority task for many countries (the most capable are US, Russia, China, and India); for this reason, we wrote this book. We define this type of observation as *hydrodynamic detection*. In this context, the book is a comprehensive review of available data and materials including original research that may have a primary consideration for the problem.

The book consists of eight chapters.

Chapter 1 is an introduction to the subject, including an historical survey, technological, experimental and theoretical aspects, processing, assessment and interpretation of optical data.

Chapter 2 outlines main oceanic phenomena and surface processes, which are potentially observable using remote sensing techniques. This chapter provides primary knowledge in hydrodynamics and ocean physics that is necessary for a better understanding of remotely sensed data.

Chapter 3 surveys physical fundamentals of ocean optics. Elements of optical theory, optical characterization of water, reflectance and propagation of optical radiance, radiative transfer, and impact of atmosphere on optical measurements—all these important subdivisions of ocean optics are briefly overviewed. This material represents an analytical summary of data and results known from the literature but they are rearranged in context with our particular subject.

Chapter 4 focuses on the optical remote sensing technologies, types of payloads and specifications of current satellite optical sensors, their parameters and ability to observe ocean surface features. A number of remarkable high-resolution satellite images of the oceans are presented in order to illustrate a multi-purpose character and great content of optical data.

Chapter 5 describes basic satellite characteristics: resolutions, orbits, image formation, methodological and practical aspects including errors formation, quantitative assessment and management. This chapter also draws up a general strategy for advanced remote sensing studies and applications.

Chapter 6 provides a comprehensive review of data processing methods, computer science and digital techniques. Indeed, data processing is one from the most important parts in remote sensing. A number of available and efficient algorithms including statistical, spectral, wavelet, correlation, and fusion techniques are discussed in detail.

Chapter 7 is an overview of our optical remote sensing studies conducted over the years. Among regular aerial optical observations of a wind-driven sea surface, wave breaking and foam and whitecap fields, we present a novel experimental material obtained from space. The focus is on analysis of multispectral satellite optical imagery of high spatial resolution (~1–4 m). The selected material may provide scientific and technological breakthroughs in the field of ocean remote sensing and advanced applications.

Chapter 8 considers the integration (synergy) concept related to *multisensor* remote sensing of ocean environments. In the concluding paragraphs, we outline some sophisticated applications and detection capabilities of high-resolution optical imagery.

The text has not attempted to be exhaustive but is balanced enough to give the reader knowledge of optical remote sensing and perspectives for further scientific research and developments. The bibliography and references are at the end of each chapter or section. The reader can find needed information much easier and get insight, if desired.

It is my hope that this book and material presented will be useful for many specialists, scientists, researchers, engineers, and students working on cross-disciplinary problems, including geosciences and remote sensing, computer science, applied physics, oceanography, satellite observation technology, and optical engineering.

I would like to gratefully acknowledge all of the people who have made possible publishing this book. Many thanks to CRC Press team and editors, especially Irma Britton, for encouragement and assistance.

Victor Raizer
Fairfax, Virginia

MATLAB® is a registered trademark of The MathWorks, Inc. For product information,

please contact:
The MathWorks, Inc.
3 Apple Hill Drive
Natick, MA 01760-2098 USA
Tel: 508-647-7000
Fax: 508-647-7001
E-mail: info@mathworks.com
Web: www.mathworks.com

Author

Victor Raizer, PhD, DSc, physicist and researcher, has multi-year experiences in the field of electromagnetic wave propagation, radiophysics, microwave radiometer/radar and optical techniques, and remote sensing of the Earth's environment. He has provided a broad spectrum of scientific research including the development of airspace observation technology, modeling, simulation, and prediction of complex remotely sensed data. He is the author of several books.

Author

Victor Raizer, PhD, has played in and research that has ... ... of experience in the field of characterization of ... ... wave propagation radial pixel infrarowaves to metric ... ... water and ... influence and ... remote sensing of the ... ... Earth development. He has provided a broad view from the ... ... of scientific research, including the development of air- ... ... space observational technology, modeling, simulation, ... and prediction of complex remotely sensed data. ... ... is the author of several books.

1 Introduction

This chapter outlines the basic principles of remote sensing technology as a powerful tool for exploration of ocean environments. The introduction contains common knowledge material which can help the reader to better understand optical remote sensing capabilities. The volume of published materials in the field of remote sensing grows dramatically and a survey of the literature would be tedious enough. Meanwhile, we offer an extended bibliography at the end of this chapter for readers who want more information and details. Selected textbooks and monographs refer to a variety of general science disciplines—hydrodynamics, ocean optics, remote sensing, observation technology, and data/image processing. The reader may gain insights into all these areas and choose the topic that he/she finds the most interesting.

1.1 DEFINITION AND OBJECTIVES

The term "remote sensing" first was used in the United States in late 1960 by Ms. Evelyn Pruitt of the US Office of Naval Research. Twenty-five years later, the official definition of this term was formed in United Nations Plenary meeting (as part of the general assembly resolutions A/RES/41/65, 95th Plenary meeting, December 3, 1986) as the following: "The term remote sensing means the sensing of the Earth's surface from space by making use of the properties of electromagnetic wave emitted, reflected or diffracted by the sensed objects, for the purpose of improving natural resource management, land use and the protection of the environment." Another, more understandable short definition is "remote sensing is the acquiring of data about an object without touching it." In context with our book, *remote sensing is a scientific technology of data acquisition, based on fundamental principles of electromagnetic radiation.*

But a real history of remote sensing began a hundred years earlier, in 1858, when Gaspard-Felix Tournachon first made a picture of Paris from a hot air balloon. Remote sensing grows from there. The next occasion happened during the US Civil War, when messenger pigeons, kites, and unmanned balloons with attached photo cameras were flown over enemy territory. The most remote sensing (aerial photography) missions were developed and dedicated to military surveillance during World Wars I and II and reached a reconnaissance climax during the Cold War. Today, remote sensing includes radio waves, microwave, and optical bands—infrared, visible light, and ultraviolet. Remote sensing satellites are used to gain information on global processes and events that occur on the Earth, planets, and solar system and beyond.

The successful implementation of remote sensing for geophysical and/or military purposes requires the development of a full-scale physics-based methodology and high level of competence in the field of information technology and management.

Thus, a remote sensing field experiment conducted *in the right place at the right time* is almost a success. From this point of view, remote sensing can be perceived as an *art of cognition*.

Remote sensors are divided into two categories: active and passive. Active remote sensors are radar, scatterometer, interferometer, altimeter, global position system (GPS) tracker, sonar, and lidar. The active sensor transmits electromagnetic waves at a certain frequency and measures the scattering or reflected signal from the object of interest. In this case, we usually obtain selective information about spatially statistical (geometrical) characteristics of the object. Passive remote sensors are microwave and infrared radiometers, visible and infrared spectrometers, photographic, digital, electro-optical, and TV cameras (without flash), and ultraviolet spectrophotometer. The passive sensor does not transmit an electromagnetic signal, but it measures spectral radiation emitted by the object itself. As a result, we obtain integral information about the thermodynamic, structural, and physical properties of the object. The difference between active and passive, including both optical and microwave sensors, is also in day/night observation capabilities. Some microwaves can pass through the atmosphere and clouds and have all-weather capability, which is their principal advantage. Optical sensors provide much better spatial resolution than microwave remote sensing systems and that is an important advantage for Earth's observations.

The entire remote sensing process involves many elements and consistent steps. The main are the following:

1. Specification of the source of electromagnetic energy
2. Characterization of electromagnetic radiation (type, frequency, polarization)
3. Interaction of electromagnetic radiation and an object (land, ocean, atmosphere, etc.)
4. Propagation electromagnetic radiation through the atmosphere
5. Receiving and recording of electromagnetic radiation by remote sensor
6. Transmission, reception, and processing of the recorded by the sensor data
7. Assessment, analysis, and classification of the collected remotely sensed data
8. Thematic (geophysical) interpretation and application is the final and most important step of the remote sensing process. As a result, the user (researcher) should be able to obtain needed information and generate scientific or commercial product.

In practice, however, such an ideal multistage scheme may not always be achieved because of unknown properties of environment or target being observed. An example is an ocean environment where many hydro-physical (or hydrodynamic) parameters and events constantly vary in space and time. This leads to acquiring multivariate remotely sensed data and certain difficulties in geophysical interpretation. In reality, optical observations due to their *documental character* provide a unique scientific basis for exploring multiscale ocean hydrodynamic phenomena, on the one hand, and creating a broad framework for analysis of objective, non-abstract data type, on the other hand.

Optical remote sensors operate with a part of electromagnetic spectrum covering the visible, near-infrared, shortwave infrared, and thermal infrared bands. The word "optical" usually means connection with natural solar radiation, i.e., sunlight which reaches the Earth's surface. In this case, an image (e.g., digital photo) is formed by recording of light, reflected from the object. It seems very easy to obtain information about ocean surface parameters using airspace optical cameras; however, in reality, this process is much more complicated than expected.

At the most basic level, we have to embrace dynamic data formed under the influence of many factors and processes. These include incident solar radiance, atmospheric effects, electromagnetic properties (reflectivity) of the surface, as well as instrument characteristics and an observation process (geometry, swath, resolution). Physical principles, experimental design, and research methodology are critical issues in remote sensing of the oceans. As a whole, advanced optical technologies may have much greater potential to make significant contributions to Earth science and remote sensing than other available techniques.

1.2   PHYSICS AND TECHNIQUES OF OPTICAL REMOTE SENSING

An understanding of the physics of ocean remote sensing requires the knowledge of mechanisms of propagation and interaction of electromagnetic radiance in the ocean-atmosphere system. These mechanisms are determined by many surface processes and phenomena related to ocean physics, fluid mechanics, nonlinear physics, and thermodynamics. The choice of the most efficient and informative range of the electromagnetic spectrum for remote sensing of ocean hydrodynamics is not an easy task because we actually deal with a *stochastic multiscale dynamic system* with distributed parameters which is real world ocean environment. Detailed study of such a complex dynamic system is always a challenge.

Traditionally, the visible and near-infrared regions of the electromagnetic spectrum have been the most commonly used in remote sensing of the Earth including the ocean as well. This is due to our feeling that our visual perception of natural scenes and/or extended bright objects, illuminated by sunlight, can be easily recorded using affordable light detectors—photographical or electro-optical. However, in practice, the situation is more complicated. The impact of environment factors on the quality of optical data becomes critical. An example is aerial photography of the ocean, which is very sensitive to variations of light conditions and atmospheric effects.

In general, several principle mechanisms of interactions between light and matter are known. They are the following: reflection (mirror, specular, and diffuse), refraction, diffraction, scattering, shadowing (self-shadowing), transmission, absorption, and reemission (fluorescence or phosphorescence). These mechanisms manifest themselves differently under different conditions or situations. The interactions may also cause joint effect leading to light interference.

Optical images of the sea surface are formed due to specular (or quasi-mirror) reflection of light from surface slopes. Diffuse reflection and volume scattering can also occur on microstructural nonuniformities (bubbles, spray, aerosol) generated by wave breaking processes or other sources. In this case, electromagnetic properties of the air-sea interface vary significantly, not only at microwave frequencies (that is well known fact), but also in visible and near-infrared ranges.

In aerial photographs and/or video of stormy seas, we often observe fuzzy, vague, indistinct patterns of explicit surface waves and blurry contours of foam/whitecap coverage areas. Such *non-robustness* of optical images is a result of many impacts as natural as well instrumental. Natural factors are associated with high spatiotemporal dynamics and stochastic behaviour of the surface wave facet slopes causing strong variations (fluctuations) of the reflected radiation. In such situations, so-called *image motion* effect can appear that produces blur. Image motion may happen when an exposure time option is inappropriate. One method to reduce such a *motion artefact* is to use image forward motion compensation. This is important at large-scale aerial photography of the sea surface (and digital video as well) when using fast aircraft at relatively low altitudes. With a slow enough flying platform, e.g., a helicopter or air balloon, image motion is not frequently a problem.

Other important factors are the relative dispositions of the sun, surface, and sensor that determines light intensity and illumination conditions. With low-flying platforms, the geometry of optical observations has a critical impact on the quality of imaging data. Uniform illumination yields, as a rule, monotonic distribution of brightness and contrast in the optical images of the ocean surface at low and moderate winds. At the same time, non-uniform illumination leads to the appearance of nonmonotonic brightness and contrast in the images. Figure 1.1 shows aerial photographs of sea surface at different lighting situations. These images have different contrast, brightness, and texture.

Under certain conditions, there occur fascinating phenomenon known as *sun glint* or *sun glitter* and *glitter pattern*. This remarkable optical event is the direct result of specular reflection of sunlight from the surface wave facets. Sun glitter actually is a

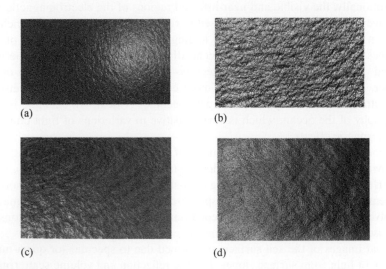

(a)   (b)

(c)   (d)

FIGURE 1.1 **(See color insert.)** Aerial photographs of sea surface: (a) sunglint texture, (b) monotonic texture, (c) variable texture, and (d) patterned texture.

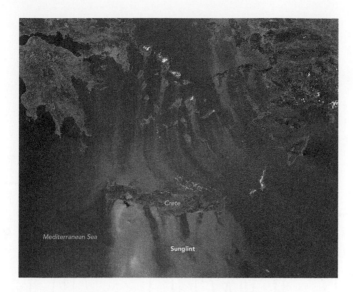

FIGURE 1.2 **(See color insert.)** Satellite image. Sunglint on the Aegean and Mediterranean (MODIS). (Image from NASA.)

random composition of innumerable bright points. The shape of the glitter pattern varies with the solar altitude. Glitter image is difficult to investigate quantitatively without special digital processing. Figure 1.2 shows low-resolution optical satellite image with sun glint which, probably, impossible to use for hydrodynamic studies.

Meanwhile, there was a great achievement when, in the early 1950s, Charles Cox and Walter Munk's use of photographs of sun glitter derived a statistical law of sea slope distribution. It is no surprise that these pioneer works and the Cox–Munk model for slope statistics of sea waves have become literary classics for remote sensing and many other applications.

Not all sunlight energy reflected from the surface reaches the sensor. This is because of absorption and scattering within the atmosphere. Figure 1.3 shows atmospheric "windows" where atmospheric transmission is highest only at selected spectral regions. The most optical sensors use visible (400–700 nm), near-infrared (720–1300 nm) and shortwave infrared (1300–3000 nm) wavelength bands. Thermal infrared systems employ the midwave infrared (3–5 µm) and the longwave infrared (8–14 µm) wavelength bands. Note that in visible range, the images are formed due to the solar radiance reflected from an object, whereas a thermal infrared sensor measures the radiance emitted from an object itself. Although thermal radiance includes reflection and emission components, the temperature of an object yields the largest contribution to thermal infrared signal. For this reason, a thermal infrared sensor is unable to detect wave hydrodynamic phenomena directly, but it enables to measure the surface temperature with a high accuracy.

Atmospheric absorption reduces the solar radiance within the absorption bands of the atmospheric gases (Figure 1.3). The attenuation of the reflected radiance is wavelength dependent. Atmospheric scattering occurs from constituent gases, small particles, aerosols, and dust in the atmosphere. The radiation scattered by

(a)

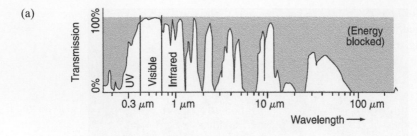

(b)

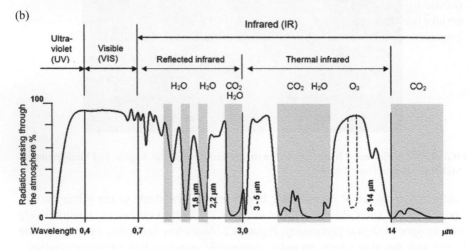

FIGURE 1.3 Atmospheric optical windows. (Based on Lillesand et al. 2015.)

the atmosphere toward the sensor can produce a *hazy appearance* of the image. Sometimes, in the high-resolution images, we can observe a so-called *adjacency effect* that is significant sources of errors. Possible causes of this effect are multiple scattering and diffusion of light in the atmosphere column within the field of view of optical sensor. Finally, atmospheric stratification and turbulence (fluctuations of the refractive index) reduce a spatial resolution of optical images of the ocean surface.

Strong atmospheric effects may lead to unacceptable level of image quality— reduction of spatial resolution, brightness, and contrast that make it difficult or even impossible to provide assessment and/or quantitative analysis of optical data. Eventually, the influence of these and other limitation factors should be minimized when registering and processing photographic or electro-optical images. If this is done right, we have a good chance to evaluate hydro-physical information from optical data.

Let's consider several optical techniques capable to provide remote sensing studies of ocean hydrodynamics. Briefly, they are the following.

Conventional and traditional photography both use single-lens film (black and white or color) cameras. These methods have been employed in the earlier 20th century but they are still in demand in many applications. In the 1960s and 1970s, many great studies were done using aerial photography and stereophotography in order to investigate geometric and statistical characteristics of sea surface waves. For this goal, aerial photographs were analyzed using analog and analytic methods of *photogrammetry*. In the early years, aerial photogrammetric methods were used in many geophysical applications, including regional oceanography, marine weather forecasting, and reconnaissance; some of them are still in practice today.

Multiband aerial imagery enables us to measure optical radiance at several wavelength bands simultaneously at visible and near-infrared ranges of the electromagnetic spectrum. In the 1970s, airspace multiband imagery was implemented using a single lens film camera equipped by different color filters. Many pioneering remote sensing studies of ocean dynamics were conducted in the USSR over the 1980s decade (the Pacific Ocean) using Carl Zeiss, Jena 6- and 4-channel multispectral aerial photo cameras MKF-6 and MSK-4 (Figure 1.4 and Table 1.1). These photographic cameras had more advanced optical characteristics than others at that time.

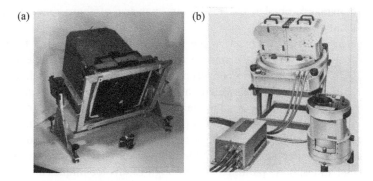

FIGURE 1.4 Aerial photo cameras MKF-6 (a) and MSK-4 (b). (Source: Carl Zeiss Jena, Deutsches Museum, Munich, Germany.)

TABLE 1.1
Specifications of the Six-Channel Multispectral Aerial Photo Camera MKF-6

	1	2	3	4	5	6
Channel #		Four Visible			Two Infrared	
Spectral band (nm)	480	540	600	660	720	840
Focus of objective, f (mm)		125				
Max. optical resolution		150 lines/mm (~2–3 m for ocean conditions)				
Max. relative aperture		1/4				
Field of view size		0.4–0.64 H (H=altitude, km)				
Size of photo frame		56 mm×81 mm				
Basic altitude		3 and 5 km				
Scale (L=H/f)		From 1:20,000 to 1:40,000				
Overlap		20%, 60%, 80%				
Product		Photo film roll containing up to 2500 frames				

Source: After Raizer (2017).

Digital photography appeared in the late 1960s. This technique was invented to obtain quick shot permanent pictures of an object and store imaging data in a digital file. The history and evolution of digital image technology is an amazing story which has been described in literature many times. Without going into detail, there are two basic types of detectors to perform digital imagery: *charge-coupled-device* (CCD) and *complementary metal-oxide semiconductor* (CMOS). Both detectors enable to capture a wider range of electromagnetic spectrum than photographic film. Conventional CCD and CMOS imaging sensors have limitations by spatial resolution (low number of pixels means low resolution) and dynamic range as well. Meanwhile, high-resolution professional CCD and/or CMOS digital cameras are used successfully for the investigations of sea surface roughness, wind-generated waves, breaking waves, foam, and whitecaps. Such "video" experiments with further data processing can be conducted from various low-cost instrument platforms (including unmanned vehicles and vessels) in open sea, offshore zones, lakes, rivers, and laboratory tanks.

Digital electro-optical imagery is the most advanced observation technology which is widely used in airspace remote sensing missions. Passive electro-optic/ infrared (EO/IR) imaging systems convert measured electromagnetic radiation (analog signal) into electronic signal which can be enhanced and displayed in digital format. Digital data can be processed in real time. Satellite EO/IR imagers (Landsat, Terra, AVHHR, SeaWiFS, COES, etc.), operated at visible, infrared, and thermal infrared regions of the electromagnetic spectrum, provide the global monitoring of the Earth including mapping of ocean colour, bio-productivity, sea surface temperature (SST) and atmospheric parameters. Although EO/IR techniques are constantly improving and evolving dramatically, optical detection of hydrodynamic features in the ocean is not really a priority task.

Multispectral (MS) and Hyperspectral (HS) remote sensing technology, also known as imaging spectroscopy, enables the measuring of electromagnetic radiation within many narrow spectral bands. MS images usually contain 3–10 spectral bands of wide bandwidth (70–400 nm) and HS images contain as many as 200 (or more) spectral bands of narrow bandwidth (5–10 nm). Eventually, HS images contain much more spectral information than MS images and therefore, HS sensors have a greater ability (and sensitivity) to capture fine details and distinguish the smallest variations of radiation. In satellite oceanography, MS and HS imagers provide measurements of ocean colour, bioactivity, pollutions, bathymetry, monitoring of coral reef health and coastal zones. Satellite MS and/or HS imaging data, which can be directly associated with ocean surface hydrodynamics are very limited.

LiDAR (Light Detection and Ranging) or simply "lidar" is an active electro-optical technique that uses light sources (laser) to measure distance between the object and sensor. Lidar also enable to deduce physical properties of the object through detection of the light-object interaction process such as scattering, absorption, reflection, and fluorescence. Lidar has many applications in geophysics and remote sensing including ocean topography missions and diagnostics of marine environment—pollutions, biochemical and (non)organic compounds, phytoplankton, and fishing activity. Lidar also provides aerosol profile measurements in the atmosphere. Additional lidar capabilities include laser radar, altimeter, and spectrometer. These advanced technologies have great potential value for study of ocean hydrodynamics and hopefully will deploy more efficiently in the near future.

1.3 OPTICAL OCEANOGRAPHY: STATE OF SCIENCE

The first known scientific investigation of the optical properties of the ocean was conducted in the early 19th century when a Russian naval officer, Otto von Kotzebue, undertook an oceanographic cruise in 1817. He made measurements of the depth of light penetration below the surface using a red piece of cloth. In later experiments, this optical instrument was replaced by a flat white disk also known as the *Secchi disk*. In the 20th century, scientists from US, Europe, and Russia used photographic methods for measuring light in the sea. In the second half of the 20th century, basic light propagation characteristics were defined and documented in many papers and books (see our Bibliography).

The appearance of electro-optical (laser-based) technology had a pronounced effect on the study of interactions of light and ocean water. The military interest for *undersea warfare systems* in both USA and former Soviet Union provided significant funding for the research in order to develop new optical methods and instruments. The most known scientific centers where these works were performed are the Scripps Institute of Oceanography, La Jolla, CA (https://scripps.ucsd.edu/) in the US and the Shirshov Institute of Oceanology, Moscow in Russia (https://ocean.ru/en/). Within decades these and the other Institutes have made significant contributions to ocean optics conducting sophisticated measurements of optical properties (including scattering, absorption, attenuation) throughout the world's oceans. In the 1970s and 1980s, numerous innovative commercially available optical instruments (infrared radiometers, visible spectrometers, scanners, altimeters, etc.) were developed and applied for oceanographic studies. However, interest to *in situ* optical observations slightly declined because the satellite remote sensing industry was getting ready for the next revolution in optical oceanography.

A modern history of satellite oceanography has begun in 1978 with the launch of the SeaSat satellite carrying a remarkable set of remote sensing instruments— altimeter, microwave radiometer and scatterometer, visible and infrared radiometer, and Synthetic Aperture Radar (SAR). The collected materials have opened a new era in the exploration of the ocean from space. In the past few decades, the following major satellites were intended for oceanographic studies:

Nimbus 7 (1978) including the CZCS, the Coastal Zone Color Scanner; Kosmos 1076 and Kosmos 1151 (1979–1980), Kosmos 1870 (1987); Topex-Poseidon (1992); SeaWiFS(1997), Terra/MODIS (1999), JASON 1 (2001), Aqua (2002), Resourcesat-1 (2003), and Meteor-3M (2007).

From these and the other airspace missions, it becomes clear to us that deep ocean processes can only be detected if they have *observable surface hydrodynamic manifestations (or signatures)*. The best examples available today are observations of oceanic internal waves. Their surface manifestations are the result of hydrodynamic modulations of surface roughness by currents induced by internal waves.

Satellite optical oceanography is based on several fundamental physical principles of electrodynamics which are as follows: spectral characterization of light, interaction (refraction, transmission) with medium, propagation (reflection, scattering, absorption, attenuation), and instrumental ability to measure total solar radiance and irradiance through photographic or digital imagery. The choice of an appropriate optical system for ocean observations depends on the task and applications. However, as the practice shows, the digital imagery is significantly greater than of digitalized analog photography or film. A typical example is computation of 2D Fourier spectra by optical images obtained at variable lighting conditions. As a whole, digital imagery provides a better result than photographic film.

Optical satellite oceanography includes many subdisciplines; among them bio-geo-optics, ocean colour, coastal waters, bathymetry, and sea surface topography are the most popular emerging research areas today. As a whole, optical remote sensing is a high performance integrated framework including a number of components (Figure 1.5). Note that an extended *multisensor* framework is much more efficient for hydrodynamic studies.

Other practical interest in ocean optics is measurement of scattering and absorption of light by seawater. Optical spectroscopy (including laser and lidar remote sensing) has fundamental importance for marine biochemistry and microbiology providing detection of sediments and organic compounds at subsurface ocean layer. The technique has been developed using a concept of "Inherent Optical Properties" (IOPs) and "Apparent Optical Properties" (AOPs) of ocean environment. Briefly, IOPs are those optical properties that depend only on medium (seawater) and are independent of the ambient light within medium. IOPs are scattering, absorption, and attenuation. AOPs are those optical properties that depend both on medium (IOPs) and an ambient light. AOPs are radiance, irradiance, and reflectance.

Radiative transfer theory provides the connection between the IOPs and the AOPs. Environmental factors such as surface roughness, underwater bottom topography,

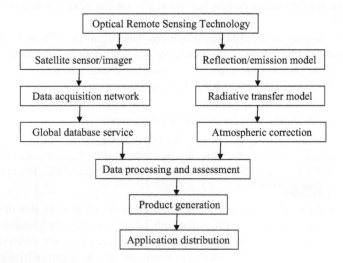

FIGURE 1.5   Basic optical remote sensing framework.

SST and sea surface salinity, the near-surface atmospheric boundary layer and other geophysical factors can be incorporated into the theory via the boundary conditions. This allows oceanographers to perform a more adequate interpretation of experimental data.

Because light propagation in pure water and natural seawater differs (due to the presence of solid microparticles in seawater), absorption and scattering spectra measured in visible/near-infrared range depend on variations of colored dissolved and particulate substances contained in the water. These substances are often grouped at the subsurface ocean layer within three main categories: phytoplankton, non-algal particles, and colored dissolved organic matter. We mention these things because seawater chemical and biological (and radioactive) pollutants may cause microstructure variations of the surface tension resulting to the change of surface roughness. Possible fine-structure hydrodynamic perturbations can be detected using a sensitive optical sensor, e.g., lidar (more detailed discussion is beyond the scope of this book).

Another significant contribution of optical remote sensing methods is monitoring of SST across the globe. SST is a fundamental geophysical variable which strongly influences climate variability and the Earth's hydrological cycle. Thermal infrared measurements of SST have a long heritage (more than 30 years). Today satellite infrared and microwave radiometers provide SST mapping on a regular basis. Example of skin SST map derived from satellite infrared and microwave data is shown in Figure 1.6. This image displays the global distribution of SST with mesoscale thermal features and anomalies including patterns of water circulation, locations of cold water upwelling near the coasts, and warm water currents such as the Gulf Stream.

Many models and algorithms have been used to compute thermal infrared SST from remotely sensed data. The problem is to achieve higher accuracy (~0.1°C) of the bulk SST measurements from satellites. This task requires the development of improved parametrical oceanographic models for the bulk skin sea temperature as well the application of advanced infrared radiometry. The existence of the ocean skin layer and day/night temperature gradients (so-called diurnal thermocline) are most important natural factors effecting variations of skin temperature. Atmospheric

FIGURE 1.6 (See color insert.) Example of global SST map derived from satellite data (Terra/MODIS). (Image from NASA.)

conditions and cloudiness have potential impact on satellite SST measurements as well. Rigorous radiative transfer models including atmospheric correction provide comprehensive numerical analysis of the ocean-atmosphere radiation properties. Finally, instrument errors and calibration biases may also affect the retrieval process and accuracy of geolocated SST maps.

In certain situations, variations of skin layer temperature can change *thermohydrodynamic* conditions of the ocean-atmosphere interface causing the generation of temperature anomaly. Temperature anomaly is characterized by increased (or decreased) value of SST (usually ~0.5°C or less) relative to ambient surface temperature. The generation of temperature anomaly can be associated with a number of natural causes: hurricane impact, tropical rain, strong current or thermal convection in a mixed ocean layer.

Many believe that the water heated by the nuclear submarine can create the so-called *thermal wake* that is detectable by an optical/infrared sensor. However, simple calculations of heat transfer from the nuclear reactor show that the amount of thermal radiation entering the water is so small that conventional thermal infrared technology unable to provide detection of such kind of wake from space. Indeed, available data demonstrate that the changes (increment) of SST in the field of the "thermal wake" do not exceed approximately 0.005°C (0.009°F).

Unfortunately, space observations of ocean hydrodynamics have no priority or preference relative to climate, meteorology, or ecology missions. The only exception is the remote sensing of coastal hydrodynamics, which is important for offshore engineering and management. One possible reason of ignoring the problem is a lack of knowledge and experiences how to use space-based technology for detection of deep ocean and/or surface hydrodynamic events/disturbances. Meanwhile, it is generally believed that it is strictly military matter but this is something that we shall overcome sooner or later...

1.4   OPTICAL DATA ACQUISITION AND ANALYSIS

Remote sensing involves many state-of-the-art methods and algorithms aimed data collection, processing, assessment, analysis, interpretation, application, and management. The entire process of capturing, measuring, and exploring measured information is known as data acquisition and analysis. An acquisition of remotely sensed data can be conducted from different platforms—ground-based, vehicles, ships, aircraft, satellites, balloons, rockets, space shuttles, etc. Therefore, there are many different data acquisition methods and analysis techniques in remote sensing dependent on sensor-platform combinations.

Effectiveness and efficiency of data acquisition system is defined by a number of factors; the most important are sensor specification, type of data, and data processing capabilities. Let's consider briefly these factors in a view of optical observations of the ocean.

Science and technology are key inputs in choosing sensor/instrument specification. In our case, three major characteristics—spectral band, resolution, and swath width—define sensor capabilities and information performance. The success of optical imagery of the ocean depends on full instrument resolution. In remote sensing, there are four types of resolution: spatial, temporal, spectral, and radiometric. Resolution parameters are defined by payload design goals and engineering trade-offs.

In the case of an ocean environment, observations at very high spatial resolution (~1 m or better) are more important than observations at improved spectral or radiometric resolutions with lower spatial resolutions. The *effective swath width* should match typical scales of hydrodynamic processes (~5–10 km). Therewith, it is necessary to acquire *statistically representative* experimental data in order to choose an appropriate sampling strategy and provide correct assessments. Nowadays, the IKONOS, QuickBird, GeoEye, WorldView, and the others similar (military) optical instruments can satisfy these criteria.

There are two categories of optical data: photographic and digital. The first is made through a photographic process using light-sensitive chemically active materials (film or plate). The second (which is actually called "image") is recorded directly using electro-optical device, digital camera, or scanning system; these data are stored in digital memory. Optical data may vary significantly depending on sensor design characteristics and properties of the object being studied.

Satellite optical observations usually provide three types of imaging data: (1) panchromatic PAN, (2) polarimetric, and (3) multispectral, MS (or hyperspectral, HS). In the past, only two types—PAN and polarimetric imagery—were acquired for a given application. Black and white PAN is sensitive to the same wavelength band as the human eye (400–700 nm) and produces a natural-looking picture. Polarimetric imagery has some advantages providing considerable contrast enhancement and equalization of image brightness especially at unstable illumination conditions.

Nowadays, MS and HS data are most commonly used in remote sensing of the Earth surfaces. It became clear that several spectral (visible/infrared) bands offer more geophysical information than just a single band. Efficient tool for analysis of MS and/or HS data is fusion. Fusion provides a better classification of land cover, vegetation, and urban areas. However, there is no experimental evidence that MS/HS data can yield more useful *hydrodynamic* information than the same PAN data. The key factor here is spatial resolution, which for PAN is much higher than for MS/HS data.

Other fundamental characteristics are spatiotemporal scales of observations. In particular, the geographic image/map scale is defined as a ratio of distance between corresponding points on the image and on the ground. In cartography and Geographic Information Systems (GIS) mapping scales have the following classification:

Small scale = 1:24,000 and smaller
Medium scale = 1:10,000–1:24,000
Large scale = 1:1,000–1:10,000
Very large scale = 1:1,000 and larger.

The larger the scale, the more detail with more accuracy appears in the image/map.

The scale transformation has become acute in remote sensing and GIS because of the need to compare and integrate data collected at different scales, resolutions, and accuracies. At the same time, spatiotemporal measurements allow the researches to investigate scaling and self-similarity of geophysical objects being observed. Unlike microwave (radar/radiometer) observations, high-resolution optical imagery provides direct *evidence* of detected surface phenomena/events. It is not necessary to make any manipulations with optical data in order to display geophysical content. An image processing is only needed to explore and enhance complex image features (signatures) in more detail.

Four different methods are commonly used to analyze spatial characteristics of optical data: (1) Fourier transforms, (2) variograms, (3) (multi)fractal, and (4) wavelet. Among them, spectral analysis based on the fast Fourier transform (FFT) is the most popular digital technique for the processing.

The computerized FFT version appeared first in 1965 under the name the *Cooley-Tukey algorithm*; the software was created by IBM for military purposes. Today, 1D and 2D digital FFT is standard code in many signal and image processing toolboxes. Specifically, the 2D FFT is widely used for detection of *harmonic spectral features* in the images. However, there are some nuances concerning the choice of FFT parameters for computing spatial spectra and further applications. An important problem is a juxtaposition of data generated at different scales and pixel resolutions. For oceanographic goals, sometimes it is interesting to obtain quasi-realistic sea surface wave spectra from digital 2D FFT spectra computed by optical data. This operation can be done using radiative transfer models of light propagation only. The resulting "retrieved" sea wave spectrum will depend on model parameters significantly.

More sophisticated information can be obtained using (multi)fractal and wavelet analyses. Fractal and wavelet algorithms appeared in the 1980s. Although these methods may seem more complicated than FFT, they are flexible and efficient enough in order to obtain remarkable results. Fractal formalism (known as a concept of fractal dimension) is used for the characterization of complexity and scaling of natural objects and dynamic systems. In particular, in remote sensing, fractal-based analysis offers better statistical treatment of geometrical fields having *contraction* and *aggregation* as natural property. An example is foam/whitecap objects visible in aerial photographs. Wavelet transform provides time-frequency multiresolution analysis of dynamic datasets that is advantageous over other techniques. Certain effort requires for elaborating fractal and/or wavelet methods in thematic image processing. More attention is also needed regarding combined analysis of *multisensor* datasets with different resolutions, statistics, and physics contents. One way or another, all these can be applied for analysis of optical data.

Figure 1.7 presents optical data acquisition chain for scientific and commerce applications. Such a scheme includes also quantitative accuracy assessment and management of databases that require certain software developments. The major concern is real-time acquisition and analysis of highly dynamic multivariate data (which are ocean data, actually). The real-time technology will determine *mainstream detection capabilities*.

Experiences show that data acquisition system for advanced applications requires extended networks including preliminary theoretical analysis—modeling, simulation, and prediction of remotely sensed data (signals, images, signatures). The comparison between measured (or retrieved) and predicted data is always an important stage in geophysical interpretation. Therefore, we believe that combined experimental and theoretical framework is being the most adequate approach in remote sensing of ocean environment.

PAYLOAD SENSORS	RAW DATA	SOFTWARE ANALYTICS	PRODUCT CREATION	CUSTOMER APPLICATION
Photo Camera	Photograph	Data Logistic	Images	Meteorology
Digital Camera	Image	Big Data Bases	GIS software	Geophysics
Optical Imager	Video	Visualization	Histograms	Hydrology
Spectrometer	Arrays	Classification	Diagrams	Oceanography
Laser Altimeter	Records	Data Mining	Statistics	Environment
Lidar	Stream	Processing	Databases	Climate
Atmospheric Sounder	Vectors	Thematic Processing	Maps	Fishing
Infrared Scanner	Raster	Assessment	Tables, Plots, Graphs	Maritime
Infrared Radiometer	Time-series	Regression	Reports, Publications	Military (ASW*)
		Clustering	Presentations	Reconnaissance
		Spectral Analysis	Books	Security
		Networking		
		Variogram		
		Wavelet		
		(Multi)fractal		
		Data Fusion		
		Modeling		
		Training		
		Management		

ASW* – Anti-Submarine Warfare

FIGURE 1.7 Data acquisition chain in optical remote sensing.

Nowadays, space-based optical systems enable the measurement of a very diverse range of scales—from several dozen cm to hundreds of km; therefore, there is a great opportunity for us to explore fundamental hydrodynamic properties of the ocean—*variability, scaling,* and *self-similarity.* It is difficult to study spatial environmental dynamics *in situ.* Advanced satellite observations can provide sophisticated information creating technological breakthrough in geosciences, remote sensing, and applied physics. Some state-of-the-art developments will be reported and discussed in the next chapters of the book.

BIBLIOGRAPHY

Aber, J. S., Marzolff, I., and Ries, J. 2010. *Small-Format Aerial Photography: Principles, Techniques and Geoscience Applications.* Elsevier, Amsterdam, The Netherlands.

Apel, J. R. 1987. *Principles of Ocean Physics (International Geophysics Series, Vol. 38).* Academic Press, London, UK.

Arst, H. 2003. *Optical Properties and Remote Sensing of Multicomponental Water Bodies.* Springer, Praxis Publishing, Chichester, UK.

Asrar, G. (Ed.). 1989. *Theory and Applications of Optical Remote Sensing.* John Wiley & Sons, New York.

Barale, V. and Gade, M. (Eds.). 2008. *Remote Sensing of the European Seas.* Springer, New York.

Barenblatt, G. I. 1996. *Scaling, Self-Similarity, and Intermediate Asymptotics.* Cambridge University Press, Cambridge, UK.

Benediktsson, J. A. and Ghamisi, P. 2015. *Spectral-Spatial Classification of Hyperspectral Remote Sensing Images.* Artech House Publishers, Boston, MA.

Bukata, R. P., Jerome, J. H., Kondratyev, K. Ya., and Pozdnyakov, D. V. 1995. *Optical Properties and Remote Sensing of Inland and Coastal Waters.* CRC Press, Boca Raton, FL.

Campbell, J. B. and Wynne, R. H. 2011. *Introduction to Remote Sensing*, 5th edition. The Guilford Press, New York.

Chuvieco, E. 2016. *Fundamentals of Satellite Remote Sensing: An Environmental Approach, Second Edition*, 2nd edition, CRC Press, Boca Raton, FL.

Dera, J. 1992. *Marine Physics.* Elsevier Science Publishing, New York.

Dowman, I., Jacobsen, K., Konecny, G., and Sandau, R. 2012. *High Resolution Optical Satellite Imagery.* Whittles Publishing, Dunbeath, UK.

Eckart, C. 1960. *Hydrodynamics of Oceans and Atmospheres.* Pergamon Press, Oxford, UK.

Egan, W. G. 2004. *Optical Remote Sensing: Science and Technology.* Marcel Dekker, New York.

Elachi, C. and van Zyl, J. 2006. *Introduction to the Physics and Techniques of Remote Sensing*, 2nd edition. John Wiley & Sons, Hoboken, NJ.

Emery, W. and Camps, A. 2017. *Introduction to Satellite Remote Sensing: Atmosphere, Ocean, Land and Cryosphere Applications.* Elsevier, Amsterdam, The Netherlands.

Gao, J. 2009. *Digital Analysis of Remotely Sensed Imagery.* McGraw-Hill Education, New York.

Goldstein, D. H. 2011. *Polarized Light*, 3rd edition. CRC Press, Boca Raton, FL.

Goody, R. M. and Yung, Y. L. 1995. *Atmospheric Radiation: Theoretical Basis*, 2nd edition. Oxford University Press, New York.

Graham, R. and Koh, A. 2002. *Digital Aerial Survey: Theory and Practice.* Whittles Publishing, Caithness, UK.

Halpern, D. (Ed.). 2000. *Satellites, Oceanography and Society.* Elsevier Science, Amsterdam, The Netherlands.

Hou, W. W. 2013. *Ocean Sensing and Monitoring: Optics and Other Methods.* SPIE Press, Bellingham, WA.

Ikeda, M. and Dobson, F. (Eds.). 1995. *Oceanographic Applications of Remote Sensing.* CRC Press, Boca Raton, FL.

Ishimaru, A. 1991. *Electromagnetic Wave Propagation, Radiation, and Scattering.* Englewood Cliffs, Prentice Hall, NJ.

Jensen, J. R. 2015. *Introductory Digital Image Processing: A Remote Sensing Perspective*, 4th edition. Pearson Education, Glenview, IL.

Jerlov, N. G. 1968. *Optical Oceanography.* Elsevier, Amsterdam, The Netherlands.

Jerlov, N. G. 1976. *Marine Optics.* Elsevier, Amsterdam, The Netherlands.

Jonasz, M. and Fournier, G. 2007. *Light Scattering by Particles in Water: Theoretical and Experimental Foundations.* Elsevier, London, UK.

Joseph, A. 2014. *Measuring Ocean Currents: Tools, Technologies, and Data.* Elsevier, San Diego, CA.

Khorram, S., van der Wiele, C. F., Koch, F. H., Nelson, S. A. C., and Potts, M. D. 2016. *Principles of Applied Remote Sensing.* Springer, New York.

Kirk, J. T. O. 2011. *Light and Photosynthesis in Aquatic Ecosystems*, 3rd edition. Cambridge University Press, Cambridge, UK.

Kramer, H. J. 2002. *Observation of the Earth and Its Environment: Survey of Missions and Sensors*, 4th edition. Springer, Berlin, Germany.

Kraus, E. B. and Businge, J. A. 1994. *Atmosphere-Ocean Interaction*, 2nd edition. Oxford University Press, New York.

Kuenzer, C. and Dech, S. (Eds.). 2013. *Thermal Infrared Remote Sensing: Sensors, Methods, Applications.* Springer, New York.

Lamb, H. 1932. *Hydrodynamics*, 6th edition. Cambridge University Press, Cambridge, UK.

Landgrebe, D. A. 2003. *Signal Theory Methods in Multispectral Remote Sensing.* John Wiley & Sons, Hoboken, NJ.

Lavender, S. and Lavender, A. 2015. *Practical Handbook of Remote Sensing.* CRC Press, Boca Raton, FL.

Lein, J. K. 2012. *Environmental Sensing: Analytical Techniques for Earth Observation.* Springer, Dordrecht, The Netherlands.

Lillesand, T., Kiefer, R. W., and Chipman, J. 2015. *Remote Sensing and Image Interpretation,* 7th edition. John Wiley & Sons, Hoboken, NJ.

Linder, W. 2009. *Digital Photogrammetry: A Practical Course,* 3rd edition, Springer-Verlag, Berlin, Heidelberg.

Liou, K.-N. 2002. *An Introduction to Atmospheric Radiation,* 2nd edition. Academic Press, San Diego, CA.

Manolakis, D., Lockwood, R., and Cooley, T. 2016. *Hyperspectral Imaging Remote Sensing: Physics, Sensors, and Algorithms.* Cambridge University Press, Cambridge, UK.

Martin, S. 2014. *An Introduction to Ocean Remote Sensing,* 2nd edition. Cambridge University Press, Cambridge, UK.

Massel, S. R. 1999. *Fluid Mechanics for Marine Ecologists.* Springer, Berlin, Germany.

Mather, P. M. and Koch, P. 2011. *Computer Processing of Remotely-Sensed Images: An Introduction,* 4th edition. John Wiley & Sons, UK.

Measures, R. M. 1992. *Laser Remote Sensing: Fundamentals and Applications.* Krieger Publishing Company, Malabar, FL.

Mishra, D. R., Ogashawara, I., and Gitelson, A. A. 2017. *Bio-optical Modeling and Remote Sensing of Inland Waters.* Elsevier, Amsterdam, The Netherlands.

Mobley, C. D. 1994. *Light and Water: Radiative Transfer in Natural Waters.* Academic Press, San Diego, CA.

Monin, A. S. and Krasitskii, V. P. 1985. *Yavleniya na poverkhnosti okeana (Phenomena on the Ocean Surface).* Gidrometeoizdat, Leningrad, USSR (in Russian).

Newman, J. N. 1977. *Marine Hydrodynamics.* The MIT Press, Cambridge, MA.

Njoku, E. G. 2014. *Encyclopedia of Remote Sensing (Encyclopedia of Earth Sciences Series).* Springer, New York.

Paine, D. P. and Kiser, J. D. 2012. *Aerial Photography and Image Interpretation,* 3rd edition. John Wiley & Sons, Hoboken, NJ.

Pedlosky, J. 2013. *Waves in the Ocean and Atmosphere: Introduction to Wave Dynamics.* Springer, Berlin, Germany.

Phillips, O. M. 1980. *The Dynamics of the Upper Ocean,* 2nd edition. Cambridge University Press, Cambridge, UK.

Prasad, S., Bruce, L. M., and Chanussot, J. (Eds.). 2011. *Optical Remote Sensing: Advances in Signal Processing and Exploitation Techniques.* Springer, Berlin, Germany.

Pu, R. 2017. *Hyperspectral Remote Sensing: Fundamentals and Practices (Remote Sensing Applications Series).* CRC Press, Boca Raton, FL.

Raizer, V. 2017. *Advances in Passive Microwave Remote Sensing of Oceans.* CRC Press, Boca Raton, FL.

Rampal, K. K. 1999. *Handbook of Aerial Photography and Interpretation.* Concept Publishing Co., New Delhi.

Rees, W. G. 2013. *Physical Principles of Remote Sensing,* 3rd edition. Cambridge University Press, Cambridge, UK.

Richardson, L. L. and LeDrew, E. F. 2006. *Remote Sensing of Aquatic Coastal Ecosystem Processes: Science and Management Applications.* Springer, New York.

Robinson, I. S. 2010. *Discovering the Ocean from Space: The Unique Applications of Satellite Oceanography.* Springer, New York.

Schott, J. R. 2009. *Fundamentals of Polarimetric Remote Sensing.* SPIE Press, Bellingham, WA.

Schowengerdt, R. A. 2007. *Remote Sensing: Models and Methods for Image Processing*, 3rd edition. Academic Press, Elsevier, San Diego, CA.

Shaw, J. A. 2017. *Optics in the Air: Observing Optical Phenomena through Airplane Windows*. SPIE Press, Bellingham, WA.

Shifrin, K. S. 1988. *Physical Optics of Ocean Water*, translated from Russian by D. Oliver. AIP Translation Series. American Institute of Physics, Woodbury, NY.

Shuleikin, V. V. 1968. *Fizika Moria (Physics of the Sea)*, 4th edition revised and expanded. Izdatel'stvo Akademii Nauk S.S.S.R., Moscow, USSR (in Russian).

Slater, P. N. 1980. *Remote Sensing: Optics and Optical Systems*. Addison-Wesley Publishing Company, Reading, MA.

Soloviev, A. and Lukas, R. 2014. *The Near-Surface Layer of the Ocean: Structure, Dynamics and Applications*, 2nd edition. Springer, Dordrecht, The Netherlands.

Spinrad, R. W., Carder, K. L., and Perry, M. J. (Eds.). 1994. *Ocean Optics*. Oxford University Press, New York.

Stamnes, K., Thomas, G. E., and Stamnes, J. J. 2017. *Radiative Transfer in the Atmosphere and Ocean*, 2nd edition. Cambridge University Press, Cambridge, UK.

Steele, J. H., Thorpe, S. A., and Turekian, K. K. (Eds.). 2010. *Elements of Physical Oceanography: A derivative of the Encyclopedia of Ocean Sciences*. Academic Press, Elsevier, San Diego, CA.

Stewart, R. H. 1985. *Methods of Satellite Oceanography (Scripps Studies in Earth and Ocean Sciences)*. University of California Press, Berkeley, CA.

Thomson, R. E. and Emery, W. J. 2014. *Data Analysis Methods in Physical Oceanography*, 3rd edition. Elsevier, Amsterdam, The Netherlands.

Träger, F. (Ed.). 2012. *Springer Handbook of Lasers and Optics*, 2nd edition. Springer, Heidelberg, Germany.

van de Hulst, H. C. 1981. *Light Scattering by Small Particles*, corrected edition. Dover Publications, Mineola, NY.

Verbula, D. L. 1995. *Satellite Remote Sensing of Natural Recourses*. CRC Press, Boca Raton, FL.

Victorov, S. 1996. *Regional Satellite Oceanography*. Taylor & Francis CRC Press, London, UK.

Walker, R. E. 1994. *Marine Light Field Statistics*. John Wiley & Sons, New York.

Wang, Y. (Ed.). 2010. *Remote Sensing of Coastal Environments*. CRC Press, Boca Raton, FL.

Watson, J. and Zielinski, O. (Eds.). 2013. *Subsea Optics and Imaging*. Woodhead Publishing, Cambridge, UK.

Wozniak, B. and Dera, J. 2007. *Light Absorption in Sea Water*. Springer, New York.

2 Ocean Hydrodynamics

2.1 INTRODUCTION

Fluid mechanics is a physical science concerned with the behaviour of fluid (liquids, gases, and plasmas) at rest and motion and the forces on them. Over the past few centuries, fluid mechanics sand subdivisions (Figure 2.1) have become the subject of greatest achievements in applied physics, environmental science, and engineering.

The fundamental advances in fluid dynamics began with the works of Daniel Bernoulli (1700–1782), Henri Pitot (1695–1771), Leonhard Euler (1707–1783), Louis Marie Henri Navier (1785–1836), and George Gabriel Stokes (1819–1903), who formulated physical principles and mathematical laws of fluid flow. In particular, the partial differential equations of fluid motion with internal friction, published in 1895 and known as the Navier–Stokes equations, revolutionized the analysis of a large number of complex phenomena in various fields of physics.

Here is a short definition of hydrodynamics: "The science of hydrodynamics is concerned with the behaviour of fluids in motion." The other, an extended definition, is: "Fluid dynamics or Hydrodynamics is that branch of science which is concerned with study of motion of fluids or that of bodies in contact with fluids." In context with both definitions, the ocean surface can be considered as a moving hydrodynamic system comprising flows of higher dimension and having a boundary with air or bodies. It means that we have to take into account a number of hydrodynamic aspects of fluid flow such as turbulence, interface phenomena, wave movements, stability, transformations, and interactions.

Actually, adaptation of classical hydrodynamics to real world ocean environment is very controversial problem because of complexity and variability of the phenomena. The terminology *ocean hydrodynamics* (OHD) introduced here seems more likely suitable for geosciences rather than for descriptive or physical oceanography. Some sophisticated natural phenomena (*hydrodynamic events*) can be studied from the distance or from space more adequately than *in situ*; therefore, OHD might be the subject matter of remote sensing. In this chapter, we will consider basic principles of OHD as an important part of advanced remote sensing studies and applications.

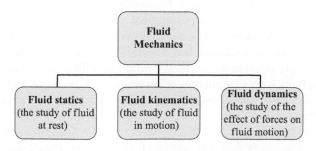

FIGURE 2.1 Branches of fluid mechanics.

2.2   ELEMENTS OF FLUID DYNAMICS

2.2.1   Fluid Classification

According to common terminology, fluids are categorized as *Newtonian and Non-Newtonian* fluids. The Newtonian fluid is defined as a fluid in which there is a linear, isotropic relation between the stresses and the velocity gradients. Newtonian fluid may be classified as "ideal" (perfect or *inviscid* flow) and "real" (actual fluid). Ideal fluid is frictionless, homogeneous, and incompressible. This concept defines the flow such as wave motion. Real fluid is viscous and compressible in nature. Viscosity, also known as an internal friction, is capable of offering resistance to shearing stress. The resistance can be relatively small (but not negligible) for fluids such as water; it is significantly larger for other fluids such as oil, kerosene, petrol, some chemical solutions, etc.

Physical properties of the fluid are characterized by (p) pressure, (ρ) density, (μ) viscosity, and (t) temperature. Pressure is defined as force per unit area. The density is a ratio of the mass (M) to the volume (V) of a substance: $\rho = M/V$. Density of water depends on temperature and pressure. There are two main types of viscosity: dynamic viscosity μ and kinematic viscosity $\nu = \mu/\rho$. The dynamic viscosity of the fluid is practically independent on pressure and depends on temperature only. The kinematic viscosity of liquid at a given pressure is a function of temperature (www.engineeringtoolbox.com). The approximation of these dependencies is $\mu = A \times 10^{B/(T-C)}$, where T is the temperature in K, μ is the dynamic viscosity in Pa·s, $A = 2.414 \times 10^{-5}$ Pa·s, $B = 247.8$ K, and $C = 140$ K. The dynamic viscosity of water is 8.90×10^{-4} Pa·s or 8.90×10^{-3} dyn·s/cm^2 or 0.890 cP at 25°C. Water has a viscosity of 0.0091 poise at 25°C or 1 centipoise at 20°C.

In our text, we consider a Newtonian fluid only, which is continuous, homogeneous, and macroscopic medium. The viscosity of the Newtonian fluid remains constant at constant temperature, pressure, and time (for a non-Newtonian fluid the viscosity varies with time; this fluid represents various plastic and polymer liquid materials).

2.2.2   Fluid Flow

Fluid dynamics is one of two branches of fluid mechanics; it studies the effect of forces on fluid motion (the other branch is fluid statics, which deals with fluids at rest). In the *American Heritage Dictionary of the English Language*, fluid dynamics is defined as "the branch of applied science that is concerned with the movement of liquids and gases." The most common liquid is water, and most hydrodynamic theories and applications deal with flow of water fluid. The movement of liquids is generally referred to as "flow," which describes the behaviour of fluid and its interactions with surrounding environment, e.g., moving surface water or moving ground water. The following are some of the important characteristics of fluid (water) flow.

There are several types of flow that occur in practice: uniform and non-uniform flow, steady and unsteady flow, laminar and turbulent flow. Figure 2.2 shows types

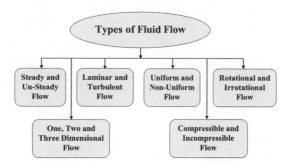

FIGURE 2.2 Types of fluid flow.

of fluid flow. Uniform flow is flow with constant section area along the flow path (the velocity at given time does not change with respect to space); non-uniform flow is flow with a variable section area (the velocity at any time changes with respect to space). Steady flow is time-independent; unsteady flow is time-dependent. Laminar flow, also known as streamline flow, is smoother, and fluid particles move very orderly along parallel layers. Turbulent flow is more chaotic, motion is locally completely random, and complex flow patterns constantly change and take form of vortexes or eddies. There is a category of flow known as the transitional flow, which is a mixture of laminar and turbulent flow (this flow mostly occurs in the pipe).

The main regime of flow, turbulent or laminar, is determined by the dimensionless *Reynolds number* (Re), which is defined as "the ratio of inertial to viscous forces." Given the characteristic velocity scale, u, and length scale, l, for a system, the Reynolds number is $Re = ul/v$, where v is the kinematic viscosity of the fluid. At low Re, the flow tends to be laminar, while at high Re, the flow tends to be turbulent (see below). The Reynolds number is also used to predict transitions from laminar to turbulent flow. Most natural flows of interest are turbulent; however, such flows can be very difficult to predict and describe in detail; distinguishing between laminar and turbulent flow is largely intuitive and often requires experimental evidence. Meanwhile, common (qualitative and understandable enough) characterization of strong turbulent flow is the following: (1) fluctuations of velocity and pressure, (2) irregular movement and stochastic behaviour, (3) vortex stretching, (4) energy dissipation, (5) diffusivity and dissipativity, and (6) Reynolds number $Re > 4000$ indicating *fully developed hydrodynamical turbulence*.

Flow patterns can be visualized using figures called "streamlines," "streaklines," "pathlines," and "timeline." These figures show trajectories of flow elements in space and time. For example, a streamline is defined as "continuous line within a fluid such that the tangent at each point is the direction of the velocity vector at that point." In the case of laminar or steady flow, the streamlines, the pathline, and the streamlines are identical; in the case of turbulent or unsteady flow, they all are quite different. Figure 2.3 illustrates these patterns schematically. The mathematical description is based on a set of differential equations which we consider below.

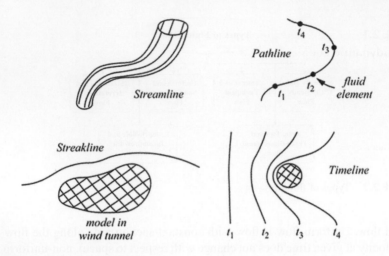

FIGURE 2.3   Fluid flow visualization.

2.2.3 BASIC EQUATIONS OF FLUID DYNAMICS

Fluid dynamics as well fluid mechanics operate with a number of classical equations known as hydrodynamic equations of motions. They are the following (Table 2.1):

- Bernoulli's equation

    It is a statement of the conservation of energy in a form useful for solving problems involving fluids. For a non-viscous, incompressible fluid in steady flow, the sum of pressure, potential and kinetic energies per unit volume is constant at any point.
- Conservation of energy

    It is a statement based on the First Law of Thermodynamics involving energy, heat transfer, and work. With certain limitations, the mechanical energy equation can be compared to the Bernoulli equation.
- Conservation of mass and pressure

    The Law of Mass Conservation is a fundamental law in fluid mechanics; it is a basis for the equation of continuity and the Bernoulli equation. The law of mass conservation states that "mass can neither be created nor destroyed." The inflows, outflows, and change in storage of mass in a system must be in balance.
- Euler equations

    In fluid dynamics, the Euler equations govern the motion of a compressible, inviscid fluid. They correspond to the Navier–Stokes equations with zero viscosity. This form of the Euler equations emphasizes the fact that they directly represent conservation of mass, momentum, and energy.
- Laplace equation

    The Laplace equation is important in fluid dynamics describing the behaviour of gravitational and fluid potentials.

TABLE 2.1
Hydrodynamic Equations of Motions

Bernoulli's equation	$\dfrac{v^2}{2} + \psi + \dfrac{p}{\rho} = 0$
Conservation of energy	$\dfrac{\partial}{\partial t}\left(\dfrac{1}{2}\rho v^2 + \rho e\right) = -\nabla \cdot \left[\rho v\left(\dfrac{1}{2}v^2 + w\right)\right], \; w = e + \dfrac{p}{\rho}$
Conservation of mass	$\nabla \cdot (\rho\vec{v}) = 0$
Euler equations	$\dfrac{\partial \rho}{\partial t} + \nabla \cdot \rho\vec{v} = 0$
Laplace equation	$\nabla^2\phi = 0 \text{ or } \nabla^2 v = 0$
Navier–Stokes equations	$\rho\left(\dfrac{\partial \vec{v}}{\partial t} + \vec{v} \cdot \nabla\vec{v}\right) = -\nabla p + \rho\vec{g} + \mu\nabla^2\vec{v} + F\,\nabla\vec{v} = 0$

$\rho = \rho(p,T)$; ρ, density; T, temperature; p, pressure; $\vec{v}$, velocity; μ, dynamic viscosity; $\vec{g}$, body acceleration (e.g., gravity); ϕ, velocity potential; e, internal energy; w, enthalpy; ψ, force potential ($\psi = \rho gh$, h indicates elevation height); F, body force.

- Navier–Stokes equations

    The motion of a non-turbulent, Newtonian fluid is governed by the Navier–Stokes equations. The equations can be used to model turbulent flow, where the fluid parameters can be interpreted as time-averaged values.
- Reynolds transport theorem

    These are basically mass conservation, energy conservation, and momentum conservation in different forms. The Reynolds theorem provides a link between the system approach and the control volume approach.

The governing equations of the fluid motion must be solved simultaneously in certain domain with specified boundary and initial conditions to obtain the solution of complete flow field. In many practical cases, simplification of basic equations is possible with certain assumption that reduces the mathematical complications. For example, quasi-analytical solutions can be obtained when (a) the flow is incompressible and the properties such as density, viscosity, and thermal conductivity are assumed to be constant and (b) the flow is inviscid (i.e., viscosity and thermal conductivity of a fluid are completely ignored) and the flow is considered to be non-turbulent. Most other cases require numerical solutions; this option is described below.

2.2.4 COMPUTATIONAL FLUID DYNAMICS

Most problems in fluid dynamics are too complex to be solved by analytic calculations.

In these cases, problems must be solved by numeric methods. This area of study is called *Computational Fluid Dynamics* (CFD), which is defined as "a branch of computer-based science that provides numerical predictions of fluid flows." CFD uses numerical techniques for modeling, simulation, and prediction of fluid flow response in problems of interest. CFD solves governing partial differential equations using finite difference (and volume) methods. Today, CFD methods and algorithms are developed for a large selection of hydrodynamic and aerodynamic applications. An important contribution of CFD is direct simulations and visualization of turbulent fluid flows (Figure 2.4). However, because turbulent flow tends to be nonlinear and chaotic, numerical solutions must take in account transition regimes and evolution of hydrodynamic flows with high accuracy. The choice of initial conditions and CFD framework plays a key role in numerical experiments. Small variations of input parameters can lead to large differences in the results and also may cause unstable solutions. To improve the performance of hydrodynamic simulations, the computational domain or computational grid is usually divided into smaller regions using high-order discretization schemes and CFD algorithms are optimized to achieve a minimum cost function.

CFD methods have many advantages. Among them the most important are: (a) relatively low cost of flow modeling and simulation, (b) better visualization and deeper insight into results, (c) testing various hydrodynamic situations with variable set of parameters, (d) practically unlimited volume of information, (e) cyclical development and reproduction capability, and (f) hydrodynamic scale-model implementation.

CFD has made remarkable progress in the past several decades especially using supercomputers which appeared first in the late 1970s. Today, powerful high-performance computers are capable of providing comprehensive numerical research of a variety of scientific and engineering problems of fluid mechanics and hydrodynamics with the commercial CFD software packages.

2.2.5 Important Results of Fluid Dynamics

There are significant fundamental achievements in fluid mechanics, hydrodynamics, and CFD which are important for us. Here are some of them:

- Direct 2D and 3D Navier–Stokes simulation of wave motion including bifurcation and breaking processes
- Simulation of hydrodynamic turbulence at different Reynolds numbers
- Simulation of the laminar–turbulent transition flow
- Simulation of *vortical* structures, their evolution, and interactions in a boundary layer
- Simulation of turbulent wakes behind moving bodies of variable geometry.

The list presented is only reflects a personal view on the problem which is vast and complex.

FIGURE 2.4 Computer fluid dynamics (CFD) simulations of turbulent fluid flows. Wake from the self-propelled submarine. (Based on Posa and Balaras 2018.)

Geophysical applications of fluid dynamics have been discussed throughout the years by many researches. Unfortunately, methods of fluid dynamics including CFD were not properly integrated into remote sensing of environment. A possible reason is difficulties in the creation of composition physics-based models in the interdisciplinary computational domain. Meanwhile, in-depth theoretical developments are of practical interest in problems of ocean remote sensing and nonacoustic detection. Today CFD-based modeling and simulation capabilities and a collected knowledge allow us to achieve success in this field. This work requires a multidisciplinary endeavor and certain resources.

2.2.6 NOTES ON THE LITERATURE

Historical developments of fluid dynamics are reviewed in books (Tokaty 1994; Darrigol 2005).

Detailed description of fundamentals of fluid dynamics can be found in standard textbooks (Lamb 1932; Birkhoff 1960; Batchelor 1967; Meyer 1971; Paterson 1983; Landau and Lifshitz 1987; Tritton 1988; Acheson 1990; Granger 1995; Faber 1995; Milne-Thomson 1996; Streeter et al. 1998; Monin and Yaglom 2007) and in more recent books (Chanson 2009; Buresti 2012; Guyon et al. 2012; Raisinghania 2013; Johnson 2016).

CFD techniques are described in books (Anderson 1995; Ferziger and Perić 2002; Zikanov 2010; Pulliam and Zingg 2014; Mueller 2015).

Descriptions of geophysical fluid dynamics are available in books (Pedlosky 1979; Salmon 1998; Kantha and Clayson 2000; McWilliams 2006; Cushman-Roisin and Beckers 2011; Imberger 2013; Vallis 2017).

Fluid dynamics with application to oceanography is discussed in books (Eckart 1960; Elder and Williams 1996).

Submarine hydrodynamics is described in book (Renilson 2018) and great illustrations of fluid motions and turbulence are available in albums (Van Dyke 1982; Samimy et al. 2003).

REFERENCES TO SECTION 2.2

Acheson, D. J. 1990. *Elementary Fluid Dynamics*. Oxford University Press, Oxford, UK.

Anderson Jr., J. D. 1995. *Computational Fluid Dynamics*. McGraw-Hill, New York.

Batchelor, G. K. 1967. *An Introduction to Fluid Dynamics*. Cambridge University Press, Cambridge, UK.

Birkhoff, G. 1960. *Hydrodynamics*, 2nd edition. Princeton University Press, Princeton, NJ.

Buresti, G. 2012. *Elements of Fluid Dynamics*. Imperial College Press, London, UK.

Chanson, H. 2009. *Applied Hydrodynamics: An Introduction to Ideal and Real Fluid Flows*. CRC Press, Boca Raton, FL.

Cushman-Roisin, B. C. and Beckers, J.-B. 2011. *Introduction to Geophysical Fluid Dynamics: Physical and Numerical Aspects*, 2nd edition. Academic Press – Elsevier, Amsterdam, The Netherlands.

Darrigol, O. 2005. *Worlds of Flow: A History of Hydrodynamics from the Bernoullis to Prandtl*. Oxford University Press, Oxford, UK.

Eckart, C. 1960. *Hydrodynamics of Oceans and Atmospheres*. Pergamon Press, Oxford, UK.

Elder, S. A. and Williams, J. 1996. *Fluid Physics for Oceanographers and Physicists. An Introduction to Incompressible Flow*, 2nd edition. Butterworth-Heinemann, Oxford, UK.

Faber, T. E. 1995. *Fluid Dynamics for Physicists*. Cambridge University Press, Cambridge, UK.

Ferziger, J. H. and Perić, M. 2002. *Computational Methods for Fluid Dynamics*, 3rd edition. Springer, Berlin, Germany.

Granger, R. A. 1995. *Fluid Mechanics*, 2nd edition. Dover Publications, New York.

Guyon, E., Hulin, J.-P., Petit, L., and Mitescu, C. D. 2012. *Physical Hydrodynamics*, 2nd edition. Oxford University Press, Oxford, UK.

Imberger, J. 2013. *Environmental Fluid Dynamics: Flow Processes, Scaling, Equations of Motion, and Solutions to Environmental Flows*. Elsevier – Academic Press, Amsterdam, The Netherlands.

Johnson, R. W. 2016. (Ed.). *The Handbook of Fluid Dynamics*, 2nd edition. CRC Press, Boca Raton, FL.

Kantha, L. H. and Clayson, C. A. 2000. *Small Scale Processes in Geophysical Fluid Flows*. Academic Press, San Diego, CA.

Lamb, H. 1932. *Hydrodynamics*, 6th edition. Cambridge University Press, Cambridge, UK.

Landau, L. D. and Lifshitz, E. M. 1987. *Fluid Mechanics (Course of Theoretical Physics, Volume 6)*, 2nd edition. Elsevier – Butterworth-Heinemann, Burlington, MA.

McWilliams, J. C. 2006. *Fundamentals of Geophysical Fluid Dynamics*, revised edition. Cambridge University Press, Cambridge, UK.

Meyer, R. E. 1971. *Introduction to Mathematical Fluid Dynamics*. Dover Publication, Mineola, New York.

Milne-Thomson, L. M. 1996. *Theoretical Hydrodynamics*, 5th edition. Dover Publications, New York.

Monin, A. S. and Yaglom, A. M. 2007. *Statistical Fluid Mechanics, Volume II: Mechanics of Turbulence*. Dover Publications, New York.

Mueller, J.-D. 2015. *Essentials of Computational Fluid Dynamics*. CRC Press, Boca Raton, FL.

Paterson, A. R. 1983. *A First Course in Fluid Dynamics*. Cambridge University Press, Cambridge, UK.

Pedlosky, J. 1979. *Geophysical Fluid Dynamics*, 2nd edition. Springer, New York.

Pulliam, T. H. and Zingg, D. W. 2014. *Fundamental Algorithms in Computational Fluid Dynamics*. Springer, Cham, Switzerland.

Raisinghania, M. D. 2013. *Fluid Dynamics: With Complete Hydrodynamics and Boundary Layer Theory*, 11th revised and enlarged edition. S. Chand & Company PVT. Ltd., New Delhi.

Renilson, M. 2018. *Submarine Hydrodynamics*, 2nd edition. Springer, Cham, Switzerland.

Salmon, R. 1998. *Lectures on Geophysical Fluid Dynamics*. Oxford University Press, New York.

Samimy, M., Breuer, K. S., Leal, L. G., and Steen, P. H. (Eds.). 2003. *A Gallery of Fluid Motion*. Cambridge University Press, Cambridge, UK.

Streeter, V. L., Wylie E. B., and Bedford, K. W. 1998. *Fluid Mechanics*, 9th revised edition. McGraw-Hill, New York.

Tokaty, G. A. 1994. *A History and Philosophy of Fluid Mechanics*. Dover Publications, New York.

Tritton, D. J. 1988. *Physical Fluid Dynamics*, 2nd edition. Oxford University Press, Oxford, UK.

Vallis, G. K. 2017. *Atmospheric and Oceanic Fluid Dynamics: Fundamentals and Large-scale Circulation*, 2nd edition. Cambridge University Press, Cambridge, UK.

Van Dyke, M. 1982. *An Album of Fluid Motion*, 14th edition. The Parabolic Press, Stanford, CA.

Zikanov. O. 2010. *Essential Computational Fluid Dynamics*. John Wiley & Sons, Hoboken, NJ.

2.3 WAVES

2.3.1 INTRODUCTION

Waves are the most fascinating manifestation of ocean dynamics, and their exploration has occupied scientists for centuries. Wave problems involve many fundamental physical aspects, which are generation, propagation, interaction, topology (geometry), scaling, and energetics. General mechanisms and periods of ocean waves are given in Table 2.2. Roughly speaking, two main categories of ocean waves can be distinguished: surface waves and internal waves (sound waves are not considered here). Origin and spatiotemporal characteristics of ocean waves are quite different; however, they have general properties described by fluid mechanics laws and mathematical unity of the subject matter. Waves are characterized by the repetitive nature of the movement, forced oscillations, and energy transmission. Basic parameters of waves are frequency, magnitude, phase and group velocity, and bandwidth; wave motions exhibit both linear and nonlinear properties and many other features known as dispersion, excitation, decay, focusing, resonance, refraction, diffraction, reflection, caustics, coupling, etc. Therefore, ocean waves can be considered and studied as a part of hydrodynamical or hydro-mechanical systems. Excellent textbooks (LeBlond and Mysak 1978; Lighthill 1978; Whitham 1999) provide a comprehensive description of the science of waves in fluids.

Wave motions and interactions are the main processes posed ocean surface dynamics. John Apel (1930–2001), who was a pioneer in satellite oceanography and

TABLE 2.2

Physical Mechanisms and Periods of Ocean Waves

Wave Type	Physical Mechanism	Period
Capillary waves	Surface tension	<0.1 s
Wind waves	Wind shear, gravity	<15 s
Rogue waves	Nonlinear mechanism	<15 s
Swell	Wind waves	<30 s
Surf beat	Wave groups	1–5 min
Seiche (standing waves)	Meteorological disturbances	2–40 min
Internal waves	Density structure instability	10 min–15 h
Tsunami	Earthquake	10 min–2 h
Tides	Gravitational action of moon and sun, earth rotation	12–24 h
Storm surges	Wind stress and atmospheric pressure variations	1–3 days

detection of ocean waves, wrote very realistically about this: "At the waves continue to increase in length and height, yet another process come to play: nonlinear, finite amplitude wave/wave interactions, in which the quasi-random gravity waves, in scattering off one another, produce both longer and shorter waves. The scattering process can be thought of as being caused by the interactions of four waves, two of which intersect to form an interference pattern that is considered to be a third, virtual wave. The virtual wave then scatters a fourth, real wave toward longer wavelengths. While weak, this process is thought to be largely responsible for the characteristic spectral shape and increasing wavelength of water waves as time goes on" (Apel 1987, pp. 172–173).

In this section, we consider ocean waves as a basic hydrodynamic factor affecting remote sensing measurements. Here we focus on three types of waves: surface waves, internal waves, and tsunami waves. These waves are perfectly detected by optical (and radar) sensors. Models and properties of ocean waves are discussed as well.

2.3.2 Surface Waves

Surface waves are defined as the *undulatory motion* of a water surface. In oceanography, surface waves are periodic deformations of the sea surface. Surface waves are generated under the influence of disturbing forces—wind, atmospheric pressure gradient, currents, tides, body impact, bottom friction, submarine landslides, earthquake, explosion, volcanic eruptions, and gravitational attraction. Restoring forces are normally driven by surface tension, gravity, and the Coriolis force. Surface waves are classified by periods or by frequency (Figure 2.5); correspondently, different types of waves have different propagation characteristics.

The problem of mathematical description of real surface waves remains complex. The early studies are summarized in classical books (e.g., Lamb 1932; Stoker 1958; Shuleikin 1968). Traditionally, wave models concern two aspects: one being hydrodynamic deterministic description and the second being spectral and/or probabilistic description. In hydrodynamic theory, surface waves are considered as

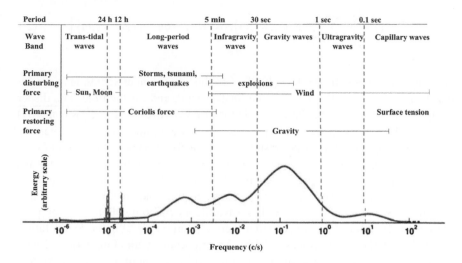

FIGURE 2.5 Classification of surface waves. (Adapted and modified from Kinsman 1984.)

monochromatic waves of finite amplitude, which are steady, regular, and periodic. In statistic theory, the wavy surface is represented by a random field of surface elevations. In both cases, it is necessary to take into account wave transformation in space and time. A wind-driven sea surface is a typical example: surface conditions and wave characteristics are changed in spatiotemporal domain under fairly random (steady or unsteady) wind forcing. The situation similar to the classical stochastic ordering in dynamic system may occur; in this case, a wave field exhibits both arbitrary and regular properties. Thus, statistical and deterministic approaches require for studying ocean surface waves.

2.3.2.1 Deterministic Theory

A history of surface wave studies goes back to the days of Airy (1845), who introduced the linear wave theory known as Airy theory, small amplitude wave, or sinusoidal wave theory. Stokes (1847) developed a perturbation technique offering higher-order wave theories, which are known as the Stokes theories of finite amplitude waves. Stokes' perturbation theories are valid only in deep water or in intermediate depths. Russell (1844) described a solitary-wave phenomenon for the case of extreme shallow water; later Boussinesq (1871) theoretically explained this phenomenon. Korteweg and de Vries (1895) proposed (KdV) equation, which describes weakly nonlinear long waves in shallow water. The KdV equation admits smooth *solitary* and *cnoidal wave* solutions. This theory was fully developed by subsequent researchers in many branches of physics, hydrodynamics, and engineering. Let's consider some elements of deterministic theory.

Surface wave theory assumes that the influences of Coriolis force, viscous friction, buoyancy, and vorticity are negligible. A velocity potential is defined by the Laplace's equation. A solution for a linearized deterministic wave system can be written as a sum of monochromatic waves

$$\xi(\vec{r},t) = \sum_{K_x} \sum_{K_y} \sum_{\omega} a(K_x, K_y, \omega) \cos\left[\vec{K} \cdot \vec{r} - \omega t + \varphi(K_x, K_y, \omega)\right], \qquad (2.1)$$

where $a(\ldots)$ is the amplitude, $\vec{K} = \{K_x, K_y\}$ is the wavenumber vector, ω is the frequency, φ is phase, $\vec{r} = \{x, y\}$ is the coordinate vector, and t is the time. The dispersion relation is given by standard formula for gravity waves $\omega(K) = \sqrt{gK \tanh(KH)}$, where $K = |\vec{K}|$ is wavenumber, H is finite depth of water, and g is the gravitational constant. Equation (2.1) describes resultant surface elevation (surface profile) from a composition of many individual harmonic waves. It is the first-order solution which provides a mathematical basis for the modeling of weakly nonlinear surface waves of the finite amplitude. There are analytic solutions for steady nonlinear waves up to fifth-order approximation for the Stokes wave theory (Fenton 1990); however, practical implementation of the higher-order theory requires direct numerical solutions (Ma 2010; Chalikov 2016; Abbasov 2018). An important result of numerical simulations is that Stokes wave trains of higher-order evolve into nonlinear chaotic system. They (Kharif et al. 2009) called this phenomenon *confined chaos*.

Main achievements of the deterministic wave theory are the following: (1) the establishment of nonlinear solutions for gravity waves on deep water, (2) computations of steep waves, (3) exact nonlinear solutions for pure capillary waves of arbitrary amplitude (known as *Crapper waves*), (4) solution of the Cauchy–Poisson problem (wave generated by oscillating sources), (5) analysis of the Kelvin ship waves, (6) concept of hydrodynamic forces of a body and wave-body interaction described by Kochin's functions; see book (Kochin et al. 1964), (7) nonlinear diffraction of water waves, (8) nonlinear theory of dispersive waves, (9) analysis of wave instability problem, and (10) analytic solitary solution of the KdV equation.

Deterministic wave theory was applied for computing 3D surface waves of finite amplitude (Figure 2.6). Such wave structures are observed in laboratory experiments and may occur at the sea surface in the form of *crescent-shaped (or shoe)* patterns (Ma 2010). They play an important role in wave bifurcation phenomena and wave breaking. The computational implementation advances the instability problem of Stokes' waves, which was a subject of many theoretical studies (Lighthill 1978; Whitham 1999). In particular, the new instability analysis involved resonant interactions of higher order. Deterministic wave models and numerical solutions have been used for estimating extreme surface wave events—rogue waves, freak waves, and long tsunami (Galiev 2015; Pelinovsky and Kharif 2016).

Deterministic wave theory is based on physical principles of classical fluid dynamics that allowed scientists and engineers to achieve remarkable results in many technical areas—hydromechanics and shipbuilding industry. However, this theory does not predict stochastic nature and a variability of ocean environment and cannot provide marine forecast. It is said that Lord Rayleigh remarked, "The basic law of the seaway is the apparent lack of any law" (Kinsman 1984).

2.3.2.2 Stochastic Theory

Wind waves are generated by turbulent airflows that form a random surface. This obvious circumstance requires more elaborate description of the surface in terms

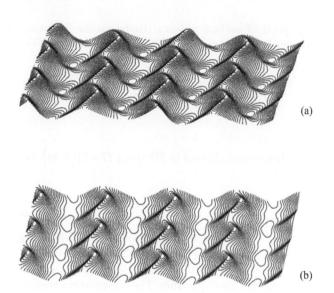

(a)

(b)

FIGURE 2.6   Three-dimensional wave patterns on Stokes waves. Schematic representation. (a) Steady (most unstable) "horse shoe" pattern, (b) unsteady "horse shoe" pattern. (Adapted from Ma 2010.)

of statistical laws rather than deterministic expressions. Statistical theory of wind waves is based on the theory of random processes. Initially, it was the random noise theory, presented by Rice (1944). As the first approximation, random waves are considered as a sum of an infinite number of sinusoidal waves with random phases. Longuet-Higgins (1952, 1975, 1983) derived statistical distributions for wave height, period, and amplitude of random waves. Yaglom (1962) developed a mathematical theory of random processes contributing in the solution of many geophysical problems including rigorous stochastic theory of sea surface waves. Detail description of wave stochastic theory can be found in the book (Ochi 1998).

The statistical theory of ocean surface waves was greatly extended by many researches (Kitaigorodskii 1973; Phillips 1980; Monin and Krasitskii 1985; Komen et al. 1996; Zakharov 1998, 1999; Newman 2017). According to the historical review (Mitsuyasu 2002), the problem included several major topics; among them surface wave generation mechanisms, energy wave spectra, and nonlinear phenomena are the most important issues of the statistical theory. Thus, surface wave generation mechanisms are known as the following: (1) surface wind stress, (2) Kelvin–Helmholtz instability due to local wind shear, (3) Miles shear instability due to the influence of a matching layer with wind profile, (4) resonance mechanism due to nonlinear interactions of gravity waves (Phillips 1980), (5) Zakharov's weak turbulence theory (Section 2.6.3). The dynamics of the surface are described by the wave kinetic equations (Hasselman 1962; Zakharov 1998, 1999), (Section 2.3.2.4) in which the listed mechanisms are taken into account in one way or another. The Hasselman-Zakharov theory also predicts self-similar wave spectra for a wind-driven sea surface (Section 2.6.3).

Statistical description of surface waves is based on the following statement. A stationary, random surface elevation $\xi(\vec{r},t)$ is represented by the Fourier–Stieltjes integral

$$\xi(\vec{r},t) = \iint\limits_{\vec{K}\ \omega} dZ_\xi(\vec{K},\omega) e^{\iota(\vec{K}\cdot\vec{r}-\omega t)}, \tag{2.2}$$

where $dZ_\xi(\vec{K},\omega)$ is differential element in 3D space $\Omega = \Omega(\vec{K},\omega)$. Fourier–Stieltjes coefficients are given by

$$dZ_\xi(\vec{K},\omega) dZ_\xi^*(\vec{K}',\omega') = \begin{cases} 0 & \text{at } \vec{K},\omega \neq \vec{K}',\omega' \\ \Phi(\vec{K},\omega) d\vec{K} d\omega & \text{at } \vec{K} = \vec{K}',\omega = \omega' \end{cases}, \tag{2.3}$$

where $\Phi(\vec{K},\omega)$ is the surface wave spectrum. The wavenumber spectrum, $\psi(\vec{K})$, and the frequency spectrum, $\psi(\omega)$, are, correspondingly

$$\psi(\vec{K}) = \int\limits_{-\infty}^{\infty} \Phi(\vec{K},\omega) d\omega, \tag{2.4}$$

$$S(\omega) = \int\limits_{-\infty}^{\infty} \Phi(\vec{K},\omega) d\vec{K}. \tag{2.5}$$

The relation between frequency spectrum and wavenumber spectrum has the form $S(\omega) = \psi(K)\dfrac{\partial K}{\partial \omega}$. Numerous measurements of wave energy spectra with average wind speeds have been conducted for the validation and possible enhancement of the available models. General form of typical frequency spectrum is given by

$$S(\omega) = A\omega^{-m} \exp\left[-B\omega^{-n}\right], \tag{2.6}$$

in which A, B, m, and n are parameters. Several most known empirical wave spectra are

$$\text{1)} \quad S(\omega) = \alpha g^2 \omega^{-5}, \alpha = 0.0081 \text{ The Phillips spectrum (1957)} \tag{2.7}$$

$$\text{2)} \quad S(\omega) = \alpha g^2 \omega^{-5} \exp\left[-0.74\left(\frac{\omega}{\omega_p}\right)^{-4}\right] \text{The Pierson – Moskowitz spectrum (1964)}$$

$$\tag{2.8}$$

$$\alpha = 0.0081, \omega_p = 0.855 g/U_{10}$$

3) $$S(\omega) = \alpha g^2 \omega^{-5} \exp\left[-\frac{5}{4}\left(\frac{\omega}{\omega_p}\right)^{-4}\right]\gamma^\delta, \delta = \exp\left[-\frac{(\omega-\omega_p)^2}{2\sigma^2\omega_p^2}\right]$$

$$\alpha = 0.076\left(\frac{U_{10}^2}{gX}\right)^{0.22}, \omega_p = 22\left(\frac{g^2}{U_{10}X}\right)^{1/3}, \gamma = 3.3, \sigma = \begin{cases} 0.07 & \omega \le \omega_p \\ 0.09 & \omega > \omega_p \end{cases} \quad (2.9)$$

The JONSWAP spectrum (1973).

Here α is constant, $\omega = 2\pi f$ is the angular frequency (f is the wave frequency in Hertz), ω_p is the peak frequency, U_{10} is the wind speed at a height of 10 m above the sea surface, parameter σ controls the width of the peak, γ is a peak enhancement factor (can vary between 3 and 7), and X is the fetch in meters. Figure 2.7 shows an example of the wave spectrum (Moskowitz 1964).

Spectral models, empirical, and theoretical approximations of surface wave spectra have been developed by many authors: Krylov, Pierson, Moskowitz, Hasselmann, Kitaigorodskii, Phillips, Mitsuyasu, Honda, Leikin, Rosenberg, Huang, Kahma, Donelan, Pierson, Zakharov, and few others. The most popular in ocean remote sensing are known from papers Apel (1994), Elfouhaily et al. (1997), and Zakharov (1999). More information can be found in books (Massel 2013; Komen et al. 1996; Young 1999).

2.3.2.3 Wind-Wave Similarity Concept

Wind is a primarily factor in generation of surface waves (in the absence of other sources). Let's consider an example related to wind fetch. Fetch is the distance that

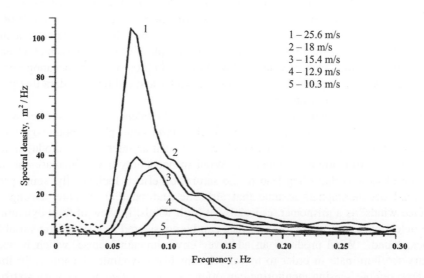

FIGURE 2.7 The JONSWAP wave spectrum at different winds. (Adapted from Moskowitz 1964.)

the wind blows long enough over the surface at the same direction. *Two idealized cases are commonly examined: fetch-limited growth* and *duration limited growth.* The first case occurs when wind of constant speed and direction blows perpendicular to straight coastline; the second case considers the development of wave field from calm to stormy sea as given by the *Beaufort wind force scale.*

Wind-wave dynamics can be investigated using scaling and similarity concept (Kitaigorodskii 1973; Komen et al. 1996; Young 1999; Janssen 2004). Dimensionless relations for wind-generated wave spectrum are given by

$$g^3 F/U^5 = f\left(U\omega/g, gX/U^2\right), g^2 m_0/U^4 = f\left(gX/U^2\right), U\omega_p/g = f\left(gX/U^2\right).$$

$$(2.10)$$

In the case of fully developed sea, the result is

$$g^3 F/U^5 = f\left(U\omega/g, gX/U^2\right), g^2 m_0/U^4 = \text{const}, U\omega_p/g = \text{const}. \quad (2.11)$$

Here $F(\omega)$ is the frequency spectrum, m_0 and ω_p are the total variance and the peak frequency, U is the wind speed, t is the duration of wind, and f denotes a general function. For gravity waves dimensionless parameters are the following: wave height $\tilde{H} = gH/U_{10}^2$, wave period $\tilde{T} = gT/U_{10}$, wave spectrum peak frequency $\tilde{\omega}_p = U_{10}\omega_p/g$, energy per unit area $\tilde{e} = g^2 m_0/U_{10}^4$, the total variance $\tilde{m}_0 = m_0 g^2/U_{10}^4$, fetch $\tilde{X} = gX/U_{10}^2$, wind duration $\tilde{t} = gt/U_{10}$, and some others (U_{10} is standard wind speed at 10 m height; water depth is not considered here).

Wind-wave analysis is based on empirical relationships between dimensionless characteristics, defined from *in situ* field and/or laboratory measurements. In particularly, the power-law relationships $\tilde{e} = A\tilde{X}^a$ and $\tilde{\omega}_p = B\tilde{X}^b$ have been established for a wide range of the dimensionless fetch, $10^2 < \tilde{X} < 10^4$. A summary of the coefficients A, a, B, and b is given in Table 2.3 (Babanin and Soloviev 1998; Hwang 2006). These data correspond to shape parameters of the Pierson–Moskowitz spectrum and the JONSWAP spectrum. Figure 2.8 shows an example of dimensionless plot. Ocean wave databases are constantly updated worldwide; in present *in situ* oceanographic measurements are combined with microwave (radar and radiometer) data that provide more representative and useful results.

There is a considerable amount of experimental evidence that the surface wave dynamics is defined not only by wind forcing (and atmospheric boundary layer action) but also by many hydrodynamic causes such as interactions, modulations, instabilities, bifurcations, currents, etc. Wind initiates only trivial increases of sea surface elevations that often lead to the saturation effect. Incidentally, some professional oceanographers assume that airborne and satellite "remote sensing of surface winds" is a misnomer, especially at high winds. In view of fluid dynamics and detection, a wind-driven sea surface can be considered as a stochastic variable "background." Wind produces an additive "environmental" noise, which is necessary to illuminate in order to reveal relevant hydrodynamic signatures. In this context, remote sensing monitoring can offer valuable and complimentary statistical oceanographic information. Numerous literature recourses are available to get

TABLE 2.3
Coefficients for Dimensionless Fetch-growth Laws

Source	A	a	B	b
JONSWAP	1.6×10^{-7}	1.00	21.98	−0.33
Bothnian Sea (unstable)	3.6×10^{-7}	1.00	19.97	−0.33
Lake Ontario	8.42×10^{-7}	0.76	11.62	−0.23
North Atlantic	1.27×10^{-7}	0.75	10.68	−0.24
Lake St. Clair	2.60×10^{-7}	0.95	17.59	−0.30
Composite stable	9.30×10^{-7}	0.77	12.00	−0.24
Composite unstable	5.40×10^{-7}	0.94	14.00	−0.28
Composite mixed	5.20×10^{-7}	0.90	13.70	−0.27
Average	7.50×10^{-7}	0.80	12.56	−0.25
Fetch-dependent growth rate	6.19×10^{-7}	0.81	11.86	−0.24

Source: After Hwang (2006).

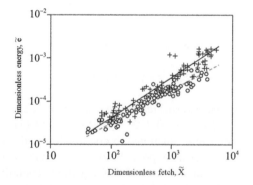

FIGURE 2.8 Dimensionless energy vs. dimensional fetch. (Adapted from Janssen 2004.)

comprehensive knowledge about wind-wave dynamics and methods of data analysis (see our notes on the literature).

2.3.2.4 Wave Kinetic Equation

The wave kinetic (WK) equation describes the dynamics of wave spectrum in weakly nonlinear and dispersive media. The WK equation has been derived in the 1960s (Hasselman 1962; Zakharov and Filonenko 1966; Zakharov 1998). The WK equation provides prediction of a wide class of wave phenomena associated, first of all, with energy cascade and transformations in sea wind-wave spectra including wave breaking as well (Section 2.6). Over the past four decades, several dozen scientific works have been published on this subject; their review is not an aim of this section. However, the problem is important for ocean remote sensing. Here is its brief description.

The classical Hasselmann's WK equation is given by

$$\frac{\partial N(\vec{K}, \vec{r}, t)}{\partial t} = S_{nl} + S_{input} + S_{diss}, \tag{2.12}$$

where $N(\vec{K}) = gF(\vec{K}, \vec{r}, t)/\omega(K)$ is the spectral density of wave action related to nonsymmetric wave spectrum $F(\vec{K}, \vec{r}, t)$ ($\vec{K}$ is the wavenumber vector and $\vec{r}$ and t are "slow" spatial and temporal variables) with dispersion relation $\omega(K) = \left[(gK + \sigma K^3) \tanh(Kh) \right]^{1/2}$, σ is the coefficient of surface tension, h is finite depth of water, g is the gravitational constant, ω is frequency, and $K = |\vec{K}|$ is the wavenumber. The right-hand side consists of three terms related to nonlinear interactions, S_{nl}, energy source, S_{input}, and energy sink (dissipation), S_{diss}. According to the Zakharov theory, these terms can be defined as follows:

$$S_{nl} = 2\pi \iiint |T_{0123}|^2 (N_1 N_2 N_3 + N_0 N_2 N_3 - N_0 N_1 N_2 - N_0 N_1 N_3)$$
$$\times \delta(\omega_0 + \omega_1 - \omega_2 - \omega_3) \delta(K_0 + K_1 - K_2 - K_3) dK_1 dK_2 dK_3, \tag{2.13}$$

$$S_{input} = \gamma_{input}(K) N(K), \tag{2.14}$$

$$S_{diss} = -\gamma_{diss}(K) N(K). \tag{2.15}$$

In (2.13) T_{0123} is the four-wave kernel (coupling coefficients) dependent on wavenumbers K_0, K_1, K_2, K_3 and $N_j = N(K_j)$, $j = 0, 1, 2, 3$. In (2.14) and (2.15) coefficients $\gamma_{input}(K)$ and $\gamma_{diss}(K)$ are some empirical functions. In contrast to S_{nl}, the knowledge of S_{input} source and S_{diss} dissipation terms is poor; both include many heuristic factors and coefficients. There are several solutions of WK equation depending on the configuration and specification of the left-hand side terms; for details see (e.g., Monin and Krasitskii 1985; Zakharov 1998; Janssen 2004). However, what is more important is that the Zakharov weak turbulent theory based on the solution of the WK equation (particularly, its stationary solution) demonstrated that the effective four-wave nonlinear interaction plays the leading role in the formation of the spectra of surface waves. One of the first numerical examples of spectral energy balance is shown in Figure 2.9.

More realistic scenarios involve more than four interacting surface waves. Multiple surface wave systems are observed and registered by high-resolution space-based optical imager. However, the theory of multi-wave resonant interactions remains open, due to the complexity of interaction kernels and rigorous mathematical solutions of the nonlinear WK equation. Note that only five-wave interaction kernel has been derived for the case of surface gravity waves (Krasitsky 1994; Krasitskii and Kozhelupova 1995). Nevertheless, we believe that the multi-wave interaction problem can be formulated and solved numerically using nonlinear Hamiltonian evolution equations for surface gravity waves. Some preliminary results have been reported

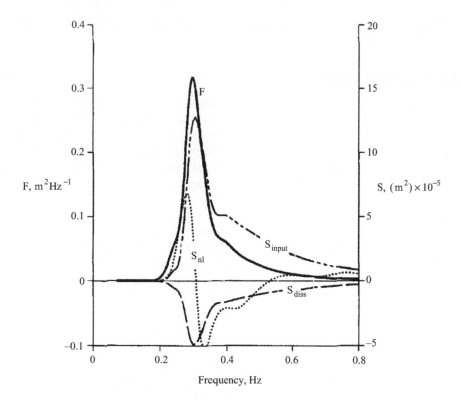

FIGURE 2.9 Calculated spectral energy balance for a JONSWAP spectrum with maximum $f_p = 0.3\,\text{Hz}$. Wind speed $U = 10\,\text{m/s}$. (Adapted from Masson and Leblond 1989.)

by Annenkov and Badulin (2001). In common case, synchronism between n-wave components must satisfy conditions: $\sum |\vec{K}_n| = 0$, $\sum \omega(|\vec{K}_n|) = 0$ (Section 7.3).

An advanced WK equation related to oceanographic forecasting systems is given by (Komen et al. 1996)

$$\left[\frac{\partial}{\partial t} + (c_g + \vec{U})\frac{\partial}{\partial \vec{r}} - \nabla\Omega\frac{\partial}{\partial \vec{K}}\right]N(\vec{K}) = S_{nl} + S_{input} + S_{diss} + S_{bt} + ..., \quad (2.16)$$

where the left-hand side represents evolution of the wave action density causing by hydro-physical and/or atmospheric processes. Here $\vec{U}$ is the surface current, $\Omega(\vec{K})$ is the Doppler-shifted frequency $\Omega(\vec{K}) = \omega(K) + \vec{K}\cdot\vec{U}$, and ∇ denotes the divergence operator. The right-hand side includes an additional term S_{bt} which is decay due to bottom friction. It is possible to add more terms here describing, e.g., shallow-water processes, swell dissipation, moving sediments, oil pollutions, etc. Although the WK Equation (2.16) represents a complicated enough simulation platform, it could be referred to the class of primitive mathematical models, which are used for estimating statistical oceanographic databases and forecasting purposes. In some cases, the WK equation may be considered as an alternative approach to the Navier–Stokes equations. But for the modeling of *local* hydrodynamic processes CFD is more preferable technique than WK theory.

2.3.3 INTERNAL WAVES

Internal waves (IWs) are waves that travel within an interior of the fluid. The books (Turner 1973; Lighthill 1978; Phillips 1980; Marchuk and Kagan 1984; Miropol'sky 2001; Vlasenko et al. 2005; Vladimirov and Bulatov 2007; Sutherland 2010; Massel 2015; Morozov 2018) provide comprehensive information about nature of IW in the stratified ocean. Some theoretical and experimental aspects of the problem will be outlined here with the emphasis on remote sensing capabilities to detect IW phenomena.

2.3.3.1 A Brief History

An old phenomenon relating to IWs is "dead water." Dead water was observed first by F. Nansen during Pridtjof Nansen's polar expedition with "Fram" in 1893–1896. A boat experienced strong resistance to forward motion in apparently calm conditions. Ekman in 1904 was first researcher who has investigated the problem of IWs mathematically. He introduced several key oceanographic terms in the context with IW phenomenon, namely, Ekman transport, Ekman spiral, Ekman layer, and several others. Until the 1960s, the research of IWs was focused on internal tides; in the 1950s and 1960s theoretical and experimental studies were developed in order to explore parameters of IWs. An important result of *in situ* measurements was the establishment of a universal frequency spectrum of IWs introduced by Garrett and Munk (1972, 1975) and known as the Garrett–Munk 79 (GM79) spectrum. The GM79 spectrum remains widely used in present in oceanography.

Remote sensing capabilities with the advent of aircraft and satellite imagery have opened a new era in the studying IWs. For example, during the Apollo–Soyuz mission in 1975 the crew provided first handmade optical photography of IW manifestations from space. John Apel, who was a pioneer in the detection of IWs reported in 1979: "At least three Apollo-Soyuz Test Project photographs have shown clear indications of oceanic internal gravity waves; the features are indicated by periodic changes in the optical reflectivity of the ocean surface overlying the waves. Two packets (or groups) of waves seen off Cádiz, Spain, have characteristics similar to internal waves seen in satellite images taken off the U.S. eastern coast. In the Andaman Sea off the Malay Peninsula, several groups were observed having wavelengths of 5 to 10 km and interpacket separations on the order of 70 to 115 km. If these are indeed surface signatures of internal waves, the waves are among the longest and fastest observed to date. Earlier shipboard measurements have shown that large amplitude internal waves exist in the area" (Apel 1979). Later from 1980s and up to now large bodies of airspace radar and optical images were obtained and analyzed to explore geophysical characteristics of IWs across the globe. Atlas (Jackson and Apel 2004) is a great collection of world satellite images of oceanic IWs.

2.3.3.2 Basic Description

Internal gravity waves, known also as tides, are waves that are formed in the water column of the oceans at different depths. They exist when the water body consists of layers of different density. This difference in water density is mostly due to a difference in water temperature, but can also be due to a difference in salinity. Often the density structure of the ocean can be approximated by two layers. The interface between layers of different densities is called pycnoline. When the density difference is due to temperature, it is called thermocline, and when it is due to salinity, it is called halocline. Large IWs are generated when ocean tidal currents interact with bottom topography, resulting in internal tides, or IWs of tidal frequency, that propagate throughout the world's oceans.

The tidally generated IWs are usually highly nonlinear and form wave groups or packets of oscillations. The amplitude of large IWs can exceed 50 m and their wavelength can be from a few hundred meters to several kilometers. Theoretically, these highly nonlinear IWs are described in terms of solitons (the term "soliton" was introduced in the 1960s, but the scientific research of solitons had started when John Scott Russell observed a large solitary wave in a canal near Edinburgh in 1834). Usually IW packet consists of several solitons.

In mathematical literature, there are a number of nonlinear equations which have *solitary* solutions. The most known in oceanography are the *Korteweg–deVries* (KdV) equation and the *Benjamin–Ono* (BO) equation. The canonical form of the KdV equation is given by

$$\frac{\partial u}{\partial t} + 6u\frac{\partial u}{\partial x} + \frac{\partial^3 u}{\partial x^3} = 0, \tag{2.17}$$

where $u = u(x,t)$ is the vertical displacement. The factor "6" in the second term is a scaling factor to make solutions easier to describe.

The canonical form of the BO equation is given by

$$\frac{\partial u}{\partial t} + u\frac{\partial u}{\partial x} + H\left[\frac{\partial^2 u}{\partial x^2}\right] = 0, \ H[f(x)] := \frac{1}{\pi}PV\int_{-\infty}^{\infty}\frac{f(y)}{y-x}dy, \tag{2.18}$$

where H is the Hilbert Transform (PV stands for Principle Value).

The KdV equation describes weakly nonlinear solitary waves in shallow water, whereas the BO equation describes solitary waves in deep stratified fluids with finite depth. However, in the case of the OB equation, wave structure is substantially more complicated (that includes shock wave solution as well) than in the case of the KdV equation having simple analytic solution in form $u(x,t) = \frac{c}{2}\operatorname{sech}^2\left(\frac{\sqrt{c}}{2}(x - x_0 - ct)\right)$, with variable parameters; for more detail see (e.g., Ostrovsky and Stepanyants 1989). There are other forms of solitary wave solutions (Grimshaw 2007).

An oscillation of a vertically displaced parcel within a statically stable environment is described by the *Brunt–Väisälä* frequency, or buoyancy frequency, which is defined as

$$N^2(z) = -\frac{g}{\rho}\frac{d\rho}{dz}, \tag{2.19}$$

where $d\rho/dz$ is the density gradient, $\rho(z)$ is the potential density depending on both temperature and salinity of water, g is the local acceleration of gravity, and z is geometric height. The Brunt–Väisälä frequency describes a competition between gravity and buoyancy; it is the principal parameter characterizing the behaviour of IWs in stratified ocean.

Miropol'sky (2001) gives the following very clear description (Figure 2.10). The density $\rho(z)$ in the upper ocean layer is almost constant. This layer is called quasi-homogeneous and its depth $h = 20–100\,m$. Below the quasi-homogeneous layer $(-h > z > -H)$ is a layer of a seasonal thermocline or a density jump layer in which density grows as a result of decrease in temperature. The lower border of this layer is $H \approx 50–200\,m$. Below the seasonal thermocline is a layer of mean thermocline with the lower border $H_0 \approx 1{,}000\,m$, where density increases continuously owing to the fall in temperature. Below the mean thermocline layer $z < -H_0$ the sea water temperature is almost constant and is at $4°C–5°C$, which causes almost constant density $\rho(z)$. In accordance with the density distribution described a variation of the Brunt–Väisälä frequency $N(z)$ occurs (Figure 2.10). The largest variations of $N(z)$ are observed in the density jump layer, where a pronounced maximum of $N(z)$ usually takes place. An average value of this maximum in the ocean is $2 \times 10^{-2}\,s^{-1}$.

2.3.3.3 Basic Mechanisms

The various mechanisms of IW generation in the ocean are known. In general, they are associated either with moving (oscillating) solid bodies, or bottom topography (bathometry), or with disturbances within the stratified fluid such as convection, flows, currents, and turbulence. Here are several important mechanisms.

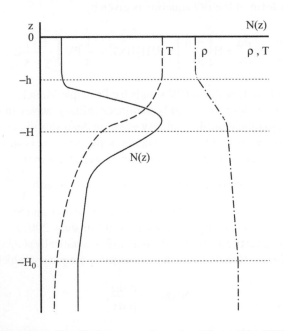

FIGURE 2.10 The Brunt–Väisälä frequency. (Based on Miropol'sky 2001.)

- *Centrifugal and gravitational forces* of moon, sun, and some other planets generate tides. The most significant of these is the principle lunar tide with approximately half-day period (called also "semi-diurnal lunar tide"). The tide effectively is a shallow IW and barotropic.
- *Flow over bottom topography* generates disturbances which usually take place when the fluid moves over the obstacle located on the ocean floor. The obstacle may represent as an isolated single hill as well periodical hill of variable amplitude. The theory provides estimates of parameters of IWs depending on configuration (amplitude and period) of the obstacle.
- *Tidal flow over bottom topography* generates IWs when oscillating barotropic flow of a density-stratified fluid moves over underwater hill called "sill." This generation mechanism is referred to as "baroclinic conversion" because it transforms the energy of barotropic tides (in which the entire ocean column oscillates) to baroclinic IWs. Through other processes, the large-scale IWs convert the energy to smaller scales, ultimately dissipating through mixing. For tidally generated IWs, the effects of Coriolis forces are crucial.
- *Wind forcing and/or atmospheric instability* can generate IWs due to air-sea interactions that may trigger oscillations of the inertial current at the mixed upper layer of the ocean. Such IWs are observed mostly in continental shelf under the influence of fluctuations of atmospheric pressure.
- *Generation by surface* (also known as *spontaneous creation*) is associated with resonant coupling of high-frequency gravity surface waves which can produce a relatively low-frequency internal wave (Brekhovskikh and Goncharov 1994).
- *Ekman layer instability* mechanism (mostly related to atmospheric IWs) supposes that instability of turbulent flow in stratified Ekman boundary layer can also generate IWs in the ocean.
- *Oscillating body* in a uniformly stratified fluid creates a cross-shaped pattern of IWs.
- The structure of the wave beams generated by a vibrating cylinder is a superposition of plane waves having different spatial structure but identical frequencies. The IW pattern is sometimes referred to as a *St. Andrew Cross* (Sutherland 2010).
- *Collapsing submerged turbulent wakes* can be efficient generators of IWs due to turbulent mixing and density fluctuations. These wave-type disturbances can propagate as in horizontal as well in vertical directions. The internal stratified layer may transmit the disturbances from the depth to the surface causing the changes in surface roughness. This event is detectable by the optical sensor under appropriate conditions.

The exact mechanisms for generation are not yet established and they are still a subject of theoretical and experimental research. In particular, the study of stochastic nonlinear dynamics of solitons, their differences and similarities is of fundamental importance for many geophysical and remote sensing applications.

2.3.3.4 Surface Manifestations

A great example of descriptive analysis of surface manifestations can be found in the atlas of satellite images (Jackson and Apel 2004) and we refer to the following summary:

- The solitons are ubiquitous in the ocean, appearing wherever the proper combination of density gradient, current flow, and bathymetry occur.
- The generation process launches undulatory internal bores on each semidiurnal tide, with significant modulations on diurnal, fortnightly, seasonal, and semiannual time scales.
- The solitons are produced via lee wave formation, shear flow instability, or scattering of barotropic modes into internal baroclinic modes at locales close to rapidly shoaling depths that protrude into the pycnocline. The exact mechanisms are not yet clear.
- They occur in packets, usually rank-ordered, with the largest, fastest, greatest-wavelength, and longest-crested oscillations appearing at the packet front, which then slowly decay to smaller-amplitude, reduced-wavelength, and shorter-crested oscillations at the rear. The orientation of packet wavefronts in the horizontal plane is controlled by refraction and to a lesser extent in narrow straits, by diffraction.
- The solitons are the leading edge of an undulatory internal tidal bore generated by tidal flow over banks, sills, and continental shelf breaks. The distance between successive bores is the internal tidal wavelength on the continental shelf. As the packet approaches, each bore is characterized by a sharp drop in pycnocline depth approximately equal to the soliton amplitude, a sequence of nonlinear oscillations that decrease in amplitude from front to rear, and a slow recovery of the depression over several hours. Wavelengths are longest at the packet front and shortest at the rear. All wavelengths increase logarithmically as time goes on, due to the higher speeds of larger solitons.
- They are dissipated by radial spreading, bottom interactions, instability, and fluid turbulence. Lifetimes are the order of a few days in the open sea and a day or soon the shelf.

Ultimately, this is one of the best descriptive evaluations of IW surface manifestations currently available.

2.3.3.5 A Concluded Note

In the early stage of the research, model calculations based on the KdV and BO equations and their combinations (or modifications) have been successfully used for description of dominant characteristics of IWs in context with the known hydrodynamic mechanisms. However, collected over the years remotely sensed data demonstrated more complicated picture than predicted. It also becomes clear that the obtained analytic and/or asymptotic theoretical solutions may not be able to explain adequately spatiotemporal dynamics and strongly nonlinear structure of real world IWs. We believe that only CFD methods enable to provide complete theoretical analysis of nonlinear solitary IWs such as shown, e.g., in optical images, Figure 1.15 (Section 1). Numerical modeling and simulations of complex scenarios will create a solid scientific basis needed for sophisticated applications.

2.3.4 TSUNAMI

The term "tsunami," also known as a sea seismic wave, is a Japanese word with the English translation "harbor wave." A tsunami is caused by sudden movements such as an earthquake, volcanic eruption, landslide, asteroid impact, or underwater explosion. A tsunami happens when the sea floor is rapidly displaced, uplifting a column of ocean water. This event forms fast moving, long surface waves of initial low amplitude (with the surface elevation <1 m) that can travel with the speed at well over 800 km/h (500 mph) for long periods of time for distance of thousands of kilometers across the open ocean.

Many books (Pelinovsky 1982, 1996; Grue and Trulsen 2006; Kundu 2007; Liu et al. 2008; Bryant 2008; Singh 2009; Levin and Nosov 2009; Constantin 2011; Joseph 2011) and book's papers (Pelinovsky 2006; Geist et al. 2017; Rabinovich et al. 2018) cover various aspects of tsunami studies including chronology of tsunami, hydrodynamic theory, analysis of experimental data, management, modeling, and new methods of tsunami warning and detection. Remote sensing (active/passive) capabilities are discussed in books (Murty et al. 2006; Marghany 2018) and also in special works (Godin 2004; Myers 2008). These and other materials demonstrate that the science of tsunami remains the greatest global challenge of our time. Below we briefly outline the main results of the theory and practice which are important for remote sensing.

2.3.4.1 Accessibility of Wave Theories to Tsunami

The problem of tsunami propagation is a special case of the general fluid mechanics problem. Although the basic governing equations have been known for over 150 years, the development of modeling tools have begun since the 1980s with the invoking numerical hydrodynamics in the 1990s. There are three major aspects of tsunami dynamics: (1) tsunami generation, (2) tsunami propagation, and (3) tsunami run-up and inundation. Today, more than ten different models are used for investigating tsunami events including the amplitude growth and collapse of the surface waves on continental shelf. The best known models are the following:

The shallow-water equations (SWE), also known as linear shallow-water wave theory or long-wave theory, describe the propagation of a long surface gravity wave of small amplitude on a homogeneous fluid layer. The SWE are derived from the Navier–Stokes equations. In canonical form, the SWE is given by

$$\xi_t + \nabla \cdot (\xi \vec{u}) = 0, \ \vec{u}_t + (\vec{u} \cdot \nabla)\vec{u} + \nabla \xi = 0, \tag{2.20}$$

where ξ is the normalized height of the free surface (i.e., water depth) and $\vec{u}$ is the horizontal velocity field. The SWE are solved numerically using finite-difference technique. The SWE is a good model for tsunami waves in the open ocean, in the continental slope, around island, and harbor. But SWE have limitations—they do not describe vertical velocity components and the behaviour of short-wave tsunami including breaking process.

Nonlinear shallow-water equations (NSWE) describe the propagation of a tsunami wave in very shallow water, where nonlinear effects and wave breaking become important. In canonical form, the 1D NSWE are given by (Liu et al. 2008)

$$\left[u(\xi+h)\right]_x + \xi_t = 0, \; u_t + uu_x + \xi_x = 0, \tag{2.21}$$

where $u(x,t)$ is the horizontal depth-averaged velocity, $\xi(x,t)$ is the free-surface elevation and $h(x) = x$ is the undisturbed water depth. The subscripts denote derivatives with respect to that variable. Here the nondimensional variables $\{u,\xi,h,t\}$ are functions of characteristic scales and the beach angle β from the horizontal. An important application of the NSWE is modeling of tsunami wave breaking in the shore (sloping beach). Various analytic approaches are considered in this context; however, the most adequate studies use 2D and 3D NSWE and high-performance computing for predicting and forecasting tsunami with interactive visualization of the result (Tiampo et al. 2008). Detailed analysis the NSWE solutions can be found in the book (Abbasov 2018).

Boussinesq (1872) and Boussinesq-type equations and models (BTE and BTM) is a family of shallow-water propagation models, describing nonlinear dispersive waves. Since pioneering work (Peregrine 1967) many efforts have been made to investigate BTE with full dispersion and nonlinearity including the application for tsunami waves as well. Here we mention only an ordinary 1D Boussinesq equation which is given by (Kundu 2007)

$$u_t + uu_x + g\xi_x - \frac{1}{3}h^2 u_{txx} = 0, \; \xi_t + \left[u(h+\xi)\right]_x = 0. \tag{2.22}$$

Here u is the velocity, ξ is the free-surface elevation and, h is the undisturbed water depth. The fully nonlinear and highly dispersive BTE are more complicated (Liu et al. 2008) and are analyzed numerically only.

In recent years, the improved BTM allow the researches to model evolution of both nondispersive and dispersive tsunami waves from weak to high nonlinearity. BTM predict tsunami wave breaking and energy dissipation with a better accuracy that corresponds to experimental data and real representation of these processes. The BTE and BTM are also successfully used in coastal engineering.

Solitary wave theory (SWT) is based on the analytic solution of the KdV Equation (2.17) under the assumption of horizontal bottom (H<<h) and 1D geometry. The solitary wave solution describes surface elevation

$$\xi(x,t) = H \operatorname{sech}^2\left(\frac{x-ct}{\Delta}\right), \; c = \left(1+\frac{H}{2h}\right)\sqrt{gh}, \; \Delta = \sqrt{4h^3/3H}, \tag{2.23}$$

where H is the wave height, h is the water depth, and c is the wave celerity. SWT has been used to model the evolution of tsunami associated with transformation of long waves into shorter waves when the front of the tsunami wave becomes sufficiently steep. In particular, the SW solutions describe profile deformations of the nearshore surface wave train (Madsen et al. 2008).

2.3.4.2 Tsunami Wave Parameters

In accordance with the linear and nondispersive wave equations, properties of tsunami waves are defined by the following main parameters:

$$\text{Phase and group velocity} \quad c = c_g = \sqrt{gh} \tag{2.24a}$$

$$\text{Energy density} \quad E = \frac{1}{2}\rho g a^2 \tag{2.24b}$$

$$\text{Energy flux} \quad F_e = E c_g \tag{2.24c}$$

$$\text{Wave period} \quad \lambda = \lambda_0 \left(\frac{h}{h_0}\right)^{1/2} \tag{2.24d}$$

$$\text{Wave amplitude} \quad a = a_0 \left(\frac{h_0}{h}\right)^{1/4} \tag{2.24e}$$

$$\text{The maximum run-up of N-waves} \quad H_{\text{Nwave}} = 3.86 (\cot\beta)^{1/2} H_b^{5/4}. \tag{2.24f}$$

Here H_{Nwave} is the maximum run-up height of tsunami above sea level (m), H_b is the wave height at shore or the toe of a beach (m), β is the slope of the seabed (degrees), ρ is the density, g is the gravitational acceleration (m/s^2), and λ_0, h_0, a_0 are initial parameters (constants). There are other practically important parameters, characterizing the tsunami resonance in the bay/harbor, coastal inundation, inland penetration, depth and velocity in shore (Bryant 2008).

2.3.4.3 A Concluding Note

Advanced research of the tsunami dynamics is associated with the 3D CFD simulations of tsunami wave breaking along a coast. It is a tsunami hazard zone. An important problem in these computations is the hydrodynamic collapse of tsunami waves on the bottom obstacles. The effect of tsunami breakdown depends on shape of the bottom and/or slope of the bottom. The tsunami wave run-up onto a shore represents strongly nonlinear process with the generation of large-scale multiple-vortex tornado. This causes very intensive turbulent fluid mixing at the air-water interface. Two-phase transition and turbulent mixing can be incorporated into the CFD simulations using a generalized Navier–Stokes solver for multi-phase flows and sediment transport model. The calculations and laboratory experiments demonstrate strong dependence of the vortex structure on the submerged obstacle/barrier geometry (Geist et al. 2017; Rabinovich et al. 2018). Although the state-of-the-art knowledge provides successfully forecast of tsunami events across the world oceans, model developments are constantly refined with every new tsunami hazard constituting a new benchmark.

Optical imagery from space gives us a unique picture of tsunami run-up onto shore (Figure 2.11). Stochastic fields of tsunami wave breaking and generations of

FIGURE 2.11 **(See color insert.)** Satellite image. Tsunami run-up onto shore, southwestern coast of Sri Lanka. December 26, 2004. (Image from Digital Globe.)

dispersed vortexes are clearly visible in the picture. Eventually, space-based tsunami early warning system should include high-resolution optical imagery which seems to be more informative than radar, altimeter, or microwave radiometer. For this goal, operational real-time monitoring is required.

2.3.5 NOTES ON THE LITERATURE

Principles of physical oceanography, marine hydrodynamics, and sea water properties are considered in many textbooks (Shuleikin 1968; Phillips 1980; Pond and Pickard 1983; Monin and Krasitskii 1985; Apel 1987; Dera 1992; Steele et al. 2010; Olbers et al. 2012; Pedlosky 2013; Thomson and Emery 2014; Knauss and Garfield 2017; Newman 2017).

Basic course of descriptive physical oceanography is available in books (Pickard and Emery 1990; Reddy 2001; Talley et al. 2011).

Mathematical theory of water waves can be found in many books (Stoker 1958; Elmore and Heald 1969; Le Mehaute 1976; Lighthill 1978; Johnson 1997; Whitham 1999; Grue and Trulsen 2006; Ostrovsky 2014).

Ocean wave mechanics is considered in books (Falnes 2002; Sundar 2016).

Linear and nonlinear water waves, interactions, and instabilities are described in books (Yuen and Lake 1982; Debnath 1994; Osborne 1991, 2010; Grue et al. 1996; Engelbrecht 1997; Infeld and Rowlands 2000; Kuznetsov et al. 2002; Voliak 2002).

Numerical models and computer simulations of ocean waves are presented in books (Kantha and Clayson 2000; Mader 2004; Lin 2008; Ma 2010; Chalikov 2016; Abbasov 2018).

Wind-generated surface waves, theory and experiments are discussed in books (LeBlond and Mysak 1978; Phillips and Hasselmann 1986; Kinsman 1984; Mei 1989; Komen et al. 1996; Young 1999; Lavrenov 2003; Janssen 2004; Holthuijsen 2007; McCormick 2007; Massel 2013; Galiev 2015).

Rogue and extreme ocean waves are studied in books (Grue and Trulsen 2006; Kundu 2007; Levin and Nosov 2016; Constantin 2011; Joseph 2011; Galiev 2015; Pelinovsky and Kharif 2016).

Coastal and nearshore hydrodynamics are discussed in books (Massel 1989; Svendsen 2006; Mani 2012).

Boundary layer, air-sea interactions, and ocean-atmosphere dynamics are reviewed in books (Kitaigorodskii 1973; Gill 1982; Fedorov and Ginsburg 1992; Kraus and Businger 1994; Soloviev and Lukas 2014).

Thermohaline structure and double-diffusive convection in the ocean are described in books (Fedorov 1978; Radko 2013).

REFERENCES TO SECTION 2.3

Abbasov, I. B. 2018. *3D Modeling of Nonlinear Wave Phenomena on Shallow Water Surfaces*. John Wiley & Sons, Hoboken, NJ.

Airy, G. B. 1845. Tides and waves. In *Encyclopedia Metropolitana*. (Eds. H. J. Rose et al.). John Joseph Griffin & Co, London. Vol. 5, pp. 241–396.

Annenkov, S. Y. and Badulin S. I. 2001. Multi-wave resonances and formation of high-amplitude waves in the ocean. In *Rogue Waves 2000: Proceedings of a Workshop Organized by Ifremer and Held in Brest, France, 29–30 November 2000 within the Brest Sea Tech Week 2000* (Eds. M. Olagnon and G. Athanasoulis). IFREMER, Plouzané, France, Vol. 32, pp. 205–214. Available on the Internet http://www.ifremer.fr/web-com/molagnon/bv/annenkov.pdf.

Apel, J. R. 1979. Observations of internal wave surface signatures in ASTP photographs. In *Apollo-Soyuz Test Project. Summary Science Report*. Scientific and Technical Information Branch NASA, Washington, Vol. II, NASA SP-412, pp. 505–509. Available on the Internet https://history.nasa.gov/astp/documents/summary%20science%20report%20sp-412%20v2.pdf.

Apel, J. R. 1987. *Principles of Ocean Physics (International Geophysics Series, Vol. 38)*. Academic Press, London, UK.

Apel, J. R. 1994. An improved model of the ocean surface wave vector spectrum and its effects on radar backscatter. *Journal of Geophysical Research*, 99(C8):16269–16291. doi:10.1029/94JC00846.

Babanin, A. V. and Soloviev, Y. P. 1998. Field investigation of transformation of the wind wave frequency spectrum with fetch and the stage of development. *Journal of Physical Oceanography*, 28(4):563–576.

Boussinesq, J. 1871. Théorie de l'intumescence liquide, appelée onde solitaire ou de translation, se propageant dans un canal rectangulaire. *Comptes Rendus de l'Académie des Sciences*, 72:755–759.

Brekhovskikh, L. M. and Goncharov, V. 1994. *Mechanics of Continua and Wave Dynamics*, 2nd edition. Springer, Berlin, Germany.

Bryant, E. 2008. *Tsunami: The Underrated Hazard*, 2nd edition. Springer, Praxis Publishing, Chichester, UK.

Chalikov, D. V. 2016. *Numerical Modeling of Sea Waves*. Springer, Switzerland.

Constantin, A. 2011. *Nonlinear Water Waves with Applications to Wave-Current Interactions and Tsunamis*. Society for Industrial and Applied Mathematics (SIAM), Philadelphia, PA.

Debnath, L. 1994. *Nonlinear Water Waves*. Academic Press, San Diego, CA.

Dera, J. 1992. *Marine Physics*. Elsevier Science Publishing, New York.

Elfouhaily, T., Chapron, B., Katsaros, K., and Vandemark, D. 1997. A unified directional spectrum for long and short wind-driven waves. *Journal of Geophysical Research*, 102(C7):15781–15796.

Elmore, W. C. and Heald, M. A. 1969. *Physics of Waves*. Dover Publication, New York.

Engelbrecht, J. 1997. *Nonlinear Wave Dynamics: Complexity and Simplicity*. Kluwer Academic Publishers, Dordrecht, Boston, London.

Falnes, J. 2002. *Ocean Waves and Oscillating Systems: Linear Interactions Including Wave-Energy Extraction*. Cambridge University Press, Cambridge, UK.

Fedorov, K. N. 1978. *The Thermohaline Finestructure of the Ocean*. Pergamon Press, Oxford, UK.

Fedorov, K. N. and Ginsburg, A. I. 1992. *The Near-Surface Layer of the Ocean (translated from Russian by M. Rosenberg)*. CRC Press, Boca Raton, FL.

Fenton, J. D. 1990. Nonlinear wave theories. In *The Sea, Vol. 9: Ocean Engineering Science* (Eds. B. Le Méhauté and D. M. Hanes). John Wiley & Sons, New York, pp. 1–17.

Galiev, S. U. 2015. *Darwin, Geodynamics and Extreme Waves*. Springer, Switzerland.

Garrett, C. and Munk, W. 1972. Space-time scales of internal waves. *Geophysical Fluid Dynamics*, 3(1):225–264. doi:10.1080/03091927208236082.

Garrett, C. and Munk, W. 1975. Space-time scales of internal waves: A progress report. *Journal of Geophysical Research*, 80(3):291–297. doi:10.1029/jc080i003p00291.

Geist, E. L., Fritz, H. M., Rabinovich, A. B., and Tanioka, Y. (Eds.). 2017. *Global Tsunami Science: Past and Future, Volumes 1*. Birkhäuser – Springer, Cham, Switzerland.

Gill, A. E. 1982. *Atmosphere-Ocean Dynamics*. Academic Press, New York.

Godin, O. A. 2004. Air-sea interaction and feasibility of tsunami detection in the open ocean. *Journal of Geophysical Research*, 109(C5) C05002. doi:10.1029/2003JC002030.

Grimshaw, R. H. J. (Ed.). 2007. *Solitary Waves in Fluids*. WIT Press Southampton Boston, Southampton, UK.

Grue, J. and Trulsen, K. (Eds.). 2006. *Waves in Geophysical Fluids: Tsunamis, Rogue Waves, Internal Waves and Internal Tides*. Springer, New York.

Grue, J., Gjevik, B., and Weber, J. E. 1996. *Waves and Nonlinear Processes in Hydrodynamics*. Kluwer Academic Publisher, Dordrecht, The Netherlands.

Hasselman, K. 1962. On the nonlinear energy transfer in a gravity wave spectrum part 1. General theory. *Journal of Fluid Mechanics*, 12(4):481–500.

Holthuijsen, L. H. 2007. *Waves in Oceanic and Coastal Waters*. Cambridge University Press, Cambridge, UK.

Hwang, P. A. 2006. Duration- and fetch-limited growth functions of wind-generated waves parameterized with three different scaling wind velocities. *Journal of Geophysical Research*, 111(C2) C02005. doi:10.1029/2005JC003180.

Infeld, E. and Rowlands, G. 2000. *Nonlinear Waves, Solitons and Chaos*, 2nd edition. Cambridge University Press, Cambridge, UK.

Jackson, C. R. and Apel, J. R. 2004. *An Atlas of Internal Solitary-Like Waves and Their Properties*, 2nd edition. Prepared under contract with Office of Naval Research. Code 322PO. Available on the Internet www. internalwaveatlas.com/Atlas_index.html.

Janssen, P. 2004. *The Interaction of Ocean Waves and Wind*. Cambridge University Press, Cambridge, UK.

Johnson, R. S. 1997. *A Modern Introduction to the Mathematical Theory of Water Waves*. Cambridge University Press, Cambridge, UK.

Joseph, A. 2011. *Tsunamis: Detection, Monitoring, and Early-Warning Technologies*. Elsevier – Academic Press, Burlington, MA.

Kantha, L. H. and Clayson, C. A. 2000. *Numerical Models of Oceans and Oceanic Processes*. Academic Press, San Diego, CA.

Kharif, C., Pelinovsky, E., and Slunyaev, A. 2009. *Rogue Waves in the Ocean*. Springer, Berlin, Germany.

Kinsman, B. 1984. *Wind Waves: Their Generation and Propagation on the Ocean Surface*. Dover Publication, New York.

Kitaigorodskii, S. A. 1973. *Physics of Air-Sea Interaction*. Israel Program of Scientific Translations, Jerusalem, Israel (translation from Russian).

Knauss, J. A. and Garfield, N. 2017. *Introduction to Physical Oceanography*, 3rd edition. Waveland Press Inc., Long Grove, IL.

Kochin, N. E., Kibel, I. A., and Roze, N. V. 1964. *Theoretical Hydromechanics. (translated from Russian by D. Boyanovitch. Edited by J. R. M. Radok)*. Interscience Publishers, New York.

Komen, G. J., Cavaleri, L., Donelan, M., Hasselmann, K., Hasselmann, S., and Janssen, P. A. E. M. 1996. *Dynamics and Modelling of Ocean Waves*. Cambridge University Press, Cambridge, UK.

Korteweg, D. J. and de Vries, G. 1895. On the change of form of long waves advancing in a rectangular canal, and on a new type of long stationary waves. XLI. *The London, Edinburgh, and Dublin Philosophical Magazine and Journal of Science, Series 5*, 39(240):422–443. doi:10.1080/14786449508620739.

Krasitsky, V. P. 1994. Five-wave kinetic equation for surface gravity waves. *Physical Oceanography*, 5(6):413–421 (translated from Russian). doi:10.1007/BF02198507.

Krasitskii (Krasitsky), V. P. and Kozhelupova, N. G. 1995. On conditions for five wave resonant interactions of surface gravity waves. *Oceanology*, 34(4):435–439 (translated from Russian).

Kraus, E. B. and Businger, J. A. 1994. *Atmosphere-Ocean Interaction*. Oxford University Press, New York.

Kundu, A. (Ed.). 2007. *Tsunami and Nonlinear Waves*. Springer, Berlin, Germany.

Kuznetsov, N., Maz'ya, V., and Vainberg, B. 2002. *Linear Water Waves: A Mathematical Approach*. Cambridge University Press, Cambridge, UK.

Lamb, H. 1932. *Hydrodynamics*, 6th edition. Cambridge University Press, Cambridge, UK.

Lavrenov, I. 2003. *Wind-Waves in Oceans: Dynamics and Numerical Simulations*. Springer, Berlin, Germany.

LeBlond, P. H. and Mysak, L. A. 1978. *Waves in the Ocean*. Elsevier Scientific Publishing, Amsterdam, The Netherlands.

Le Mehaute, B. 1976. *An Introduction to Hydrodynamics and Water Waves*. Springer, New York.

Levin, B. and Nosov, M. 2016. *Physics of Tsunamis,* 2nd edition. Springer, Switzerland.

Lighthill, J. 1978. *Waves in Fluids*. Cambridge University Press, Cambridge, UK.

Lin, P. 2008. *Numerical Modeling of Water Waves*. Taylor & Francis, London, New York.

Liu, P. L.-F., Yeh, H., and Costas, S. (Eds.). 2008. *Advanced Numerical Models For Simulating Tsunami Waves and Runup*. World Scientific Publishing, Singapore.

Longuet-Higgins, M. S. 1952. On the statistical distribution of the height of sea waves. *Journal of Marine Research*, 11(3):245–266.

Longuet-Higgins, M. S. 1975. On the joint distribution of the periods and amplitudes of sea waves. *Journal of Geophysical Research*, 80(18):2688–2694. doi:10.1029/JC080i018p02688.

Longuet-Higgins, M. S. 1983. On the joint distribution of wave periods and amplitudes in a random wave field. *Proceedings of the Royal Society A: Mathematical, Physical and Engineering Sciences*, 389(1797):241–258. doi:10.1098/rspa.1983.0107.

Ma, Q. 2010. *Advances in Numerical Simulation of Nonlinear Water Waves*. World Scientific, Singapore.

Mader, C. L. 2004. *Numerical Modeling of Water Waves*, 2nd edition. CRC Press, Boca Raton, FL.

Madsen, P. A., Fuhrman, D. R., and Schäffer, H. A. 2008. On the solitary wave paradigm for tsunamis. *Journal of Geophysical Research*, 113(C12) C12012. doi:10.1029/2008JC004932.

Mani, J. S. 2012. *Coastal Hydrodynamics*. PHI Learning Private Limited, New Delhi.

Marchuk, G. I. and Kagan, B. A. 1984. *Ocean Tides: Mathematical Models and Numerical Experiments*. Pergamon Press, Oxford, UK.

Marghany, M. 2018. *Advanced Remote Sensing Technology for Tsunami Modelling and Forecasting*. CRC Press, Boca Raton, FL.

Massel, S. R. 1989. *Hydrodynamics of Coastal Zones*. Elsevier Science Publisher, Amsterdam, The Netherlands.

Massel, S. R. 2013. *Ocean Surface Waves: Their Physics and Prediction*, 2nd edition. World Scientific Publishing, Singapore.

Massel, S. R. 2015. *Internal Gravity Waves in the Shallow Seas*. Springer, Switzerland.

Masson D. and Leblond, P. H. 1989. Spectral evolution of wind-generated surface gravity waves in a dispersed ice field. *Journal of Fluid Mechanics*, 202(5):43–81. doi:10.1017/S0022112089001096.

McCormick, M. E. 2007. *Ocean Wave Energy Conversion*. Dover Publications, Mineola, NY.

Mei, C. C. 1989. *The Applied Dynamics of Ocean Surface Waves*. World Scientific Publishing, Singapore.

Miropol'sky, Y. Z. 2001. *Dynamics of Internal Gravity Waves in the Ocean (translated and edited from Russian by O. D. Shishkina)*. Springer, New York.

Mitsuyasu, H. 2002. A historical note on the study of ocean surface waves. *Journal of Oceanography*, 58:109–120.

Monin, A. S. and Krasitskii, V. P. 1985. *Yavleniya na poverkhnosti okeana (Phenomena on the Ocean Surface)*. Gidrometeoizdat, Leningrad, USSR (in Russian).

Morozov, E. G. 2018. *Oceanic Internal Tides: Observations, Analysis and Modeling: A Global View*. Springer, Cham, Switzerland.

Moskowitz, L. 1964. Estimates of the power spectrums for fully developed seas for wind speeds of 20 to 40 knots. *Journal of Geophysical Research*, 69(24):5161–5179. doi:10.1029/JZ069i024p05161.

Murty, T. S., Aswathanarayana, U., and Nirupama, N. (Eds.). 2006. *The Indian Ocean Tsunami*. CRC Press, Boca Raton, FL.

Myers, R. G. 2008. *Potential for Tsunami Detection and Early-Warning Using Space-Based Passive Microwave Radiometry*. Master's Thesis. Massachusetts Institute of Technology. MIT, Boston, MA. Available on the Internet dspace.mit.edu/handle/1721.1/42913.

Newman, J. N. 2017. *Marine Hydrodynamics*, 40th anniversary edition. The MIT Press, Cambridge, MA.

Ochi, M. K. 1998. *Ocean Waves: The Stochastic Approach*. Cambridge University Press, Cambridge, UK.

Olbers, D., Willebrand, J., and Eden, C. 2012. *Ocean Dynamics*. Springer, Berlin, Germany.

Osborne, A. R. (Ed.). 1991. *Nonlinear Topics in Ocean Physics*. North-Holland Elsevier Science Publisher, Amsterdam, The Netherlands.

Osborne, A. R. 2010. *Nonlinear Ocean Waves and the Inverse Scattering Transform*. Academic Press – Elsevier, Burlington, MA.

Ostrovsky, L. 2014. *Asymptotic Perturbation Theory of Waves*. Imperial College Press, London, UK.

Ostrovsky, L. A. and Stepanyants, Y. A. 1989. Do internal solitions exist in the ocean? *Reviews of Geophysics*, 27(3):293–310. doi:10.1029/rg027i003p00293.

Pedlosky, J. 2013. *Waves in the Ocean and Atmosphere: Introduction to Wave Dynamics*. Springer, Berlin, Germany.

Pelinovsky, E. N. 1982. *Nonlinear Dynamics of Tsunami Waves*. Institute of Applied Physics, Gorky, USSR (in Russian).

Pelinovsky, E. N. 1996. *Tsunami Wave Hydrodynamics*. Institute of Applied Physics, Nizhny Novgorod, Russia, former Gorky (in Russian).

Pelinovsky, E. 2006. Hydrodynamics of tsunami waves. In *Waves in Geophysical Fluids: Tsunamis, Rogue Waves, Internal Waves and Internal Tides* (Eds. J. Grue and K. Trulsen). Springer, New York, pp. 1–48.

Pelinovsky, E. and Kharif, C. (Eds.). 2016. *Extreme Ocean Waves*, 2nd edition. Springer, Cham, Switzerland.

Peregrine, D. H. 1967. Long waves on a beach. *Journal of Fluid Mechanics*, 27(4):815–827. doi:10.1017/S0022112067002605.

Phillips, O. M. 1980. *The Dynamics of the Upper Ocean*, 2nd edition. Cambridge University Press, Cambridge, UK.

Phillips, O. M. and Hasselmann, K. 1986. *Wave Dynamics and Radio Probing of the Ocean Surface*. Plenum Press, New York.

Pickard, G. L. and Emery, W. J. 1990. *Descriptive Physical Oceanography: An Introduction*, 5th enlarge edition. Pergamon Press, Oxford, UK.

Pond, S. and Pickard, G. L. 1983. *Introductory Dynamical Oceanography*, 2nd edition. Butterworth-Heinemann, Oxford, UK.

Rabinovich, A. B., Fritz, H. M., Tanioka, Y., and Geist, E. L. (Eds.). 2018. *Global Tsunami Science: Past and Future, Volumes 2*. Birkhäuser – Springer, Cham, Switzerland.

Radko, T. 2013. *Double-Diffusive Convection*. Cambridge University Press, Cambridge, UK.

Rice, S. O. 1944. Mathematical analysis of random noise. *Bell System Technical Journal*, 23(3):282–332. doi:10.1002/j.1538-7305.1944.tb00874.x.

Reddy, M. P. M. 2001. *Descriptive Physical Oceanography*. A. A. Balkema Publishers, Lisse, The Netherlands.

Russell, J. S. 1844. Report on waves. *Report of the 14th Meeting of the British Association for the Advancement of Science, York, September 1844 (London 1845)*, Plates XLVII-LVII, pp. 311–390.

Shuleikin, V. V. 1968. *Fizika Moria (Physics of the Sea)*, 4th edition revised and expanded. Izdatel'stvo Akademii Nauk S.S.S.R., Moscow, USSR (in Russian).

Singh, J. 2009. *Tsunamis: Threats and Management*. I. K. International Publishing House Pvt. Ltd., New Delhi.

Soloviev, A. and Lukas, R. 2014. *The Near-Surface Layer of the Ocean: Structure, Dynamics and Applications (Atmospheric and Oceanographic Sciences Library, Vol. 48)*, 2nd edition. Springer, Dordrecht, Heidelberg.

Steele, J. H., Thorpe, S. A., and Turekian, K. K. 2010. *Elements of Physical Oceanography: A Derivative of the Encyclopedia of Ocean Sciences*. Academic Press, London, UK.

Stoker, J. J. 1958. *Water Waves: The Mathematical Theory with Applications*. A Willey-interscience Publication, New York.

Stokes, G. G. 1847. On the theory of oscillatory waves. *Transactions of the Cambridge Philosophical Society*, 8:441–455.

Sundar, V. 2016. *Ocean Wave Mechanics: Applications in Marine Structures*. Ane Books Pvt. Ltd. – John Wiley & Sons, Chichester, UK.

Sutherland, B. R. 2010. *Internal Gravity Waves*. Cambridge University Press, Cambridge, UK.

Svendsen I. A. 2006. *Introduction to Nearshore Hydrodynamics*. World Scientific Publishing, Singapore.

Talley, L. D., Pickard, G. L., Emery, W. J., and Swift, J. H. 2011. *Descriptive Physical Oceanography: An Introduction*, 6th edition. Elsevier – Academic Press, London, UK.

Thomson, R. E. and Emery, W. J. 2014. *Data Analysis Methods in Physical Oceanography*, 3rd edition. Elsevier, Amsterdam, The Netherlands.

Tiampo, K. F., Weatherley, D. K., and Weinstein, S. A. (Eds.). 2008. *Earthquakes: Simulations, Sources and Tsunamis*. Birkhäuser Verlag, Basel, Switzerland.

Turner, J. S. 1973. *Buoyancy Effects in Fluids*. Cambridge University Press, Cambridge, UK.

Vladimirov Y. V. and Bulatov V. V. 2007. *Internal Gravity Waves: Theory and Applications*. Nauka, Moscow, Russia.

Vlasenko, V., Stashchuk, N., and Hutter, K. 2005. *Baroclinic Tides: Theoretical Modeling and Observational Evidence*. Cambridge University Press, Cambridge, UK.

Voliak, K. I. 2002. *Selected Papers. Nonlinear Waves in the Ocean*. Nauka, Moscow (in Russian and English).

Whitham, G. B. 1999. *Linear and Nonlinear Waves*. John Wiley & Sons, New York.

Young, I. R. 1999. *Wind Generated Ocean Waves*. Elsevier, Oxford, UK.

Yuen, H. C. and Lake, B. M. 1982. Nonlinear dynamics of deep-water gravity waves. In *Advances in Applied Mechanics* (Ed. C.-S. Yih). Academic Press, New York, Vol. 22, pp. 67–229.

Zakharov, V. E. (Ed.). 1998. *Nonlinear Waves and Weak Turbulence*. American Mathematical Society, Providence, RI.

Zakharov, V. E. 1999. Statistical theory of gravity and capillary waves on the surface of a finite-depth fluid. *European Journal of Mechanics, B/Fluids*, 18(3):327–344.

Zakharov, V. E. and Filonenko, N. N. 1966. Energy spectrum for stochastic oscillations of the surface of liquid. *Doclady Akademii Nauk SSSR*, 170(6):1292–1295 (in Russian). English translation: *Soviet Physics–Doklady*, Vol. 11, pp. 881–884, 1967.

2.4 CURRENTS AND CIRCULATION

The combined action of horizontal and vertical movements leads to planet-wide circulation.

Ocean circulation is the large-scale movement of waters in the ocean basins. Two basic types of currents—surface and deep-water currents—characterize the flow of ocean waters across the globe. Surface circulation is driving by wind, and deep ocean circulation is driving by the cooling and sinking waters in the polar region. Both the atmosphere and the oceans transport heat from low latitudes near the equator to high latitudes near the poles. Cool air and water currents turn from high latitudes back towards the equator. In the most general sense, a current is a region of water that moves more rapidly than its surroundings. Ocean currents are caused by: (1) astronomical forces which are responsible for tide-induced currents, (2) global wind pattern, (3) the rotation of the Earth known as Coriolis effect, and (4) continental deflections—the shape of the ocean basins. The Coriolis effect influences the surface ocean as well as deeper ocean water layers, which are created by slight differences in temperature and salinity. The global pattern of winds together with the Coriolis effect and Ekman transport produce large-scale currents in the world ocean. Ocean surface currents tend to form ring-like circulation systems called Gyres (or a swirling vortex). Gyres are characterized by circulation at the scale of the ocean basin.

Mathematical models of the general circulation in oceans (also known as the primitive equations) follow from the classic nonlinear hydrodynamical equations of a rotating fluid with the use of Boussinesq, hydrostatic, incompressibility, turbulent viscosity, and diffusivity approximations; books (Marchuk and Sarkisyan 1988; Kantha and Clayson 2000; Miller 2007) provides detailed analysis of numerical methods and models. Ocean general circulation models (OGCMs) are a particular version of general circulation model (GCM) and a GCM is a type of climate model. The OGCM describes physical and thermodynamical processes in oceans. Namely, OGCM predicts the evolution of ocean horizontal and vertical velocity, temperature, and salinity fields globally over the full depth of the ocean. In physical oceanography, there is a number of OGCMs which are constantly updated and improved. Comprehensive and practical information about currents and circulation in the oceans can be found in the books (Neumann 1968; Pond and Pickard 1983; Teramoto 1993; Griffies 2004; van Aken 2007; Steele et al. 2009; Huang 2010; Samelson 2011; Siedler et al. 2013; Joseph 2014). List of available OGCMs is in the website: https://en.wikipedia.org/wiki/List_of_ocean_circulation_models.

2.4.1 SURFACE CURRENTS

Surface current is horizontal, streamlike movements of water restricted to the upper 400m of the ocean. In oceanography, several types of surface currents are distinguished: coastal currents (upwelling, downwelling), longshore currents, tidal currents, rip currents, warm currents, and cold currents. There are five major gyres—one in each major ocean basin: (1) North Atlantic Gyre, (2) South Atlantic Gyre, (3) North Pacific Gyre, (4) South Pacific Gyre, and (5) Indian Ocean Gyre. They are shown in Figure 2.12.

The surface currents have a strong effect on Earth's weather and climate. Areas near the equator receive more direct solar radiation than areas near the poles. However, these areas do not constantly get warmer and warmer, because the ocean currents and winds transport the heat from the lower latitudes near the equator to higher latitudes near the poles.

In oceanography and in fluid dynamics in general, there are two methods which are used to specify ocean surface currents—Lagrangian and Eulerian methods. Lagrangian method also called float method, determines the fluid flow properties by tracking the motion of each fluid particle trajectory. Lagrangian method is rarely used in fluid mechanics. Eulerian method also called flow method, describes the fluid flow properties of interest at every fixed point of space and time. Pressure, velocity, acceleration, and all other flow properties are described as fields within the control volume.

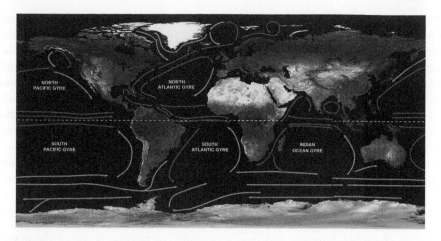

FIGURE 2.12 (**See color insert.**) Five major ocean gyres. (Image from NOAA.)

The Eulerian method is usually preferred in fluid mechanics. In coastal oceanography, both these methods are used to measure surface currents at various depths.

Because of the shortcomings of purely Lagrangian and purely Eulerian descriptions, a succeeded technique has been developed that combines the best features of both the Lagrangian and the Eulerian viewpoints. Such a technique is known as the arbitrary Lagrangian–Eulerian (ALE) description (e.g., Souli and Benson 2010). At present, the ALE method is incorporated into advanced CFD algorithms to take into account wave-current and turbulence-current interactions. The ALE is a frontier physics-based technique and can provide more accurate modeling and analysis of both *in situ* and satellite measurements of ocean surface currents.

2.4.2 Deep Ocean Circulation

In physical oceanography, deep ocean circulation is divided on two parts—thermohaline and the wind-driven. In some way or another, the energy from the sun radiation is responsible for these circulations. Correspondingly, thermohaline circulation is due to changes in water density caused by weather and wind-driven circulation is due to wind stress (Pickard and Emery 1990).

Thermohaline circulation involves the transport of heat and salt; it is caused by changes in water density which varies globally due to differences in sea temperature and salinity. Cool water can become denser than the underlying water, causing it to sink. Water gets denser in higher latitudes due to (1) the cooling of the atmosphere and (2) the increased salt levels, which result from the freezing of surface water. The earth's rotation also influences deep ocean circulation and currents. Deep ocean currents move very slowly, usually with a speed of around 2–3 cm/s.

Wind-driven circulation is principle in the upper ocean layer of few hundreds of meters. The main force driving the oceanic circulation is the wind. The wind sets motion of water as a current, the Coriolis effect, Ekman transport, and the configuration of the ocean basin. The mechanism is associated with wind blowing across the sea surface causing it to move. Part of the wind momentum is transferred to the ocean, setting up currents. In 1905 Eckman has formulated the equations of the motions. The deflection of wind-driven currents is known as Eckman motion. Because the surface layer of the ocean transfers some of its momentum to deeper water, and because these deeper layers are subject to Coriolis deflection too, successively deeper layers of the ocean are deflected to the right of the upper layers. The wind-driven currents thus form a spiral known as the Eckman Spiral. The Ekman spiral indicates that each moving layer is deflected to the right of the overlying layer's movement; hence, the direction of water movement changes with increasing depth. The average motion of the Eckman layer is 90° to the right of the wind direction in the Northern Hemisphere (to the left in the Southern). In general, wind-driven currents dominate the upper ocean, and thermohaline circulation drives the movement of the deep ocean.

2.4.3 Ocean Currents from Space

Mapping of surface currents, and their space-time variability, is one of the major challenges in physical oceanography. Satellite imagery helps to obtain valuable data. There are three basic remote sensing techniques which are capable to provide measurements of surface currents from space: thermal infrared imagery, synthetic aperture radar (SAR), and altimeter.

For example, ocean surface currents that are warmer than the surrounding water have thermal signatures that are observed by Advanced Very High Resolution Radiometer (AVHRR) operated at the visible and infrared bands. Moreover, ocean surface currents with a sufficiently strong thermal contrast can also be detected through the measurements of sea surface temperature (SST) using AVHRR or Special Sensor Microwave/Imager (SSM/I). The MODIS (onboard Aqua MODISA) and Terra (MODIST) satellites products can be used to retrieve coastal surface currents as well.

Depending on wind conditions, variations of coastal currents can be mapped using satellite SAR. The most known SAR measurements are conducted from ERS-2, Envisat, TerraSAR-X (TSX), TanDEM-X (TDX) and COSMO-SkyMed (CSK). Principles of SAR observations of ocean currents are based on the measurements of the phase difference between SAR image scenes obtained from an interferogram that is directly proportional to the surface current. The technique has been proposed and developed in the 1980s.

Satellite altimeter provides precise measurements of the sea surface height. Surface currents are detectable by variations of sea surface slopes. Small-scale surface features—like small eddies, generated by the large-scale currents (e.g., by the Gulf Stream) can be identified by altimeter data as well. Altimeter data are also used for tide modeling. Launched in 1992, TOPEX/Poseidon altimeters provide global data acquisitions which are employed for mapping basin-wide current variations and validation of OGCMs. The ocean surface currents also can be retrieved using Doppler anomaly from Sentinel-1 SAR.

Satellite remote sensing capabilities to detect and compute ocean surface currents are discussed in works (Klemas 2012; Joseph 2014; Isern-Fontanet et al. 2017; Liu et al. 2017a,b). Among them, SAR has become the most efficient operational tool to monitor ocean surface currents across the globe.

REFERENCES TO SECTION 2.4

Farazmanda, M. and Haller, G. 2012. Erratum and addendum to "A variational theory of hyperbolic Lagrangian coherent structures". *Physica D: Nonlinear Phenomena*, 241(4):439–441. doi:10.1016/j.physd.2011.09.013.

Griffies, S. M. 2004. *Fundamentals of Ocean Climate Models*. Princeton University Press, Princeton, NJ.

Huang, R. X. 2010. *Ocean Circulation: Wind-Driven and Thermohaline Processes*. Cambridge University Press, Cambridge, UK.

Isern-Fontanet, J., Ballabrera-Poy, J., Turiel, A., and García-Ladona, E. 2017. Remote sensing of ocean surface currents: A review of what is being observed and what is being assimilated. *Nonlinear Processes in Geophysics*, 24(4):613–643. doi:10.5194/npg-24-613-2017.

Joseph, A. 2014. *Measuring Ocean Currents: Tools, Technologies, and Data*. Elsevier, San Diego, CA.

Kantha, L. H. and Clayson, C. A. 2000. *Numerical Models of Oceans and Oceanic Processes*. Academic Press, San Diego, CA.

Klemas, V. 2012. Remote sensing of coastal and ocean currents: An overview. *Journal of Coastal Research*, 28(3):576–586. doi:10.2112/JCOASTRES-D-11-00197.1.

Liu, J., Emery, W. J., Wu, X., Li, M., Li, C., and Zhang, L. 2017a. Computing coastal ocean surface currents from MODIS and VIIRS satellite imagery. *Remote Sensing*, 9(10) 1083. doi:10.3390/rs9101083.

Liu, J., Emery, W. J., Wu, X., Li, M., Li, C., and Zhang, L. 2017b. Computing ocean surface currents from GOCI ocean color satellite imagery. *IEEE Transaction on Geoscience and Remote Sensing*, 55(12):7113–7125. doi:10.1109/TGRS.2017.2741924.

Marchuk, G. I. and Sarkisyan, A. S. 1988. *Mathematical Modelling of Ocean Circulation*. Springer, Berlin, Germany.

Miller, R. N. 2007. *Numerical Modeling of Ocean Circulation*. Cambridge University Press, Cambridge, UK.

Neumann, G. 1968. *Ocean Currents (Elsevier Oceanography Series, 4)*. Elsevier Publishing Company, Amsterdam, The Netherlands.

Pickard, G. L. and Emery, W. J. 1990. *Descriptive Physical Oceanography: An Introduction*, 5th enlarge edition. Pergamon Press, Oxford, UK.

Pond, S. and Pickard, G. L. 1983. *Introductory Dynamical Oceanography*, 2nd edition. Butterworth-Heinemann, Oxford, UK.

Robinson, I. S. 2010. *Discovering the Ocean from Space: The Unique Applications of Satellite Oceanography*. Springer, New York.

Samelson, R. M. 2011. *The Theory of Large-Scale Ocean Circulation*. Cambridge University Press, New York.

Shadden, S., Lekien, F., and Marsden, J. 2005. Definition and properties of Lagrangian coherent structures from finite-time Lyapunov exponents in two-dimensional aperiodic flows. *Physica D: Nonlinear Phenomena*, 212(3–4):271–304. doi:10.1016/j.physd.2005.10.007.

Siedler, G., Griffies, S. M., Gould, J., and Church, J. A. (Eds.). 2013. *Ocean Circulation and Climate: A 21st Century Perspective*. Elsevier – Academic Press, London, UK.

Souli, M. and Benson, D. J. (Eds.). 2010. *Arbitrary Lagrangian Eulerian and Fluid-Structure Interaction: Numerical Simulation*. ISTE – John Wiley & Sons, Hoboken, NJ.

Steele, J. H., Thorpe, S. A., and Turekian, K. K. (Eds.). 2009. *Ocean Currents*. Elsevier – Academic Press, London, UK.

Teramoto, T. (Ed.). 1993. *Deep Ocean Circulation: Physical and Chemical Aspects*. Elsveier Science Publisher, Amsterdam, The Netherlands.

van Aken, H. M. 2007. *The Oceanic Thermohaline Circulation: An Introduction*. Springer, New York.

2.5 INTERACTIONS

In fluid mechanics, the study of motion of the fluids is mainly addressed with reference to their transport properties, dispersion, mixing, and various aspects of the interactions between flows and waves. As a whole, there is a two-way interaction: fluid motion generates waves, but waves also force fluid motion. Because waves and flows often have very different spatial and temporal scales, and because they involve very different physical processes, the study of their interactions poses a considerable conceptual and mathematical challenge. Fundamental problem known as *wave-mean flow interaction* is concerned with coupled dynamics of waves and mean flow. It is an essential topic of geophysical fluid dynamics and one of the major driving forces of atmospheric and oceanic circulations (Bühler 2014). This topic also covers a wide range of nonlinear phenomena in oceanic science including surface hydrodynamic interactions, bifurcations, and subharmonic wave generations (Craik 1985). Here we refer to some aspects of the interactions occurring in the ocean.

2.5.1 WAVE-WAVE INTERACTIONS

The classical description of wave-wave interactions with the energy transfer was given first by Phillips (1960, 1961). His theory of nonlinear four-wave interactions was further extended to a random sea surface independently by Hasselmann (1962, 1963a,b) and Zakharov (1968). A review of these and other works can be found in books (Phillips 1980; Phillips and Hasselmann 1986; Zakharov 1998). This theory results to the six-fold integral (known also as the Boltzmann collusion integral) and the WK equation (Section 2.3.2.4). This integral describes the rate of change of wave action density due to resonant interactions between four wave vectors. Thus, nonlinear wave-wave interactions play a central role in the development of wind-generated surface waves that is important aspect of ocean remote sensing (Phillips and Hasselmann 1986).

Nonlinear wave-wave interactions can be separated into three types: "weak," "strong" and "very strong." Nonlinear interactions of wave components are characterized in terms of their evolution of spatiotemporal scales. The first type is first-order nonlinear effect for surface waves of finite amplitude with relatively small slope.

The interaction leads to a slow change of wave parameters and small perturbations. This process is characterized by a long duration of interactions. The second type is characterized by small time and small spatial scales of interactions. In this case, different types of instabilities occur (Section 2.7). The strong interaction is more rapid process leading, e.g., to the wave breaking phenomena. The steepness is another parameter that characterizes the types of interactions. Note that the suggested classification of interaction is arbitrary to some extent; it represents an attempt to organize a complicated set of wave phenomena that may not be completely described by the theory and/or tested in real world experiment (Phillips and Hasselmann 1986).

In the case of second-order resonant interactions, three pairs $\{\vec{K}_i\}$ with linearized dispersion relation for gravity-capillary waves in deep water $\{\omega_i(K) = \omega_i(K_i)\}, \vec{K}_i \neq 0$ for i = 1, 2, 3, satisfy a triad relationship:

$$\vec{K}_1 \pm \vec{K}_2 \pm \vec{K}_3 = 0, \omega_1 \pm \omega_2 \pm \omega_3 = 0, \omega = \pm\sqrt{gK + (\sigma/\rho)K^3}, \qquad (2.25)$$

where $\vec{K}_i$ and ω_i are the i-th wave number and i-th wave frequency. There are no nontrivial solutions of Equation (2.25). The resonance cannot occur at this order, and only the effect of the perturbation of the wave profile can be seen (Phillips 1980). There are two simple alternatives: if triplets of waves satisfying Equation (2.25) exist, then triad resonances, $\vec{K}_1 + \vec{K}_2 = \vec{K}_3, \omega_1 + \omega_2 = \omega_3$, are expected. If there are no such triplets, one has to look for resonant quartets.

The interaction of the three wave components $\{\vec{K}_1, \vec{K}_2, \vec{K}_3\}$ at the quadratic and cubic orders generate the components with the numbers $\{\vec{K}_1 \pm \vec{K}_2 \pm \vec{K}_3\}$. For resonance among a tetrad of wave components, the conditions of synchronism must be or near satisfied,

$$\vec{K}_1 \pm \vec{K}_2 \pm \vec{K}_3 \pm \vec{K}_4 = 0, \omega_1 \pm \omega_2 \pm \omega_3 \pm \omega_4 = 0, \omega = \pm\sqrt{gK}. \qquad (2.26)$$

The nontrivial solution of Equation (2.16) exists for four-wave interactions:

$$\vec{K}_1 + \vec{K}_2 = \vec{K}_3 + \vec{K}_4, \omega_1 + \omega_2 = \omega_3 + \omega_4, \omega = \sqrt{gK}. \qquad (2.27)$$

Figure 2.13 (Phillips 1980) shows resonant quartet graphically. This diagram shows that a pair of wave numbers K_3 and K_4 resonates with another pair of K_1 and K_2 only when they lie on the same contour of γ. There is infinite number of possible configurations of resonant four wavenumbers (Longuet-Higgins 1962).

Weak turbulence theory (Section 2.6.3) relies on these resonant equations and allows one to obtain analytical statistical solution the WK equation. In the case of an out-of-equilibrium and stationary forced system, it leads to the Kolmogorov-Zakharov (KZ) power spectra. For surface gravity waves with dispersion relation $\omega = \sqrt{gK}$, spectral solutions can be found using dimensional analysis.

An important practical case is four-wave resonant interaction with two coincident wave components $\vec{K}_3 = \vec{K}_4$; this condition leads to relationship

$$\vec{K}_1 + \vec{K}_2 = 2\vec{K}_3, \omega_1 + \omega_2 = 2\omega_3, \omega = \sqrt{gK}. \qquad (2.28)$$

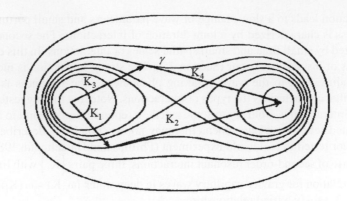

FIGURE 2.13   Phillips's "figure eight" four-wave interaction diagram. (After Phillips 1980.)

These particular conditions were tested and investigated in a number of laboratory experiments beginning from (Phillips 1980) and up to the present (e.g., Aubourg et al. 2017). During the years, resonant interactions have been a subject of many works; books (Yuen and Lake 1982; Mahrenholtz and Markiewicz 1999; Kartashova 2011) and theoretical papers published mostly by Russian authors, e.g., Zaslavskiy (1996) and Zakharov (1998) provide an extended analysis of surface wave interactions. Quasi-resonant surface wave phenomena in the ocean were also observed and investigated in the 1980s using airborne radar and optical imagery; some original data were published by Voliak (2002); see also Chapter 7.

2.5.2 WAVE-CURRENT INTERACTION

As mentioned above, in fluid mechanics, wave-current interaction is defined as the interaction between surface waves and a mean flow. According to Peregrine (1976), there are following major types of wave-current interactions:

- interaction of waves with a slow large-scale current
- interaction of waves with strong small-scale current (in contrast with the case above)
- the combined wave-current motion for depth-varying current below the free surface
- interaction of waves with turbulence
- interaction of ship waves and surface current producing a ship wake.

Kinematic effect of wave-current coupling is described by the following equations (Peregrine 1976):

$$\frac{\partial \vec{K}}{\partial t} + \nabla \omega = 0, \tag{2.29}$$

$$\nabla \times \vec{K} = 0, \tag{2.30}$$

$$\omega_r = \left(\omega - \vec{K} \cdot \vec{U}\right)^2 = gK \tanh(Kh), \tag{2.31}$$

where $\vec{K}$ and ω are the wave vector and the angular frequency at the absolute celerity of the waves, ω_r is the relative angular frequency of waves on uniform current, h is the water depth, and $\vec{U}$ is the current field. Formula (2.31) is the linear dispersion relation for waves propagating over an unsteady but slowly varying current field $\vec{U}$. This relation describes the Doppler shift of wave frequency due to effect of moving fluid flow, i.e., current.

Further research is associated with the solution of Equations (2.29) and (2.30) which are combined to form

$$\left[\frac{\partial}{\partial t} + \left(\vec{U} + \vec{c}_g\right) \cdot \nabla\right]\omega = \vec{K} \cdot \frac{\partial \vec{U}}{\partial t} + \frac{\partial \omega_r}{\partial h}\frac{\partial h}{\partial t}, \tag{2.32}$$

where $c_g = \omega/K$ is the phase velocity $\left(K = |\vec{K}|\right)$. Note that phase velocity is not a vector, e.g., the phase velocity along a line in the direction of a unit vector $\vec{e}$ is $\omega/\left(\vec{K} \cdot \vec{e}\right)$.

There are different modifications of Equation (2.32) resulting to different solutions of the problem. For example, linear and nonlinear dynamics of waves on currents are considered for depth-integrated and depth-varying cases that seem to be realistic enough approaches. In many practical situations, averaged equations of motion and spectral models are used as well. Finally, advanced studies and applications provide CFD-based numerical simulations of wave-current interactions using full set of hydrodynamic equations.

Wave-current interactions are known to contribute to the formation of extreme wave events (freak waves, tsunami) with disastrous effect. These and other problems of wave-current interactions are discussed in the books (Huang 2009; Constantin 2011).

2.5.3 SURFACE–INTERNAL WAVE INTERACTION

There is a numerous literature on the problem of interaction between internal and surface waves. The study is mostly associated with SAR imagery of IWs and theoretical analysis is based on a coupled KdV–Schrödinger system and/or Hamiltonian formalism. Surface–internal wave interactions are considered in context with WK equation as well taking into account a spatial modulation of the wave spectrum by underlined current in the presence of IWs or solitons. This particular "spectral statistical model" can be used for interpretation of remotely sensed data.

Along with resonant interactions between surface waves, resonant interactions between surface and IWs may also cause changes in the surface wavenumber spectrum. For example, effects of amplification or depression of spectral components in the wave spectrum can occur at certain conditions. However, rigorous theory of resonant surface-internal wave interactions is more complicated (and limited) than the theory of surface wave-wave interactions. This is due to the differences in spatiotemporal scales of internal and surface waves.

It is known (e.g., Miropol'sky 2001) that nonlinear interactions between two surface gravity waves can generate internal wave in stratified upper ocean layer, especially at resonant triad

$$\vec{K} = \vec{K}_1 - \vec{K}_2, \, \omega(\vec{K}) = \omega(\vec{K}_1) - \omega(\vec{K}_2), \qquad (2.33)$$

where $\{\vec{K}_{1,2}, \omega(\vec{K}_{1,2})\}$ are wave vectors and angular frequencies of a pair of surface waves and $\{\vec{K}, \omega(\vec{K})\}$ are wave vector and frequency of an internal wave. Appropriate conditions (2.33) may occur at $\omega(K) \ll \omega(K_1)$; $\omega(K_2)$ and $\vec{K} = \vec{K}_1 - \vec{K}_2, \, \omega(\vec{K}) = \omega(\vec{K}_1) - \omega(\vec{K}_2), |\vec{K}_1| \approx |\vec{K}_2|$. At strong ocean stratification, the resonant triad (in statistical sense) plays an important role in the formation of modulation mechanism causing narrow-band variations (transformations) in the surface wavenumber spectrum.

In remote sensing applications, specifically related to interpretation of SAR data (and passive microwave imagery as well), it is convenient to introduce spectral density perturbation function $f(\vec{k}, \vec{r}, t)$ associated with current field $\vec{U}(\vec{K}, \vec{r}, t)$, induced by a number of internal solitons

$$\vec{U}(\vec{K}, \vec{r}, t) = \sum_{i=1}^{n} U_i(\vec{r}, t), \, \vec{U}_i(\vec{r}, t) = U_{0i} \operatorname{sech}^2(\vec{K}_i \cdot \vec{r} - c_{gi}t - \psi_i), \qquad (2.34)$$

$$f(\vec{k}, \vec{r}, t) = \frac{F(\vec{k}, \vec{r}, t) - F_0(\vec{k}, \vec{r}, t)}{F_0(\vec{k})}, \qquad (2.35)$$

where $\vec{U}_i(\vec{r})$ is the current velocity induced by i-th soliton with wave number vector $\vec{K}_i$, peak U_{0i}, phase speed c_{gi}, and phase ψ_i; n is the number of participating solitons; $F(\vec{k}, \vec{r}, t)$ and $F_0(\vec{k}, \vec{r}, t)$ are perturbed and unperturbed surface wave spectra. The spectral density of wave action is $N(\vec{k}, \vec{r}, t) = \frac{\omega_r}{k} F(\vec{k}, \vec{r}, t)$ with local dispersion relation $\omega = \omega_r + \vec{k} \cdot \vec{U}(\vec{r}, t)$, where the intrinsic frequency is given by $\omega_r = \sqrt{gk + (\sigma/\rho)k^3}$.

The WK Equation (2.12) yields for $f(\vec{k}, \vec{r}, t)$

$$\frac{\partial}{\partial t} f(\vec{k}, \vec{r}, t) = -\gamma_{input} f(\vec{k}, \vec{r}, t) + \frac{\partial \vec{U}}{\partial \vec{r}} G(\omega, k, F_0), \qquad (2.36)$$

where γ_{input} is a coefficient associated with the wind source term in the WK Equation (2.12) or (2.16) and $G(\omega, k, F_0)$ is some function. In (2.36) variations of the current field are assumed to be small and the induced perturbations of the wave spectrum are also expected to be small, $f \ll 1$. In the simplified case of stationary field of current, Equation (2.36) at 1D geometry is reduced to proportional relationship $f \sim \frac{1}{\gamma_{input}} \frac{\partial U}{\partial x}$.

Even though the simplicity of the expression, it may provide good qualitative agreement with remote sensing observations. In particular, internal wave packets interacting with wind-generated surface waves usually create periodic-like signatures (due to modulation of surface roughness by current) that can be perfectly registered in high-resolution active/passive microwave images. Model of surface currents (2.34), incorporated into the observation model yields similar oscillation features in the images.

REFERENCES TO SECTION 2.5

Aubourg, Q., Campagne, A., Peureux, C., Ardhuin, F., Sommeria, J., Viboud, S., and Mordant, N. 2017. Three-wave and four-wave interactions in gravity wave turbulence. *Physical Review Fluids*, 2(11):114802-1–114802-19.

Bühler, O. 2014. *Waves and Mean Flows (Cambridge Monographs on Mechanics)*, 2nd edition. Cambridge University Press, Cambridge, UK.

Constantin, A. 2011. *Nonlinear Water Waves with Applications to Wave-Current Interactions and Tsunamis.* Society for Industrial and Applied Mathematics (SIAM), Philadelphia, PA.

Craik, A. D. D. 1985. *Wave Interactions and Fluid Flows.* Cambridge University Press, Cambridge, UK.

Hasselmann, K. 1962. On the nonlinear energy transfer in a gravity spectrum, Part I: General theory. *Journal of Fluid Mechanics*, 12(4):481–500. doi:10.1017/S0022112062000373.

Hasselmann, K. 1963a. On the nonlinear energy transfer in a gravity wave spectrum, Part II, Conservation theorems, wave particle analogy, irreversibility, *Journal of Fluid Mechanics*, 15(2):273–281. doi:10.1017/S0022112063000239.

Hasselmann, K. 1963b. On the nonlinear energy transfer in a gravity wave spectrum, Part 3, Conservation theorems, wave-particle analogy, irreversibility. *Journal of Fluid Mechanics*, 15(3):385–398. doi:10.1017/S002211206300032X.

Huang, H. 2009. *Dynamics of Surface Waves in Coastal Waters: Wave-Current-Bottom Interactions.* Higher Education Press, Beijing – Springer, Dordrecht, The Netherlands.

Kartashova, E. 2011. *Nonlinear Resonance Analysis: Theory, Computation, Applications.* Cambridge University Press, Cambridge, UK.

Longuet-Higgins, M. S. 1962. Resonant interactions between two trains of gravity waves. *Journal of Fluid Mechanics*, 12(3):321–332. doi:10.1017/S0022112062000233.

Mahrenholtz, O. and Markiewicz, M. (Eds.). 1999. *Nonlinear Water Wave Interaction (Advances in Fluid Mechanics).* WIT Press/Computational Mechanics Publications, Southampton, UK.

Miropol'sky, Y. Z. 2001. *Dynamics of Internal Gravity Waves in the Ocean (translated from Russian by O. D. Shishkina).* Springer, New York.

Peregrine, D. H. 1976. Interaction of water waves and currents. *Advances in Applied Mechanics*, 16:9–117. doi:10.1016/S0065-2156(08)70087-5.

Phillips, O. M. 1960. On the dynamics of unsteady gravity waves of finite amplitude, Part I, The elementary interactions. *Journal of Fluid Mechanics*, 9(2):193–217. doi:10.1017/S0022112060001043.

Phillips, O. M. 1961. On the dynamics of unsteady gravity waves of finite amplitude, Part II, Local properties of random wave field. *Journal of Fluid Mechanics*, 11(1):143–155. doi:10.1017/S0022112061000913.

Phillips, O. M. 1980. *The Dynamics of the Upper Ocean*. Cambridge University Press, Cambridge, UK.

Phillips, O. M. and Hasselmann, K. 1986. *Wave Dynamics and Radio Probing of the Ocean Surface*. Plenum Press, New York.

Voliak, K. I. 2002. *Selected Papers. Nonlinear Waves in the Ocean*. Nauka, Moscow (in Russian and English).

Yuen, H. C. and Lake, B. M. 1982. Nonlinear dynamics of deep-water gravity waves. In *Advances in Applied Mechanics* (Ed. C.-S. Yih). Academic Press, New York, Vol. 22, pp. 67–229.

Zakharov, V. E., 1968. Stability of periodic waves of finite amplitude on the surface of a deep fluid. *Journal of Applied Mechanics and Technical Physics*, 9(2):86–94. doi:10.1007/BF00913182.

Zakharov, V. E. (Ed.). 1998. *Nonlinear Waves and Weak Turbulence*. American Mathematical Society, Providence, RI.

Zaslavskiy, M. M. 1996. On the role of four-wave interactions in formation of space–time spectrum of surface waves. *Izvestiya, Atmospheric and Oceanic Physics*, 31(4):522–528 (translated from Russian).

2.6 TURBULENCE

2.6.1 INTRODUCTION

In fluid dynamics turbulent flow is characterized by irregular (chaotic) movement of particles of the fluid. Turbulent flow is a nonlinear dynamic system with an extremely large number of degrees of freedom. Unlike laminar flow, mixing, intermittent and irregularity in turbulent flow are very high; therefore, the behaviour of turbulent flow is complicated to predict and describe mathematically. In fact, the nature of turbulence has been called "the last unsolved problem in classical physics."

The following characteristics of turbulence are known: irregularity, diffusivity, large Reynolds numbers, 3D velocity fluctuations, dissipation, continuum, "turbulent flows are flows." A concept of continuum was formulated by (Tennekees and Lumley 1972): "Turbulence is a continuum phenomenon, governed by the equations of fluid mechanics. Even the smallest scales occurring in a turbulent flow are ordinarily far larger than any molecular length scale." Turbulence is also characterized by strong vortex generation mechanism (vortex stretching) and spectral energy cascade.

Various methods are employed to model the turbulent flows: statistical, spectral, diffusional, semi-empirical theories, and CFD methods. In particular, the most adequate numerical modeling is based on solving the system of Navier–Stokes equations. In present, widely accepted models of turbulent flow are the following: (1) Direct numerical simulation (DNS), (2) Standard k–ε model, (3) the Reynolds averaged Navier–Stokes (RANS), (4) Reynolds stress model, and (5) Large eddy simulation (LES). Fundamentals and models of turbulence as well as numerical data are described in many books (e.g., Monin and Yaglom 1971, 1975; Pope 2000;

Belotserkovskii et al. 2005; Schiestel 2008; Cardy et al. 2008; Durbin and Pettersson Reif 2011; Rebollo and Lewandowski 2014). More specifically, the phenomenon of oceanic turbulence, methods of its observation and analysis are discussed in books (Monin and Ozmidov 1985; Baumert et al. 2005; Thorpe 2007). In this section we consider only three categories of turbulent flow, which are closely related to ocean surface dynamics, namely, weak turbulence, strong turbulence, and turbulent wake. These phenomena are of a great importance in OHD and remote sensing.

2.6.2 KOLMOGOROV LAW

Statistical theory of turbulence was initially proposed by Taylor in 1935, von Kármán in 1937 and developed by Kolmogorov in 1941 and Obukhov in 1941, and later by Batchelor in 1953; for references see, e.g., books (Monin and Yaglom 1975; Frisch 1995; Lesieur 2008). Today, this theory is the most frequently used in studying environmental turbulence. The theory is based on the three following hypotheses: (1) turbulence is homogeneous and isotropic at small scales, (2) on a small scale the statistical properties depend only on the kinematic viscosity ν and the dissipated energy or dissipative rate ε, and (3) in the inertial range ("small scales, but not too small") the statistical properties depend only on the dissipated energy. This theory was formulated using the second order of the *longitudinal structure function* of the velocity differences in a turbulent flow, which is

$$S_2(\vec{r}) = <\left[\left(u(\vec{x}+\vec{r})-u(\vec{x})\right)\hat{r}\right]^2> = C_2\varepsilon^{2/3}r^{2/3} \qquad (2.37)$$

with scale $(\varepsilon r)^{2/3}$, where $\vec{x}, \vec{x}+\vec{r}$ are points in a turbulent flow field, u is the component of the velocity in the direction $\vec{r}$, $r = |\vec{r}|$ is the length, $(\hat{r} = |\vec{r}|/r$ is the unit vector), ε is the mean rate of energy dissipation, C_2 is a constant, and brackets denote an average. Dimension analysis gives the energy spectrum of turbulence

$$E(k) = C_k\varepsilon^{2/3}k^{-5/3}, 1/\ell \ll k \ll 1/\eta, \qquad (2.38)$$

which is the 1D spectrum of velocity, but the similar law applies for the 3D spectrum with constant $(18/55)C_k$. This expression is known as Kolmogorov or Kolmogorov–Obukhov spectrum or the "five-thirds law." In (2.38) $C_k = 1.5$ is the universal Kolmogorov constant, ℓ and η are lengths at large scale and at small scale, respectively, $1/\ell \ll k \ll 1/\eta$ is the "inertial range," and k is the wavenumber. The length, η, time, τ, and velocity, u, scales of the smallest eddies of turbulence are defined as

$$\eta \equiv \left(\frac{\nu^3}{\varepsilon}\right)^{\frac{1}{4}}, \tau \equiv \left(\frac{\nu}{\varepsilon}\right)^{\frac{1}{2}}, u \equiv (\nu\varepsilon)^{\frac{1}{4}}. \qquad (2.39)$$

The Kolmogorov theory postulates that the small-scale turbulence is in equilibrium (independent of large-scale) and is controlled solely with ε and ν. The "five-thirds law" provides predictions of turbulence properties that agree well with many geophysical experiments when the effect of intermittency is negligible.

Later it was found (e.g., Landau and Lifshitz 1987; Frisch 1995) that this scaling result, based on (2.37) is generalized by the structure function of any order

$$S_n(\vec{r}) = <\left[(u(\vec{x}+\vec{r}) - u(\vec{x}))\hat{r} \right]^n > \propto C_n r^{\varsigma(n)}.$$ (2.40)

At scaling exponent $\varsigma(n) = n/3$ the Kolmogorov scaling law is valid, with no corrections. Generalized statistical theory of turbulence with structure function (2.40) has been considered by many authors at different turbulent regimes and scaling exponents.

The Kolmogorov's theory has led to many activities in the rigorous mathematical calculations and measurements of turbulence. In particular, on this basis, it was created so-called *weak wave turbulence (WT) theory* dealing with stochastic nonlinear wave fields in incompressible turbulent fluid. WT theory describes cascade processes and energy dissipation similar to the Kolmogorov's scaling law. Mathematical formulations and details of the WT theory can be found in books (Zakharov et al. 1992; Nazarenko 2011).

2.6.3 WEAK TURBULENCE

Initially WT theory was developed in the 1960s to describe the transfer of energy among turbulent, weakly nonlinear, dispersive waves in plasma and fluid (Kadomtsev 1965; Zakharov et al. 1992). Most hydrodynamic studies and applications were made assuming that fluctuations of velocity in turbulent flow are small (i.e., flow contains small eddies) and the statistics of weak turbulence is quasi-Gaussian. This, as a whole, corresponds to the Kolmogorov's scaling theory.

Mathematically, the WT theory is based on the solution of the nonlinear Schrödinger equation under the assumption of weak nonlinearity and small wave amplitudes. Important cases have been considered in the statistical dynamics of sea surface waves (gravity, capillary) as well as geophysical (Rossby) waves. Under certain conditions, there is the *stationary* analytic solution of the WK equation with wave resonant interactions (Hasselmann equation, Section 2). This yields theoretical spectra and fluxes of weak turbulence E(k) known as Kolmogorov–Zakharov (KZ) spectra. Here are several examples of analytical spectra (Zakharov et al. 1992):

- four-wave interactions, gravity waves on a deep water, direct energy cascade

$$E(k) \propto \varepsilon^{1/3} g^{1/2} k^{-5/2}, \omega = \sqrt{gk}$$ (2.41)

- four-wave interactions, gravity waves, inverse energy cascade

$$E(k) \propto q^{1/3} g^{2/3} k^{-7/3}, \omega = \sqrt{gk}$$ (2.42)

- five-wave interactions, gravity waves

$$E(k) \propto \varepsilon^{1/4} g^{5/8} k^{-21/8}, \omega = \sqrt{gk}$$ (2.43)

- three-wave interactions, capillary waves

$$E(k) \propto \varepsilon^{1/2}\sigma^{1/4}k^{-7/4}, \omega = \sqrt{gk + \sigma k^3} \sim \sqrt{\sigma k^3}, \qquad (2.44)$$

where ε is the energy flux to high wave numbers and q is the wave action flux to small wave numbers, σ is the surface tension, g is the gravitational constant, ω is frequency, and k is wavenumber.

One of the central discoveries in WT theory is the establishments of a family of power-law energy spectra $E(k) \propto \varepsilon^{1/(N-1)}k^{-p}$, where the exponent of the energy flux depends on the N-wave interaction process which is itself fixed by both the wave dispersion relation and the dominant nonlinear interaction. This out-of-equilibrium solution takes place in presence of the energy sources and sinks separated by a large inertial range of scales. The exponent $p > 0$ depends on the scaling properties of the interaction coefficient and type of the solution.

Experimental evidence of KZ spectra have been found in many physical applications of weak turbulence including ocean surface dynamics (e.g., Nazarenko 2011; Shrira and Nazarenko 2013). The WT theory is a very general framework by which the statistical properties of a large number of interacting waves can be studied. In particular, WT theory has found further application in the Zakharov's analytic theory of a wind-driven sea surface and ocean wave forecasting.

2.6.4 Strong Turbulence

A principal point is a division of different turbulent states in a weak and a strong one. Unlike WT definition which is formulated mathematically using the WK equation and its analytical analysis, the definition of strong turbulence (ST) is purely descriptive. Actually, it is assumed that ST is any turbulence which is not weak. It is well known that WT theory breaks down when wave amplitudes are large enough and/or velocity fluctuations in turbulent flow are not small. This is a case of *fully developed turbulence* which is characterized by extremely irregular variations of velocity at each point. Several factors—stochastic motions (*vortex shedding*), interactions, intermittency, and strong nonlinearity are typical for ST flow.

Transition between WT and ST can be investigated using the *complex Ginzburg–Landau equation* (CGLE) introduced in 1950, which plays an important role in describing many dynamic physical systems. The CGLE is given by

$$\frac{\partial A}{\partial T} = (1 + \iota\alpha)\Delta A + A - (1 + \iota\beta)|A|^2 A, \qquad (2.45)$$

where $A(X,T)$ is a complex-valued function of (X) space and (T) time variables and α, $\beta \geq -1$ are parameters known as linear dispersion and nonlinear dispersion, respectively. The CGLE possesses a family of time-periodic solutions:

$$A(X,T) = r_q e^{\iota q X + \iota\omega_q T + \iota\varphi_0} \qquad (2.46)$$

with parameters $q, \varphi_0, r_q, \omega_q$. The amplitude A describes slow modulations in space and time of the underlying bifurcating spatially periodic pattern including hydrodynamic stability and intermittency problems (i.e., property peculiar to ST). The CGLE is a flexible mathematical model platform for studying *complex turbulent flows* especially at 2D or 3D spaces. In some case, the CGLE reduces to the nonlinear *Schrödinger* wave equation which can also yield some turbulent solutions but for WT. Other possibility to model ST is DNS using the Navier–Stokes equations for viscous incompressible fluid with the corresponding space and time-dependent parameters. Available models for this were listed in Section 2.6.1.

At present, the development of the theory of nonlinear dynamic systems makes considerable progress in the understanding of turbulence. In particular, the chaos theory and (multi)fractal formalism provide asymptotical description of stochastic motions similar to ST that is an important attribute of the most geophysical and hydrodynamic systems. An important hydrodynamic example of ST is a free-surface water flow that includes such natural events as wave breaking or splitting at the sea surface, turbulent jet, wake, and so-called strong Langmuir turbulence.

Turbulence-wave interactions in ocean boundary layer (or wall- or near-wall bounded turbulent flows) produce sometimes 2D surface quasi-symmetric chaotic patterns—*coherent structures*, of variable geometry known as Lagrangian coherent structures (LCS), *hairpin* vortices, *cat's eye* eddies and *horseshoe* vortices (Figure 2.14), which can also be categorized as ST. LCS as well as the near-wall turbulence (or shear flow), created directly by a body force are studied using space and time-dependent Navier–Stokes DNS only.

Very ST at the air-sea interface is an essentially two-phase region involving significant gas motion and violent flows of bubbles and droplets. For example, a steep breaking water wave crest produces high-velocity turbulent two-phase dispersed flow (known as "whitecap") which is the major contributor to ocean microwave emissivity at high winds. Ocean whitecap and foam are perfectly registered in optical images and can be investigated using digital processing (Chapter 7).

 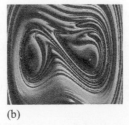 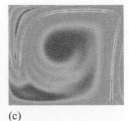

(a) (b) (c)

FIGURE 2.14 (**See color insert.**) Lagrangian coherent structures: (a) Hairpin-shape vortices. Mediterranean Sea, Shear Wall Spiral Eddies. (Image from NASA). Obtained by astronaut Paul Scully-Power, Space Shuttle mission STS-41-G, 1984. (Based on Robinson 2010), (b) Cat's eye eddy, computer generation. (Based on Farazmanda and Haller 2012), (c) Horseshoe vortex, computer generations. (Based on Shadden et al. 2005.)

2.6.5 Wake Turbulence (Turbulent Wake)

According to fluid mechanics classification both phenomena in the title are defined as turbulent flow. However, the words "wake turbulence" and "turbulent wake" may have different meaning. Wake turbulence mostly refers to a disturbance in the atmosphere (e.g., wake around aircraft) and turbulent wake is usually connected with fluid flow. We use the second definition. In our consideration, the occurrence of turbulent wake is associated with a body, moving through a fluid with enough velocity; the flow separates from the body and forms a wake region (or a vapor-gas cavity that is not considered here).

Several types of wakes are known: the Kelvin wake, a von Kármán vortex street, wave breaking wake, as well the Bernoulli hump, and some other specific wakes. Hydrodynamics and parameters of turbulent wake are discussed in many books (e.g., Birkhoff and Zarantonello 1957; Holt 1968; Townsend 1976; Green 1995; Mathieu and Scott 2000; Piquet 2001; Biswas and Eswaran 2002; Bernard and Wallace 2002; Davidson 2015). But the most data are concerned with laboratory and model studies (e.g., Voropayev and Afanasyev 1994).

A typical case is the turbulent wake induced by a circular cylinder. Evolution of the wake is characterized by the Reynolds number: at $1 < \text{Re} < 40$ two symmetric vortexes appear and at $40 < \text{Re} < 9500$ several vortexes appear and form a staggered parallel pattern known as *von Kármán vortex street*. At higher Reynolds number this pattern becomes more irregular causing the turbulent mixing that leads to "fully established turbulent flow" (at $\text{Re} \approx 10^8$). More familiar type of wake is the Kelvin wake which is a V-shaped wave pattern behind the ship on the surface. The Kelvin wake pattern is perfectly observed in optical images and its geometry is well studied. Radar also enables the detection of ship wake signatures but their specification requires additional analysis. Figure 2.15 illustrates different types of turbulent wakes.

In real ocean, a turbulent wake has more complicated structure and configuration dependent on parameters of a moving body (size, velocity, depth) and environmental conditions—stratification of the upper ocean layer, boundary layer properties, thermocline depth, ambient turbulence, bathometry, etc.

Stratified turbulent flows (wake), induced by a moving body are characterized by the following dimensionless parameters:

- the Reynolds numbers, $\text{Re} = UD/\nu$
- the internal Froude number $\text{Fr} = U/ND$
- the buoyancy (Brunt–Väisälä) frequency $N = [g\Delta\rho/\rho_0 H]^{1/2}$ with ρ_0 as the mean density at mid-depth
- the Strouhal number $\text{St} = f(D / U)$ with f as the oscillation frequency
- the Roshko number $R_0 = fD^2/\nu = \text{St} \cdot \text{Re}$ (introduced in 1952),

where D is diameter, U is speed, and H is depth of a moving body.

Hydrodynamic calculations, utilizing CFD techniques attempt to predict the structure and the behaviour of turbulent wakes from a moving submerged body of different shape. As it follows from available data (e.g., from the publications in the Proceedings of Symposium on Naval Hydrodynamics), the results of numerical simulations are highly dependent on overall model specification. Discussion of

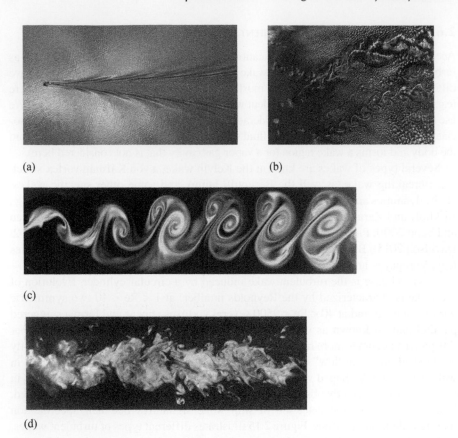

FIGURE 2.15 Turbulent wakes: (a) Kelvin wake pattern (aerial view), (b) von Kármán vortex street. (Image from NOAA), (c) Kármán vortex street behind a bluff body, laboratory experiment. (Adapted from Afanasyev and Korabel 2006), (d) photograph of turbulent wake of a cylinder. (Adapted from Van Dyke 1982.)

these works is beyond the scope of this book. However, the structure and configuration of the generated wake at representative environmental conditions may be described schematically. Obviously, turbulent flow behind a body and hydrodynamic perturbations are changed with the distance and time. Detailed analysis shows that induced wake pattern behind a body can be divided into several connected regions (Figure 2.16). More realistic wake scenario includes the developments of instabilities and spatiotemporal *intermittent* regime.

There is an important conclusion (Mathieu and Scott 2000) that "in the turbulent regime, although oscillations may still discernible win the wake, the flow is steady in statistical sense. Separation produces shear layers that divide the fast-moving fluid outside the wake from slowly moving fluid behind the body." It is also assumed that the wake becomes self-similar at certain distance from a body. This can occur in the late stage of the transition from turbulent regime to collapse regime.

Some specific situation is associated with strong density stratification in the ocean. Collapsing submerged wake generates a set of turbulent spots which

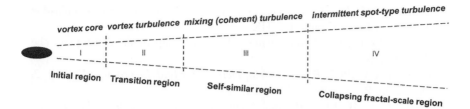

FIGURE 2.16 Schematic structure of induced turbulent wake. (I)—initial stage where the vortex occur very closely to the body, (II)—turbulence induced by the vortex (or a number of vortexes), (III)—turbulent mixing and generation of coherent (e.g., LCS) structures, (IV)—the collapsing wake and generation of (fractal-like) turbulent spots.

become pancake-shaped causing the fine structure of turbulence. Double-diffusive and thermohaline processes may trigger the generation and evolution of turbulent spots (due to the development of secondary instabilities in stratified fluid) which can reach somehow the near-surface boundary layer and effect on surface roughness. In this case, the spot-type signatures can be detected by a sophisticated remote sensor.

REFERENCES TO SECTION 2.6

Afanasyev, Y. D. and Korabel, V. N. 2006. Wakes and vortex streets generated by translating force and force doublet: Laboratory experiments. *Journal of Fluid Mechanics*, 553(1):119–141. doi:10.1017/S0022112006008986.

Batchelor, G. K. 1953. *The Theory of Homogeneous Turbulence (Cambridge Science Classics)*. Cambridge University Press, New York.

Baumert, H. Z., Simpson, J., and Sündermann, J. (Eds.). 2005. *Marine Turbulence: Theories, Observations, and Models*. Cambridge University Press, Cambridge, UK.

Belotserkovskii, O. M., Oparin, A. M., and Chechetkin, V. M. 2005. *Turbulence: New Approaches*. Cambridge International Science Publishing, Cambridge, UK.

Bernard, P. S. and Wallace, J. M. 2002. *Turbulent Flow: Analysis, Measurement, and Prediction*. John Wiley & Sons, Hoboken, NJ.

Birkhoff, G. and Zarantonello, E. H. 1957. *Jets, Wakes, and Cavities*. Academic Press, New York.

Biswas, G. and Eswaran, V. (Eds.). 2002. *Turbulent Flows: Fundamentals, Experiments and Modeling*. CRC Press, Boca Raton, FL.

Cardy, J., Falkovich, G., and Gawedzki, K. 2008. *Non-equilibrium Statistical Mechanics and Turbulence* (Eds. S. Nazarenko and O. V. Zaboronski). Cambridge University Press, Cambridge, UK.

Davidson, P. A. 2015. *Turbulence: An Introduction for Scientists and Engineers*, 2nd edition. Oxford University Press, Oxford, UK.

Durbin, P. A. and Pettersson Reif, B. A. 2011. *Statistical Theory and Modeling for Turbulent Flows*, 2nd edition. John Wiley & Sons, Chichester, UK.

Frisch, U. 1995. *Turbulence: The Legacy of A. N. Kolmogorov*. Cambridge University Press, Cambridge, UK.

Green, S. (Ed.). 1995. *Fluid Vortices. Volume 2*. Springer, Dordrecht, The Netherlands.

Holt, M. (Ed.). 1968. *Basic Developments in Fluid Dynamics, Volume 2*. Academic Press, New York.

Kadomtsev, B. B. 1965. *Plasma Turbulence (translated from Russian by L. C. Ronson and edited by M. G. Rusbridge)*. Academic Press, London, New York.

Landau, L. D. and Lifshitz, E. M. 1987. *Fluid Mechanics (Course of Theoretical Physics, Volume 6, translated from Russian by J. B. Sykes and W. H. Reid)*, 2nd edition. Elsevier – Butterworth-Heinemann, Burlington, MA.

Lesieur, M. 2008. *Turbulence in Fluids*, 4th edition. Springer, Dordrecht, The Netherlands.

Mathieu, J. and Scott, J. 2000. *An Introduction to Turbulent Flow*. Cambridge University Press, Cambridge, UK.

Monin, A. S. and Ozmidov, R. V. 1985. *Turbulence in the Ocean*. D. Reidel Publishing Company, Dordrecht, Holland.

Monin, A. S. and Yaglom, A. M. 1971. *Statistical Fluid Mechanics, Volume 1: Mechanics of Turbulence (English translation, edited by J. L. Lumley)*. The MIT Press, Cambridge, MA.

Monin, A. S. and Yaglom, A. M. 1975. *Statistical Fluid Mechanics, Volume 2: Mechanics of Turbulence (English translation, edited by J. L. Lumley)*. The MIT Press, Cambridge, MA.

Nazarenko, S. 2011. *Wave Turbulence*. Springer, Berlin, Germany.

Piquet, J. 2001. *Turbulent Flows: Models and Physics*, 2nd printing. Springer, Berlin, Germany.

Pope, S. B. 2000. *Turbulent Flows*. Cambridge University Press, Cambridge, UK.

Rebollo, T. C. and Lewandowski, R. 2014. *Mathematical and Numerical Foundations of Turbulence Models and Applications*. Birkhäuser – Springer, New York.

Schiestel, R. 2008. *Modeling and Simulation of Turbulent Flows*. ISTE – John Wiley & Sons, Hoboken, NJ.

Shrira, V. and Nazarenko, S. (Eds.). 2013. *Advances in Wave Turbulence*. World Scientific Publishing, Singapore.

Tennekes, H. and Lumley, J. L. 1972. *A First Course in Turbulence*. The MIT Press, Cambridge, MA.

Thorpe, S. A. 2007. *An Introduction to Ocean Turbulence*. Cambridge University Press, Cambridge, UK.

Townsend, A. A. 1976. *The Structure of Turbulent Shear Flow*, 2nd edition. Cambridge University Press, New York.

Voropayev, S. I. and Afanasyev, Y. D. 1994. *Vortex Structures in a Stratified Fluid: Order from Chaos*. Chapman & Hall/CRC Press, Boca Raton, FL.

Zakharov, V. E., L'vov, V. S., and Falkovich, G. 1992. *Kolmogorov Spectra of Turbulence 1: Wave Turbulence*. Springer, Berlin, Germany.

2.7 INSTABILITIES

Stability is an important aspect of any dynamical system. Helmholtz established the science of hydrodynamic stability in 1868. Since then, hydrodynamic stability has become the main tool in analysis of small perturbations in sheared flow including turbulence as well (Section 2.6).Hydrodynamic instabilities have been studied extensively; see books (Zakharov 1968; Chandrasekhar 1981; Riahi 1996; Moiseev et al. 1999; Schmid and Henningson 2001; Drazin 2002; Drazin and Reid 2004; Rahman 2005; Manneville 2010; Charru 2011; Sengupta 2012; Yaglom and Frisch 2012; Gaissinski and Rovenski 2018). One of the first experiments demonstrating instability of surface waves was conducted by Su and Green (1984).

In classical hydrodynamic theory, the Navier–Stokes equations of motion globally describe fluid flow at any Reynolds number. It is assumed that flow instability is formally a linear concept, applicable only for infinitesimal perturbations to a steady or periodic solution to the governing equations. The concept of stability of steady flow has been clearly formulated by Landau and Lifshitz (1987): "Yet not every solution of the equations of motion, even if it is exact, can actually occur in Nature. The flows that occur in Nature must not only obey the equations of fluid dynamics, but also be stable. For the flow to be stable it is necessary that small perturbations, if they arise, should decrease with time. If, on the contrary, the small perturbations which inevitably occur in the flow tend to increase with time, then the flow is absolutely unstable. Such a flow unstable with respect to infinitely small perturbations cannot exist."

The following types of instabilities are distinguished: (1) *primarily instability* becomes possible if basic flow state changed to another flow state under the influence of its critical instability parameters; (2) *secondary instability* occurs when the flow state has been changed already due to primarily instability and it is changed again due to the influence of critical instability parameters; and (3) *tertiary instability* is a result of the consecutive actions of preliminary and secondary instabilities.

The analysis of secondary instability is of practical interest in plasma and hydrodynamics (Moiseev et al. 1999; Rahman 2005). In this situation, a self-similar transition in the hydrodynamic system (for example, shear flow) from one state to other state can occur. The secondary instability is also a possible wave generation mechanism that can trigger and amplify certain harmonics in the surface wave spectrum (Craik 1985). The concept of secondary hydrodynamic instability has also been applied in the development of passive microwave remote sensing technique associated with the detection of deep ocean events (Cherny and Raizer 1998).

At the same time, the following types of hydrodynamic instabilities, observed in real world are known:

- *Kelvin–Helmholtz instability* at the interface between two fluids moving with different velocities.
- *Rayleigh–Taylor instability* at interface between two fluids of different density.
- *Benjamin–Feir instability* or modulation instability for nonlinear Stokes waves on the water surface.

- *Taylor–Couette instability* related to convection rolls, vortexes, and/or spiraling eddies.
- *Benard instability* for many varieties of ocean-atmosphere convection.
- *Baroclinic instability* in stratified shear flows.

Practically, all listed hydrodynamic instabilities, one way or another, play an important role in the dynamics of ocean boundary layer causing the transition of wave motions to turbulence accomplishing by the change of the surface wave spectrum. Due to surface instabilities the redistribution of wave energy occurs as spontaneously as well through multiple cascades that can lead to the excitation (or amplification) or suppression of certain surface wave components or redistribution of wave group structure. Instability-induced amplification mechanism may trigger multiple excitation of the wave number spectrum at certain intervals of spatial frequencies. The possible causes are associated with the following processes:

- kinematics of long-short surface waves and wave-wave resonant interactions
- surface wave-current interactions
- double-diffusive convection
- oscillations of ocean boundary layer parameters (wind speed, drag coefficient, roughness coefficient)
- acoustical action or/and underwater sound effects ("parametric excitation")
- underwater explosions (UDEX).

Mathematical treatment and stability analysis have attracted the attention of many scientists and both linear and weakly nonlinear stability theories has become a mature topic in fluid mechanics and CFD.

REFERENCES TO SECTION 2.7

Chandrasekhar, S. 1981. *Hydrodynamic and Hydromagnetic Stability*. Dover Publications, New York.

Charru, F. 2011. *Hydrodynamic Instabilities (Cambridge Texts in Applied Mathematics)*. Cambridge University Press, Cambridge, UK.

Cherny, I. V. and Raizer, V. Y. 1998. *Passive Microwave Remote Sensing of Oceans*. John Wiley & Sons, Chichester, UK.

Craik, A. D. D. 1985. *Wave Interactions and Fluid Flows*. Cambridge University Press, Cambridge, UK.

Drazin, P. G. 2002. *Introduction to Hydrodynamic Stability (Cambridge Texts in Applied Mathematics)*. Cambridge University Press, Cambridge, UK.

Drazin, P. G. and Reid, W. H. 2004. *Hydrodynamic Stability*, 2nd edition. Cambridge University Press, Cambridge, UK.

Gaissinski, I. and Rovenski, V. 2018. *Modeling in Fluid Mechanics: Instabilities and Turbulence*. CRC Press, Boca Raton, FL.

Landau, L. D. and Lifshitz, E. M. 1987. *Fluid Mechanics (Course of Theoretical Physics, Volume 6)*, 2nd edition. Elsevier – Butterworth-Heinemann, Burlington, MA.

Manneville, P. 2010. *Instabilities, Chaos and Turbulence (ICP Fluid Mechanics)*, 2nd edition. Imperial College Press, London, UK.

Moiseev, S. S., Pungin, V. G., and Oraevsky, V. N. 1999. *Non-Linear Instabilities in Plasmas and Hydrodynamics*. CRC Press, Boca Raton, FL.

Rahman, M. 2005. *Instability of Flows (Advances in Fluid Mechanics)*. WIT Press Computational Mechanics, Southampton, UK.

Riahi, D. N. 1996. *Mathematical Modeling and Simulation in Hydrodynamic Stability*. World Scientific Pub Co Inc., Singapore.

Schmid, P. J. and Henningson, D. S. 2001. *Stability and Transition in Shear Flows (Applied Mathematical Sciences, Volume 142)*. Springer, Dordrecht, The Netherlands.

Sengupta, T. K. 2012. *Instabilities of Flows and Transition to Turbulence*. CRC Press, Boca Raton, FL.

Su, M.-Y. and Green, A. W. 1984. Coupled two- and three-dimensional instabilities of surface gravity waves. *Physics of Fluids*, 27(1):2595–2597.

Yaglom, A. M. and Frisch, U. (Eds.). 2012. *Hydrodynamic Instability and Transition to Turbulence (Fluid Mechanics and Its Applications 100)*. Springer, Dordrecht, The Netherlands.

Zakharov, V. E. 1968. Stability of periodic waves of finite amplitude on the surface of a deep water. *Journal of Applied Mechanics and Technical Physics*, 9(2):190–194 (translated from Russian). doi:10.1007/BF00913182.

2.8 WAVE BREAKING

2.8.1 INTRODUCTION

Wave breaking is known in fluid mechanics as a hydrodynamic transformation process occurring due to local gravitational instability of the wave system resulting to loss of its equilibrium. Wave breaking is a strongly nonlinear 3D phenomenon. Wave breaking is an important mechanism of turbulence production, mixing, and hydrodynamic forces. Wave breaking leads to the formation of disperse media at the ocean-atmosphere interface known as foam, whitecap, bubbles, spray, and *indissoluble* two-phase turbulent flows. Wave breaking and foam/whitecap are important contributors to the sea surface microwave emission (and backscatter) at high winds.

Wave breaking dynamics is a subject of long-term theoretical and experimental studies which are discussed in many books (Phillips 1980; Phillips and Hasselmann 1986; Bortkovskii 1987; Su 1987; Banner and Peregrine 1993; Monahan and MacNiocaill 1986; Massel 2007; Sharkov 2007; Babanin 2011; Soloviev and Lukas 2014). Considerable progress in the study of wave breaking characteristics has been made in hydrodynamics of surfzone because of importance of coastal engineering (Svendsen 2006). In this context, numerical solutions of the Navier–Stokes equations have demonstrated realistic results (e.g., Lemos 1992; Lubin and Glockner 2015). At the same time, wave breaking hydrodynamics and phase transitions (whitecapping) in the open ocean have not been fully investigated and understood. Many researchers continue to explore various quantities of wave breaking events through laboratory measurements, field observations, and numerical modeling and simulations.

Historically, Phillips (1985) proposed a framework based on so-called breaking rate distribution, $\Lambda(c)dc$, which is the average length of the breaking crest per unit area traveling with velocities in the range $(c, c + \Delta c)$. The Phillips's $\Lambda(c)$ concept is used in oceanography for estimating the total energy dissipation rate due to wave breaking $E = b\rho g^{-1} \int c^5 \Lambda(c)dc$, the momentum flux from the waves to currents $M = b\rho g^{-1} \int c^4 \Lambda(c)dc$, and active whitecap fraction $W = T_{phil} \int c\Lambda(c)dc$, where g is gravity, ρ is the density of water, b is a numerical constant or "breaking parameter," and T_{phil} is an average bubble persistence time introduced by Phillips (1985). Detailed review and interpretation of the $\Lambda(c)$ concept are given by Banner et al. (2014). Formally, the calculated by $\Lambda(c)$-distribution whitecap area fraction W can be incorporated into remote sensing models. However, it becomes clear that such a phenomenological single-parameter approach has certain limitations because it does not describe a variety of geometrical and structural properties of real world foam/whitecap environments.

Meanwhile, large-scale wave breaking processes in the ocean are perfectly registered practically by any conventional passive optical and/or microwave sensor. In optical panchromatic images, foam and whitecap are revealed as distinct bright (white) objects of variable geometry and texture (Figure 2.17). Optical observations provide unique material for investigating foam/whitecap geometrical parameters and coverage statistics (Chapter 7).

2.8.2 Hydrodynamic Characterization

There are several classical physical theories (and equations) which can be used to evaluate pre-breaking evolution of nonlinear gravity waves on a water surface. The most known are the following: (1) the Stokes theory for limited deep-water waves, (2) the KdV equation for solitary-wave limit, (3) the Boussinesq equation for

FIGURE 2.17   Aerial photograph of foam and whitecap.

cnoidal waves, (4) the nonlinear Schrödinger equation for wave collapse, (5) the 2D Kadomtsev–Petviashvili (KP) equation for nonlinear wave motions (1970), and (6) numerical solutions of the Navier–Stokes equations which predict spatiotemporal nonlinear evolution of surface waves up to the wave breakpoint.

The criteria for individual wave breaking event in deep water have been formulated first by Stokes in 1847 and later by Michell in 1893. They are the following (Massel 2007):

- The particle velocity of fluid at the wave crest equals the phase velocity
- The crest of the wave attains a sharp point with an angle of 120°
- The ratio of wave height to wavelength is approximately 1/7 (= 0.142)
- Particle acceleration at the crest of the wave equals 0.5 g.

Since that, many criteria have been proposed to predict the onset of wave breaking. They can be classified into three categories: (1) geometric criteria based on local wave shape, (2) kinematic criteria based on particle and phase velocities, and (3) dynamic criteria based on acceleration at the wave crest. These wave breaking criteria were analyzed and tested in a number of laboratory experiments (e.g., Wu and Nepf 2002; Perlin et al. 2013).

In oceanography, four types of breaking waves (excluding micro-breaking) are divided (Figure 2.18):

- Spilling breakers—wave crests spill forward, creating foam and turbulent water flow
- Plunging breakers—wave crests form spectacular open curl; crests fall forward with considerable force

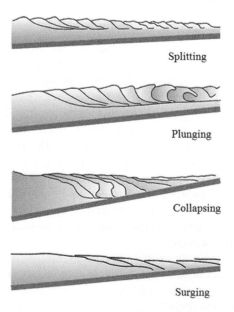

Splitting

Plunging

Collapsing

Surging

FIGURE 2.18   Four types of breaking waves. (Based on Holthuijsen 2007.)

- Collapsing breakers—wave fronts form steep faces that collapse as waves move forward
- Surging breakers—long, relatively low waves whose front faces and crests remain relatively unbroken as waves slide up and down.

This classification is based on the changes of the surface wave geometry during the breaking process. All four types of breakers can occur in shallow water, but splitting breakers are the most common in deep water. In the open ocean at strong gales, more fascinating picture is observed (especially from aircraft)—we call this type as "cascade wave breaking," which is a 3D stochastic process of the entire wave's top collapse, accompanied by intensive whitecapping and bubble/spray production. This leads to a strong and previously unexplored source of local deep ocean mixing and turbulence. Conventional theories may not be valid in such situations and, perhaps, more complicated models, involving two-phase flow dynamics are required. A possible variant can be considered on the basis of conservation equations for volume, momentum, and buoyancy which can also characterize the phenomenon known as "wave breaking in dense plumes."

2.8.3 Spectral Characterization

Wave breaking is a central problem in statistical spectral analysis of a wind-driven sea surface. The basic factor considered here is the nonlinear transformation of wave spectra through the energy dissipation during wave breaking process known as "spectral dissipation." Pioneering studies in this field were carried out by Kitaigorodskii, Longuet-Higgins, Hasselmann, Phillips, Zakharov, Melville, and later by many other researchers. The corresponding descriptions and references can be found in the books (Massel 2007; Babanin 2011).

In statistical theory of a wind-driven sea surface, spatiotemporal evolution of the surface wave spectra is described by the Hasselmann WK Equation (2.12), $\partial N / \partial t = S_{nl} + S_{input} + S_{diss}$, with dissipation terms due to wave breaking $S_{diss}(K) \neq 0$. The full-scale solution of the WK equation requires the knowledge of the right-hand site terms, in particularly, the spectral dissipation source $S_{diss}(K)$. It is generally assumed that $S_{diss}(K)$ is a function of the spectral density of wave action $S_{diss} = -\gamma_{diss}(K)N(K)$ or surface wave spectrum $S_{diss}(K) \propto \{F(K)\}^n$, $n \geq 1$. The most known parameterizations of $S_{diss}(K)$ are based on the following schemes: (1) Phillips $\Lambda(c)$-rate, (2) the Longuet-Higgins probability model, (3) wave breaking threshold, (4) Zakharov's WT theory, and the others; see review (Young and Babanin 2006; Ardhuin et al. 2010). An improvement of the physical basis of the dissipation term in the WK equation can be done, in our opinion, using self-similar scaling of wave breaking fields that will characterize their global spatially-statistical properties much better than semi-implicit schemes.

In conclusion, we may assume that wave breaking (the term is used in a wide sense) in the open ocean is one from the most complex dynamic event in nature. The conventional wave theories, perhaps, are incapable to describe fully this hydro-mechanical process resembling critical phenomena in phase transition physics. Stochastic and/or spectral models seem to be more appropriate for practice and can find certain applications in oceanography, climatology, and remote sensing. However, their adequate implementation requires the overall knowledge of spectral-energetic characteristics of ocean environment at variable conditions. Needed global information can be obtained using space-based observation technologies only.

REFERENCES TO SECTION 2.8

Ardhuin, F., Rogers, E., Babanin, A. V., Filipot, J.-F., Magne, R., Roland, A., van der Westhuysen, A., Queffeulou, P., Lefevre, J.-M., Aouf, L., and Collard, F. 2010. Semi-empirical dissipation source functions for ocean waves. Part I: Definitions, calibration and validations. *Journal of Physical Oceanography*, 40(9):1917–1941. doi:10.1175/2010JPO4324.1.

Babanin, A. 2011. *Breaking and Dissipation of Ocean Surface Waves*. Cambridge University Press, Cambridge, UK.

Banner, M. L. and Peregrine, D. H. 1993. Wave breaking in deep water. *Annual Review of Fluid Mechanics*, 25:373–397.

Banner, M. L., Zappa, C. J., and Gemmrich, J. R. 2014. A note on the Phillips spectral framework for ocean whitecaps. *Journal of Physical Oceanography*, 44(7):1727–1734.

Bortkovskii, R. S. 1987. *Air-Sea Exchange of Heat and Moisture during Storms*. D. Reidel, Dordrecht, The Netherlands.

Lemos, C. M. 1992. *Wave Breaking: A Numerical Study*. Springer, Berlin, Germany.

Lubin, P. and Glockner, S. 2015. Numerical simulations of three-dimensional plunging breaking waves: Generation and evolution of aerated vortex filaments. *Journal of Fluid Mechanics*, 767:364–393.

Massel, S. R. 2007. *Ocean Waves Breaking and Marine Aerosol Fluxes*. Springer, New York.

Monahan, E. C. and MacNiocaill, G. 1986. *Oceanic Whitecaps*. D. Reidel, Dordrecht, The Netherlands.

Perlin, M., Choi, W., and Tian, Z. 2013. Breaking waves in deep and intermediate waters. *Annual Review of Fluid Mechanics*, 45:115–145. doi:10.1146/annurev-fluid-011212-140721.

Phillips, O. M. 1980. *The Dynamics of the Upper Ocean*. Cambridge University Press, Cambridge, UK.

Phillips, O. M. 1985. Spectral and statistical properties of the equilibrium range in wind-generated gravity waves. *Journal of Fluid Mechanics*, 156:505–531.

Phillips, O. M. and Hasselmann, K. (Eds.). 1986. *Wave Dynamics and Radio Probing of the Ocean Surface*. Premium Press, New York.

Sharkov, E. A. 2007. *Breaking Ocean Waves: Geometry, Structure and Remote Sensing*. Praxis Publishing, Chichester, UK.

Soloviev, A. and Lukas, R. 2014. *The Near-Surface Layer of the Ocean: Structure, Dynamics and Applications (Atmospheric and Oceanographic Sciences Library)*, 2nd edition. Springer, Dordrecht, Heidelberg.

Su, M.-Y. 1987. Deep-water wave breaking: Experiments and field measure-
    ments. In *Nonlinear Wave Interactions in Fluids. The Winter Annual
    Meeting of the American Society of Mechanical Engineers.* The American
    Society of Mechanical Engineers, pp. 23–36.
Svendsen, I. A. 2006. *Introduction to Nearshore Hydrodynamics.* World
    Scientific Publishing, Singapore.
Wu, C. H. and Nepf, H. M. 2002. Breaking criteria and energy losses for
    three-dimensional wave breaking. *Journal Geophysical Research,*
    107(C10):41-1–41-18. doi:10.1029/2001JC001077.
Young, I. R. and Babanin, A. V. 2006. Spectral distribution of energy dissipa-
    tion of wind-generated waves due to dominant wave breaking. *Journal of
    Physical Oceanography,* 36(3):376–394. doi:10.1175/JPO2859.1.

2.9 HYDRODYNAMICS OF EXPLOSION

This section provides a brief introduction to the basic surface phenomena associated with underwater explosions. Underwater explosions are generally referred to as
UNDEX. In our opinion, UNDEX are of a great interest in OHD and remote sensing
since this fascinating and complex physics event can be perfectly registered and
explored by satellite optical sensors.

Systematic investigations of UNDEX have been undertaken since the 1940s
and surveyed in the classical book (Cole 1948, translated in Russian as well). The
most important contributions to the problem were made after World War II in the
1950s–1960s in US, UK, and Soviet Union. The first exact similarity solutions of
the problem of intense point-source explosions were given in the late 1940s independently by Sedov and Taylor, known as *Sedov–Taylor Spherical Blast Wave*
(Sedov 1993). This concept plays a key role in the development of blast wave theory. Many scientific results including adaptation of the Sedov's theory for UNDEX
are discussed in the books (Yakovlev 1961; Zamyshlyaev and Yakovlev 1973).
Moreover, the surface phenomena are investigated in a number of scientific papers
(e.g., Kranzer and Keller 1959; van Dorn 1961; Collins and Holt 1968; Ballhaus and
Holt 1974; Falade and Holt 1978; Costanzo 2011). Excellent reviews are given by
Holt (1977) and Kedrinskii (1978); some data are available in books (Teller et al.
1968; Le Mehaute and Wang Shen 1996; Kedrinskii 2005). Selected material is
presented below.

2.9.1 BASIC CHARACTERIZATION

The primarily features of UNDEX include: the explosion or detonation phase, the formation of the shock wave and its effects, the secondary loading effect known as bulk
cavitation, the effects of the expanding and contracting gas bubble, and shock wave
refraction effects. These features represent complex, integral combination of many
interconnected physical phenomena which are studied and described at some level of
detail depending on the type of the explosion—chemical (conventional) or nuclear.

As a destructive factor, UNDEX produce a localized disturbance of the water
surface in the form of crater and generate a group of surface waves referred as to

Explosively Generated Water Waves (EGWW) by the terminology suggested in book (Le Mehaute and Wang Shen 1996). The characteristics of EGWW depend on a number of parameters which are: (a) the yield in TNT, (b) depth of burst in feet, and (c) the efficiency of the wave generation process. Correspondently, UNDEX can be classified on three following categories: (1) deep-water explosion when the explosion crater on the water surface is small compared with to the water depth, (2) intermediate depth explosion, and (3) shallow-water explosions when water depth is small compared to the explosion crater size. Different categories of UNDEX cause different EGWW. The most critical UNDEX is a very shallow-water explosion when EGWW and wave propagation process are affected by the bathometry (bottom slope).

2.9.2 SEQUENCE OF EVENTS

UNDEX causes a complex sequence of physical phenomena. Very briefly, the following phases are known (at shallow water):

- detonation
- formation of underwater compressive shock waves
- creation of "slick" or "crack" on the surface (also known as "cavern")
- reflection of shock wave from the air-water boundary and water bottom
- part of the shock wave passes through the surface into the air and may form a condensation cloud
- creation of bulk cavitation due to reflected shock wave
- generation, expansion, contraction, and migration of the gas bubble
- formation of high-velocity jet over the surface known as *sultan* in Russian literature (Kedrinskii 1978). Sultans are referred to as "spray dome" or "splashes" or "plume" in English. Sultans are defined as a throwing out of water (vertical column)
- the radioactive contents of the bubble are vented through this hollow column and may form a cauliflower-shaped cloud at the top
- vertical column of water and spray (plume) falls back to the surface and creates gigantic wave (or droplet's cloud) of mist completely surrounding the column at its base
- creation of so-called base surge which is a dense cloud of small water droplets, moving rapidly outward from the column
- generation of EGWW; they move outward from the center of the explosion across the ocean.

2.9.3 SURFACE DISPLACEMENT

UNDEX create large-amplitude EGWW which may propagate on long distance and cause enormous damage to shorelines and harbors. It was shown (Kranzer and Keller 1959; Falade and Holt 1978) that in the far field the EGWW can be estimated using linear hydrodynamic theory because of maximum of wave slope is small. The general linear solution for the surface displacement $\eta(r,t)$ induced by the EGWW is given by (Kranzer and Keller 1959)

$$\eta(r,t) = -\frac{1}{\rho g^{1/2}} \lim_{y \to 0} \int_0^\infty s^{3/2} \overline{I}(s) [\tanh(sh)]^{1/2} \sec h(sh)$$

$$\times \cosh(y+h)\sin[(gs\tanh(sh))]^{1/2} tJ_0(rs)ds. \qquad (2.47)$$

Here $\overline{I}(s) = \int_0^\infty I(r)J_0(sr)rdr$ is the zero-order Hankel transform of the impulse distri-

bution given by $I(r) = \int_0^{t_1} P(r,t)dt$, where P is the pressure and t_1 is the cutoff time,

ρ is the water density, g is the gravity constant, y is the vertical distance measured from the undisturbed surface, h is the depth of ocean (= 100 dimensionless units), s = σ/h, is an auxiliary variable, and r is the radial distance from the point of explosion. Figure 2.19 illustrates calculated by (2.47) dependencies of $\eta(r,t)$. According to this formula, typical impulse-type displacement (wave height) can reach $\eta \approx$ 2–3 m over the range r = 20 km (estimates are based on linear theory). But in reality, amplitude of EGWW is larger.

2.9.4 BUBBLE MOTION

UNDEX create a bubble with very high internal temperature and pressure. The bubble contains detonation gas products or water vapor. In the case of nuclear explosions, the bubble also contains most of the radioactive debris. Furthermore, bubble motion displaces a great amount of water and produces surface waves of very large amplitude. The simplest theories of underwater bubble motion are based on solutions of the equations governing unsteady incompressible flow with spherical symmetry. The kinetic energy of the bubble is calculated on the assumption that it moves as an

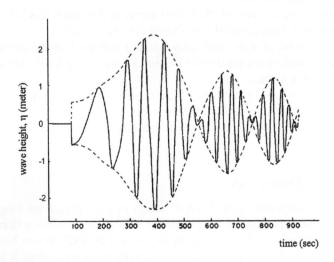

FIGURE 2.19 UNDEX EGWW. The wave height $\eta(r,t)$ in meters as a function of time in seconds. (Adapted from Kranzer and Keller 1959.)

oscillating spherical vortex. The expression for the potential of a spherical vortex of constant radius moving in a uniform stream is given by Lamb (1932).

2.9.5 CONCLUDED NOTE

UNDEX phenomena are summarized schematically in the Figure 2.20. In the case of nuclear explosion, the following citation gives us very laconic description: "… In an underwater burst, approximately one-half of all the energy is contained in the primary shock wave, which also causes the most destruction. Typical of the underwater explosion is the formation of a large bubble around the center of the burst, which undergoes pulsating motions that subside with time. The secondary waves emitted as a result of the pulsations have much less effect than the primary shock waves. For a 20-kiloton nuclear explosion at shallow depth, the radius of strong destructive action, causing ships to sink, is ~0.5 km. In an underwater nuclear explosion, an enormous column of water vapor and spray rises into the air above the water. Intense surface waves also arise, which spread for many km: the height of the wave crest reaches 3 m at a distance of 3 km from the epicenter of a 20-kiloton underwater burst…" (Raizer, Yu. P. *The Great Soviet Encyclopedia* 1979). Available on the Internet https://encyclopedia2.thefreedictionary.com/Explosion.

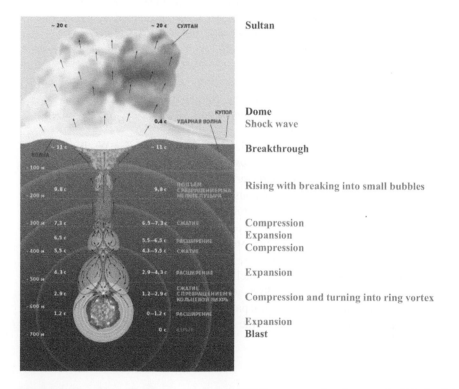

FIGURE 2.20 **(See color insert.)** UNDEX 30 KT summary, time in seconds, "c", depth in meters, "m" (original in Russian).

REFERENCES TO SECTION 2.9

Ballhaus, W. F. and Holt, M. C. 1974. Interaction between the ocean surface and underwater spherical blast waves. *Physics of Fluids*, 17(6):1068–1079.

Cole, R. H. 1948. *Underwater Explosions*. Princeton University Press, Princeton, NJ (published by Dover Publications Inc., 1965).

Collins, R. and Holt, M. 1968. Intense explosions at the ocean surface. *Physics of Fluids*, 11(4):701–713. doi:10.1063/1.1691988.

Costanzo, F. A. 2011. Underwater explosion phenomena and shock physics. In: *Structural Dynamics, Volume 3: Proceedings of the 28 IMAC, A Conference on Structural Dynamics, 2010. Part 1* (Ed. T. Proulx). Springer, New York, Dordrecht, pp. 917–937. doi:10.1007/978-1-4419-9834-7_82.

Falade, A. and Holt, M. 1978. Surface waves generated by shallow underwater explosions. *Physics of Fluids*, 21(10):1709–1716.

Holt, M. 1977. Underwater explosions. *Annual Review of Fluid Mechanics*, 9:187–214. doi:10.1146/annurev.fl.09.010177.001155.

Kedrinskii, V. K. 1978. Surface effects from an underwater explosion (review). *Journal of Applied Mechanics and Technical Physics*, 19(4):474–491 (translated from Russian). doi:10.1007/BF00859396.

Kedrinskii, V. K. 2005. *Hydrodynamics of Explosion: Experiments and Models (translated from Russian by S. Yu. Knyaseva)*. Springer, Berlin, Germany.

Kranzer, H. C. and Keller, J. B. 1959. Water waves produced by explosions. *Journal of Applied Physics*, 30(3):398–407. doi:10.1063/1.1735176.

Lamb, H. 1932. *Hydrodynamics*, 6th edition. Cambridge University Press, Cambridge, UK (reprinted 1945 by Dover, New York).

Le Mehaute, B. and Shen, W. 1996. *Water Waves Generated by Underwater Explosion*. World Scientific Publishing, Singapore.

Sedov, L. I. 1993. *Similarity and Dimensional Methods in Mechanics*, 10th edition. CRC Press, Boca Raton, FL.

Teller, E., Tallen, W. K., Higgins, G. H., and Johnson, G. W. 1968. *The Constructive Uses of Nuclear Explosives*. McGraw-Hill, New York.

van Dorn, W. G. 1961. Some characteristics of surface gravity waves in the sea produced by nuclear explosions. *Journal of Geophysical Research*, 6(11):3845–3862. doi:10.1029/JZ066i011p03845.

Yakovlev, Y. S. 1961. *Hydrodynamics of Explosion*. (Яковлев, Ю. С. *Гидродинамика взрыва*. Судпромгиз, Ленинград, 1961, in Russian).

Zamyshlyaev, V. B. and Yakovlev, Y. S. 1973. *Dynamic Loads in Underwater Explosion*. Naval Intelligence Support Center Washington, D. C. (Замышляев, Б. В., Яковлев Ю. С. *Динамические нагрузки при подводном взрыве*. Ленинград, Судостроение, 1967, in Russian). English translation available on the Internet www.dtic.mil/dtic/tr/fulltext/u2/757183.pdf.

2.10   SUMMARY

Ocean Hydrodynamics (OHD) plays a dominant role in the development of detection technology. For this reason, we have paid more attention to the major problems and investigations related to OHD. These investigations can be branched out in several directions: (1) dynamics of ocean waves, (2) ocean turbulence, (3) interactions phenomena, and (4) event-induced hydrodynamic impacts. We have also reviewed a number of theoretical and experimental data that enable some important aspects of OHD to be tested in more detail especially in context with satellite-based optical monitoring. Among them the most important for us are nonlinear dynamics of mid- and large-scale (~10–40 m) surface waves, their multiple (resonant) interactions, specifications of hydrodynamic instabilities, effects of (sub)surface turbulence, and the generation of OHD signatures induced by underwater sources/events.

Many attempts have been made during the years to establish the validity of theoretical models, approaches, and semi-empirical approximations that can explain the complexity of wave motions in the oceans. At least but not last, the OHD *background framework* has been created anyway and an incredible progress in understanding of dynamical nature of ocean environment has been made by many scientists. However, the collected experiences may not be enough to make determinative description and/ or prediction of stochastic OHD processes and fields observed in real world remote sensing experiments. In particular, *nonlinearity, randomization*, and *fractionation* in the surface wave environment, contributions from wind action, boundary-layer, and wave breaking processes, as well the spatial *structurization or deformation* of the surface under the influences of different (induced) factors are not fully investigated and/or described. Those are just a short list of critical issues arising in remote sensing of the oceans. In this context, methods, and capabilities of CFD, operated with fundamental equations of fluid mechanics seem to be the most promising theoretical tool in the study of OHD. Direct numerical modeling and simulations of a variety of hydrodynamic scenes, scenarios, processes, phenomena, and events provide comprehensive research to accomplish data prediction needed for sophisticated applications and further development.

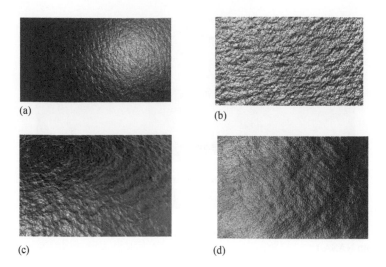

FIGURE 1.1 Aerial photographs of sea surface: (a) sunglint texture, (b) monotonic texture, (c) variable texture, and (d) patterned texture.

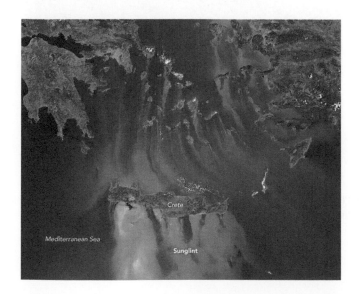

FIGURE 1.2 Satellite image. Sunglint on the Aegean and Mediterranean (MODIS). (Image from NASA.)

FIGURE 1.6 Example of global SST map derived from satellite data (Terra/MODIS). (Image from NASA.)

FIGURE 2.11 Satellite image. Tsunami run-up onto shore, southwestern coast of Sri Lanka. December 26, 2004. (Image from Digital Globe.)

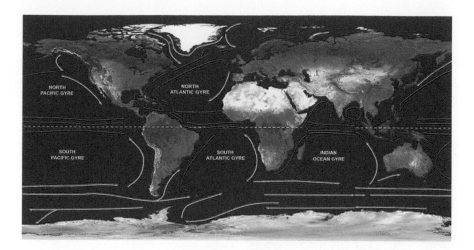

FIGURE 2.12 Five major ocean gyres. (Image from NOAA.)

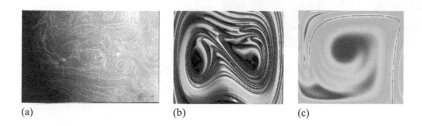

(a)   (b)   (c)

FIGURE 2.14 Lagrangian coherent structures: (a) Hairpin-shape vortices. Mediterranean Sea, Shear Wall Spiral Eddies. (Image from NASA). Obtained by astronaut Paul Scully-Power, Space Shuttle mission STS-41-G, 1984. (Based on Robinson 2010), (b) Cat's eye eddy, computer generation. (Based on Farazmanda and Haller 2012), (c) Horseshoe vortex, computer generations. (Based on Shadden et al. 2005.)

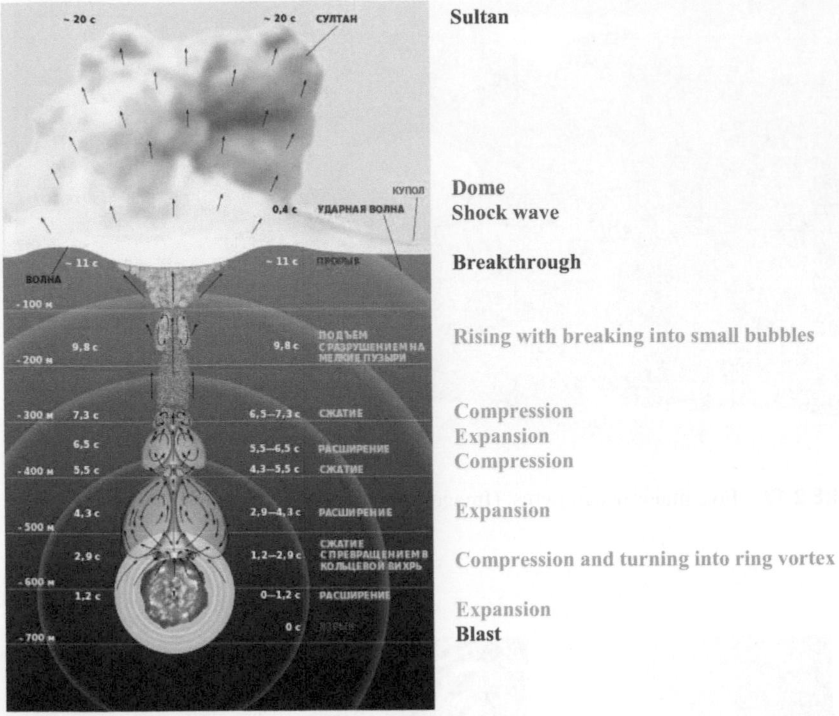
Sultan
Dome
Shock wave
Breakthrough
Rising with breaking into small bubbles
Compression
Expansion
Compression
Expansion
Compression and turning into ring vortex
Expansion
Blast

FIGURE 2.20 UNDEX 30 KT summary, time in seconds, "c", depth in meters, "m" (original in Russian).

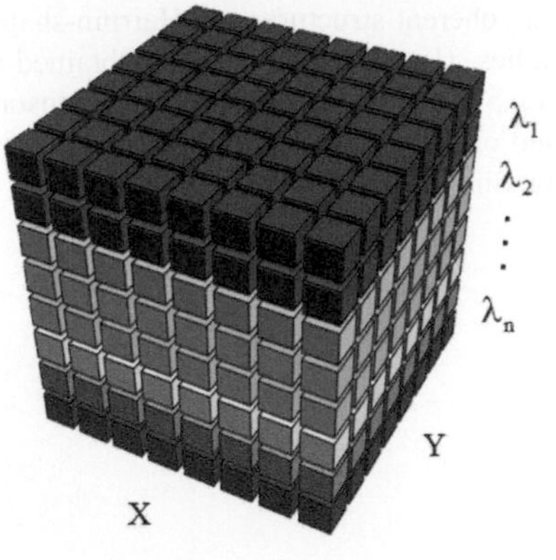
λ1
λ2
λn
Y
X

FIGURE 4.2 Hyperspectral data cube.

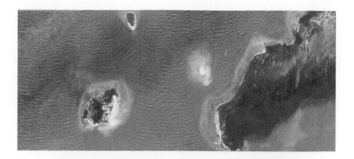

FIGURE 4.3   Satellite image. Wave refraction. (Image from Digital Globe.)

FIGURE 4.5   Satellite image. Internal waves. The island of Trinidad in the southeastern Caribbean Sea. (Image from NASA.)

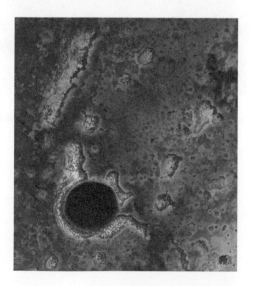

FIGURE 4.7   Satellite image. Corals around Belize Great Blue Hole. (Image from Digital Globe.)

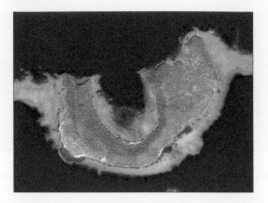

FIGURE 4.8 Satellite image. Coral reefs around the island. Matangi Island, South Pacific. (Image from Digital Globe.)

FIGURE 4.9 Satellite image. Bioproductivity. Black Sea (Aqua MODIS). (Image from NASA.)

FIGURE 4.10 Satellite image. Phytoplankton bloom. Northeast of the Falkland Islands (Terra MODIS). (Image from NASA.)

FIGURE 4.11 Satellite image. Algal bloom. Central Baltic Sea. (Image from Copernicus Sentinel-2A.)

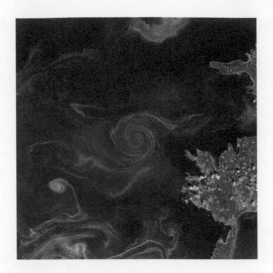

FIGURE 4.12 Satellite image. Algal bloom at coastal waters (Landsat 8). (Image from NASA.)

FIGURE 4.13 Satellite image. Oil pollutions. The Gulf of Mexico (Terra MODIS). (Image from NASA.)

FIGURE 4.14 Satellite image. Drilling slicks. The Gulf of Mexico. (Image from Digital Globe.)

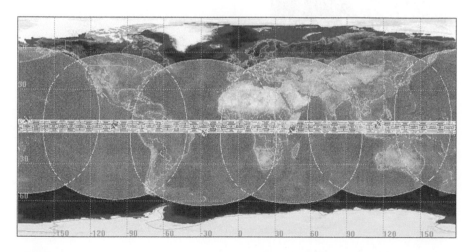

FIGURE 5.7 Tsunamisat Constellation. (After Myers 2008; Myers et al. 2008.)

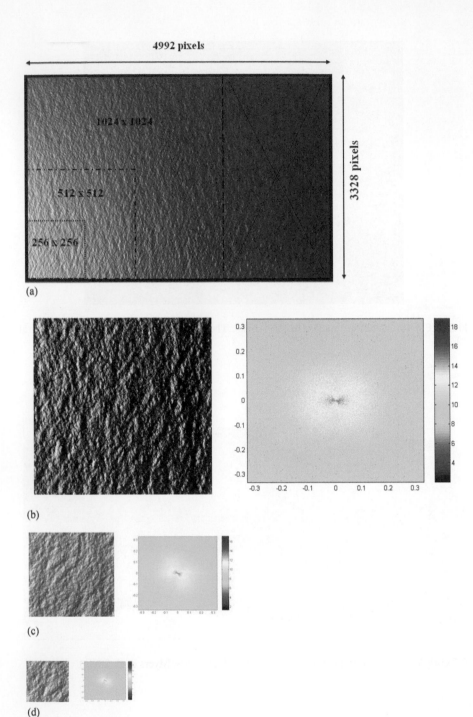

FIGURE 7.1 Variants of 2D FFT digital processing of sea aerial photography. (a) Basic image, (b) fragment 1024×1024, (c) fragment 512×512, (d) fragment 256×256.

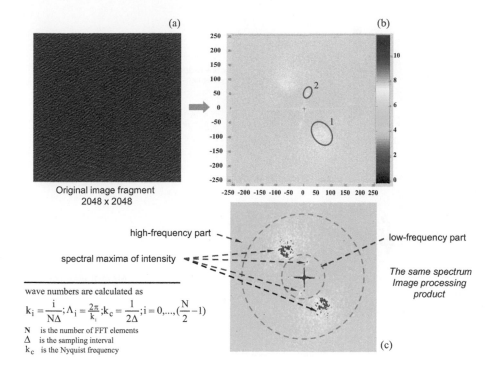

(a)

(b)

Original image fragment
2048 x 2048

high-frequency part

low-frequency part

spectral maxima of intensity

The same spectrum
Image processing
product

wave numbers are calculated as

$$k_i = \frac{i}{N\Delta}; \Lambda_i = \frac{2\pi}{k_i}; k_c = \frac{1}{2\Delta}; i = 0,...,(\frac{N}{2}-1)$$

N   is the number of FFT elements
Δ   is the sampling interval
k_c is the Nyquist frequency

(c)

FIGURE 7.3   Enhancement and extraction of spectral features from 2D FFT. (a) Original IKONOS image, (b) 2D FFT spectrum 512×512, (c) enhanced spectrum and extraction of spectral maxima.

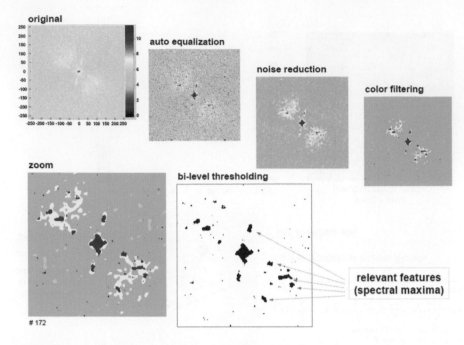

FIGURE 7.4 Extraction and display of spectral peaks during image processing.

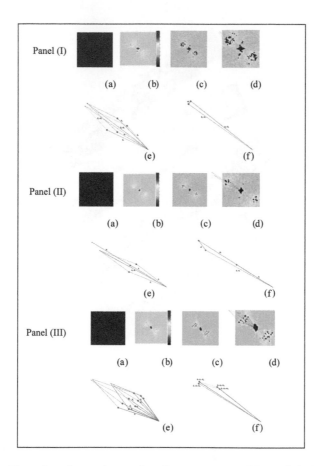

FIGURE 7.8 Examples of wave interaction diagrams extracted from digital 2D FFT spectra. Panel (I): four-wave interaction, Equation 7.1. Panel (II): four-wave interaction with two coincident waves, Equation 7.2. Panel (III): multiple-wave interaction, Equation 7.3. (a) original image (IKONOS-PAN), (b) 2D FFT spectrum, (c) enhanced spectrum, (d) spectral maxima with zoom, (e) wave vector diagram, (f) wave synchronism.

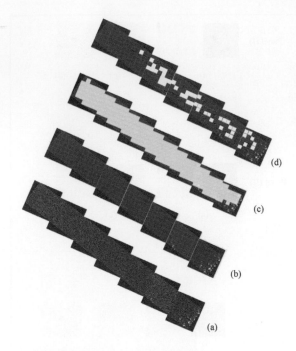

(d)

(c)

(b)

(a)

FIGURE 7.14 Optical spectral portrait implementation (see text).

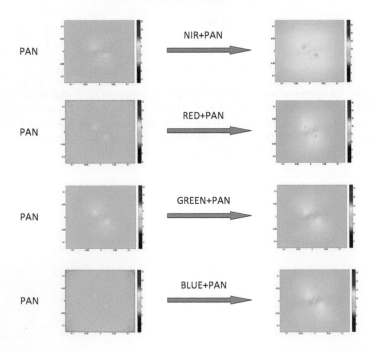

FIGURE 7.15 Examples of fused spatial spectra generated from multispectral optical images.

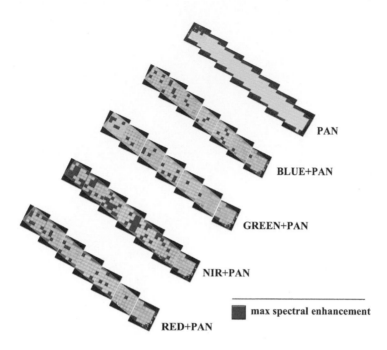

PAN

BLUE+PAN

GREEN+PAN

NIR+PAN

RED+PAN

■ max spectral enhancement

FIGURE 7.16   Realizations of fused spectral portrait.

FIGURE 5.16 Realization of band spectral portion.

3 Fundamentals of Ocean Optics

3.1 INTRODUCTION

Optics is a part of the physics which studies properties of light. As a special case of electromagnetic phenomena, the science of optics is generally divided on three areas associated with three theoretical approaches: (1) geometrical optics, which study of light as rays describing reflection and refraction; (2) physical optics, which study of light as electromagnetic waves describing diffraction, interference, and polarization; and (3) quantum optics, which describe light as an integer number of particles known as quantized photons. At the same time, there is a phenomenological description of the behaviour of light based on radiative transfer theory operated with light intensity. The connection between electromagnetic and radiative transfer theories has not been yet fully established mathematically. Since the behaviour of light depends on properties of matter, optics is divided into a number of branches (subdivisions) describing characteristics of light in geophysical systems of different types. Figure 3.1 illustrates the relationship between different optical subdivisions.

Traditional ocean optics (sometimes called marine or hydrologic optics) is concerned with the propagation, interaction, and distribution of visible light in aquatic media. Propagation characteristics such as scattering, absorption, and attenuation, vary with angle, depth, and wavelengths depending on physical, chemical, and biological conditions of the upper ocean layer. Recent advancements in ocean remote sensing have demonstrated the need for the incorporation of both electromagnetic and radiative transfer parts of theory into optical observation technologies. Radiative transfer creates a basis for study of light propagation in ocean medium including subsurface active mixed layer, whereas electromagnetic approach provides the framework for study of the interactions of optical radiance with the surface offering possibilities for remote sensing of hydrodynamic phenomena. Thus, ocean optics involve several fundamental and applied problems which are known as bio-optics, color optics, underwater optics, and also "observational optics" which is capable of detecting various sea surface phenomena. Those are wind waves, wave breaking and foam/whitecap fields, wind stress, currents, internal wave manifestations (slicks, rip currents), *suloy*, roughness patterns and turbulence, vortexes, eddies, wakes, oil pollutions, atmospheric boundary layer convective cells, and sea ice. Hydrodynamic transport processes involving advection, dispersion, and mixing can also be inferred from optical data.

This chapter considers major elements of ocean optics with an emphasis on remote sensing. Although a large number of excellent books was written on ocean optics (see, e.g., Bibliography, Chapter 1), just of few of them discusses quite frankly the relationships between optical observations and ocean hydrodynamics. The material in this chapter provides more insight than others sources. This may help the

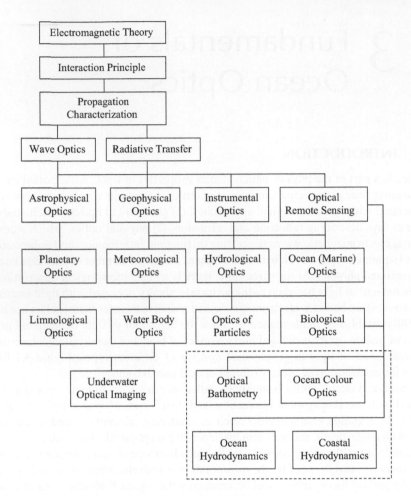

FIGURE 3.1   Optical subdivisions. (Updated and modified from Kirk (2011).)

readers to realize further progress in this field. Several classical books are devoded to ocean optics (Jerlov 1968, 1976; Monin 1983; Shifrin 1988; Kirk 2011; Spinrad et al. 1994; Walker 1994).

3.2 OPTICAL FIELDS AND QUANTITIES

3.2.1 MAXWELL EQUATIONS

The set of partial differential equations for an electromagnetic field was proposed by James Clerk Maxwell in the middle of the 19th century as a result of generalization of studies of electricity and magnetism. The inhomogeneous, macroscopic Maxwell equations (in Gaussian units) are

$$\nabla \times \vec{E}(\vec{r},t) = -\frac{1}{c}\frac{\partial \vec{B}(\vec{r},t)}{\partial t}, \qquad \text{Faraday's Law,} \qquad (3.1)$$

$$\nabla \times \vec{H}(\vec{r},t) = \frac{1}{c}\frac{\partial \vec{D}(\vec{r},t)}{\partial t} + \frac{4\pi}{c}\vec{j}(\vec{r},t), \qquad \text{Maxwell – Ampere's Law,} \qquad (3.2)$$

$$\nabla \cdot \vec{D} = \rho(\vec{r},t), \qquad \text{Gauss's Law for electric field,} \qquad (3.3)$$

$$\nabla \cdot \vec{B} = 0, \qquad \text{Gauss's Law for magnetic field,} \qquad (3.4)$$

which are supplemented by material equations (for simple medium)

$$\vec{j} = \sigma\vec{E}, \qquad (3.5)$$

$$\vec{D} = \vec{E} + 4\pi\vec{P}, \qquad (3.6)$$

$$\vec{B} = \vec{H} + 4\pi\vec{M}. \qquad (3.7)$$

where $\vec{E}$ is the electric field, $\vec{H}$ is the magnetic field, $\vec{B}$ is the magnetic induction, $\vec{D}$ is the electric displacement, $\vec{P}$ is the polarization field, $\vec{M}$ is the magnetization field, $\vec{j}$ is the electric current density, ρ is the electric charge density, σ is the conductivity, ε is the permittivity, μ is the permeability, t denotes time, $\vec{r}$ denotes coordinate vector, and c denotes the velocity of light; the $\nabla \times$ symbol denotes the curl operator and the ∇ symbol denotes divergence operator.

In the linear approximation, the polarization field $\vec{P}$, the magnetization field $\vec{M}$, the electric susceptibility χ_e, and magnetic susceptibility χ_m are defined as

$$\vec{P} = \chi_e\vec{E}, \quad \vec{M} = \chi_m\vec{H}, \quad \varepsilon = 1 + 4\pi\chi_e, \quad \mu = 1 + 4\pi\chi_m, \qquad (3.8)$$

so that

$$\vec{D} = \varepsilon\vec{E}, \quad \vec{B} = \mu\vec{H}. \qquad (3.9)$$

Maxwell has also established purely theoretically that *light is an electromagnetic disturbance in the form of waves propagated through the luminiferous aether* with speed $c = 1/\sqrt{\varepsilon_0\mu_0} = 2.9979 \times 10^8$ m/s, where $\varepsilon_0 = (1/36\pi) \times 10^{-9}$ F/m (farad per meter) and $\mu_0 = 4\pi \times 10^{-7}$ H/m (henry per meter) are permittivity and permeability of free space, correspondently.

Maxwell equations cover all aspects of electrodynamics and optics and the solutions are enormously varied depending on the issues of subject matter.

3.2.2   The Wave Equation, the Velocity of Light, and the Propagation Constant

It is considered the case when there are no electric charge ($\rho = 0$) and currents $\left(\left|\vec{j}\right| = 0\right)$. The vector wave equations for inhomogeneous medium are given by (Born and Wolf 1999)

$$\nabla^2 \vec{E} - \frac{\varepsilon\mu}{c^2}\frac{\partial^2 \vec{E}}{\partial^2 t} + \left(\text{grad}\ln\mu\right) \times \text{curl}\vec{E} + \text{grad}\left(\vec{E}\cdot\text{grad}\ln\varepsilon\right) = 0, \tag{3.10}$$

$$\nabla^2 \vec{H} - \frac{\varepsilon\mu}{c^2}\frac{\partial^2 \vec{H}}{\partial^2 t} + \left(\text{grad}\ln\varepsilon\right) \times \text{curl}\vec{H} + \text{grad}\left(\vec{H}\cdot\text{grad}\ln\mu\right) = 0, \tag{3.11}$$

For homogeneous medium Maxwell equations reduce to

$$\nabla \times \vec{E}(\vec{r},t) = -\frac{\mu}{c}\frac{\partial \vec{H}(\vec{r},t)}{\partial t}, \tag{3.12}$$

$$\nabla \times \vec{H}(\vec{r},t) = \frac{\varepsilon}{c}\frac{\partial \vec{E}(\vec{r},t)}{\partial t}, \tag{3.13}$$

$$\nabla \cdot \vec{E}(\vec{r},t) = 0, \tag{3.14}$$

$$\nabla \cdot \vec{H}(\vec{r},t) = 0, \tag{3.15}$$

and wave equations ($\text{grad}\ln\varepsilon = \text{grad}\ln\mu = 0$) become

$$\nabla^2 \vec{E} - \frac{\varepsilon\mu}{c^2}\frac{\partial^2 \vec{E}}{\partial^2 t} = 0, \tag{3.16}$$

$$\nabla^2 \vec{H} - \frac{\varepsilon\mu}{c^2}\frac{\partial^2 \vec{H}}{\partial^2 t} = 0. \tag{3.17}$$

These equations for electric $\vec{E}(\vec{r},t)$ and magnetic $\vec{H}(\vec{r},t)$ fields are particularly special and important in classical optics.

The basic solutions of the wave Equations (3.16) and (3.17) are written in the form of two uniform plane waves

$$\vec{E}(\vec{r},t) = \vec{E}_0 \exp\left[\iota\left(\vec{k}\vec{r} - \omega t\right)\right], \tag{3.18}$$

$$\vec{H}(\vec{r},t) = \vec{H}_0 \exp\left[\iota\left(\vec{k}\vec{r} - \omega t\right)\right], \tag{3.19}$$

where $\vec{E}_0$ and $\vec{H}_0$ are complex amplitudes of the electric and magnetic fields; $\vec{k}$ is the complex wave vector, ω is the angular frequency of the wave, and $\iota = (-1)^{1/2}$ is the imaginary unit.

For a source-free, linear, isotropic, homogeneous, *lossy* medium ($\rho = 0$, $|\vec{j}| \neq 0$, $\sigma > 0$), the wave equation has the form

$$\nabla^2 \vec{E} - \gamma^2 \vec{E} = 0, \tag{3.20}$$

$$\nabla^2 \vec{H} - \gamma^2 \vec{H} = 0, \tag{3.21}$$

where γ is called the *propagation constant* of the medium which is

$$\gamma = \alpha + \iota\beta = \sqrt{\iota\omega\mu(\sigma + \iota\omega\varepsilon)} = \iota\omega\sqrt{\mu\varepsilon}\left(1 + \frac{\sigma}{\iota\omega\varepsilon}\right)^{1/2} \quad \text{or} \tag{3.22}$$

$$\alpha = \omega\sqrt{\frac{\mu\varepsilon}{2}\left[\sqrt{1 + \left(\frac{\sigma}{\omega\varepsilon}\right)^2} - 1\right]}, \tag{3.23}$$

$$\beta = \omega\sqrt{\frac{\mu\varepsilon}{2}\left[\sqrt{1 + \left(\frac{\sigma}{\omega\varepsilon}\right)^2} + 1\right]}, \tag{3.24}$$

where α is the attenuation constant and β is the phase constant (here we set relative constants $\mu = \mu_r\mu_0$, $\varepsilon = \varepsilon_r\varepsilon_0$, μ_r, and ε_r are permittivity and permeability of media). Equation (3.20) or (3.21) is also known as homogeneous vector *Helmholtz equation* or *phasor* vector wave equation (symbol ∇^2 denotes the vector Laplacian operator). The $\vec{k} - \omega$ relationship gives the velocity of light in the medium $\upsilon = \dfrac{\omega}{k} = \dfrac{c}{\sqrt{\mu_r\varepsilon_r}}$, $\left(k = |\vec{k}|\right)$ and the index of refraction $n = \dfrac{c}{\upsilon} = \sqrt{\mu_r\varepsilon_r}$ of the medium. In general, an index of refraction is complex number and $n > 1$ ($\upsilon < c$); it is an optical property of particular environmental medium and a function of frequency $n(\omega)$.

Solutions of the vector wave Equations (3.10) and (3.11) for random inhomogeneous media require rigorous calculations involving Green's function and numerical methods. The description is beyond of scope of this book and we refer the reader to the corresponding literature on electrodynamics and optics. Meanwhile, direct numerical simulations of electromagnetic fields based on Maxwell equations, are of practical interest for computing wave propagation characteristics in rough terrestrial media. In ocean remote sensing, numerical modeling and simulations also make sense to develop in order to analyze effects of scattering and emission from surface waves and/or hydrodynamic disturbances of complex geometry, e.g., strongly nonlinear surface waves on steep slopes. Moreover, numerical experiments may allow us to reveal much better the contributions associated with bifurcations and breaking of short gravity surface waves.

3.2.3 From Maxwell Equations to Geometrical Optics

The geometrical optics (ray) approximation follows from Maxwell equations at neglecting the electromagnetic wavelength (known also as high-frequency approximation or zero-wavelength approximation). In this case, the solution of Maxwell equations (3.12)–(3.15) for electric and magnetic fields is written in the form of time-harmonic plane waves

$$\vec{E}(\vec{r},t) = \vec{E}_0(\vec{r})e^{-\iota\omega t}, \vec{E}_0(\vec{r}) = \vec{e}_0(\vec{r})e^{\iota k_0 S(\vec{r})},\qquad(3.25)$$

$$\vec{H}(\vec{r},t) = \vec{H}_0(\vec{r})e^{-\iota\omega t}, \vec{H}_0(\vec{r}) = \vec{h}_0(\vec{r})e^{\iota k_0 S(\vec{r})},\qquad(3.26)$$

where $\vec{e}_0(\vec{r})$ and $\vec{h}_0(\vec{r})$ are constant complex vectors of position, $S(\vec{r})$ is a real scalar function of position known as optical path, and $k_0 = \dfrac{\omega}{c} = \dfrac{2\pi}{\lambda_0}$ is the wave number and λ_0 is the wavelength in free space. Substituting (3.25) and (3.26) into Equations (3.12)–(3.15), at large $k_0 \to \infty$ (assuming λ_0 is small relative to all gradients of $\vec{e}_0, \vec{h}_0, S$), we obtain

$$\mu\frac{\partial\vec{H}(\vec{r},t)}{\partial t} = \nabla\times\vec{E}(\vec{r},t) = \nabla\times\vec{e}_0(\vec{r})e^{\iota k_0 S(\vec{r})}e^{-\iota\omega t} + \iota k_0\nabla S(\vec{r})\times\vec{E}(\vec{r},t)$$

$$\approx \iota k_0\nabla S(\vec{r})\times\vec{E}(\vec{r},t),\qquad(3.27)$$

$$\varepsilon\frac{\partial\vec{E}(\vec{r},t)}{\partial t} = \nabla\times\vec{H}(\vec{r},t) = \nabla\times\vec{h}_0(\vec{r})e^{\iota k_0 S(\vec{r})}e^{-\iota\omega t} + \iota k_0\nabla S(\vec{r})\times\vec{H}(\vec{r},t)$$

$$\approx \iota k_0\nabla S(\vec{r})\times\vec{H}(\vec{r},t).\qquad(3.28)$$

The highest order term in k_0 leaves just the gradient of S, and the time-harmonic term gives:

$$\nabla S(\vec{r})\times\vec{e}_0(\vec{r}) - \mu\vec{h}_0(\vec{r}) = 0, \vec{h}_0(\vec{r})\cdot\nabla S(\vec{r}) = 0,\qquad(3.29)$$

$$\nabla S(\vec{r})\times\vec{h}_0(\vec{r}) + \varepsilon\vec{e}_0(\vec{r}) = 0, \vec{e}_0(\vec{r})\cdot\nabla S(\vec{r}) = 0.\qquad(3.30)$$

From (3.29) and (3.30) it follows

$$\left[\nabla S(\vec{r})\right]^2 - n^2(\vec{r}) = 0 \text{ or } \left(\frac{\partial S}{\partial x}\right)^2 + \left(\frac{\partial S}{\partial y}\right)^2 + \left(\frac{\partial S}{\partial z}\right)^2 = n^2(x,y,z),\qquad(3.31)$$

which is known as the *Eikonal* equation; it is the basic equation of geometrical optics and the function $S(\vec{r})$ is called eikonal. The ray can formally be defined as a line the tangent to which at each point coincides with the vector grad S. The gradient function S is directed along the normal to the surface S = constant. Ray approximation illustrates Figure 3.2. The plane perpendicular to light rays is the wavefront.

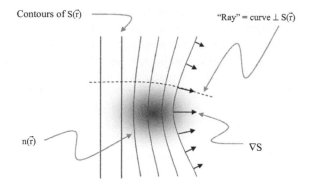

FIGURE 3.2 Geometrical optics. Ray approximation.

The theory of geometrical optic defines several important laws:

- Light rays in homogeneous media propagate in straight line path
- Law of reflection: the angle of reflection is equal to angle of incidence, i.e., $\theta_r = \theta_i$
- Specular and diffuse reflection from rough surfaces
- Snell's law of refraction: light passes through a boundary between two different isotropic media of different complex refractive indexes n_1 and n_2, so that $n_1 \sin\theta_1 = n_2 \sin\theta_2$ (θ_1 and θ_2 are angles of incidence and refraction, correspondingly)
- Ray trajectories in a motionless inhomogeneous medium are determined by the ray equation

$$\frac{d}{ds}\left[n(\vec{r})\frac{dr}{ds}\right] = \nabla n(\vec{r}), \qquad (3.32)$$

where s is the distance along ray trails, $n(\vec{r})$ the spatial distribution of the index of refraction, and $\vec{r}(s)$ radius vector (ray trajectory).

The principles and methods of geometrical optics are used in many applications involving communication, remote sensing, imaging, holography as well optical instrument design (e.g., Kravtsov 2005). Geometrical optics is a general-theory approach which is considered at optical observations of the ocean surface.

3.2.4 Light Propagation in Scattering Media

Scattering is one of the key problems in remote sensing of environment. Scattering medium is characterized by an ensemble of nonuniformities (also called *scatterers*) which force to deviate initial straight trajectory of light paths when they pass through the medium or its boundary. In optics, scattering (or multiple scattering) refers to interactions of light and material media containing particles (their groups or clusters). Light scattering is observed in the case that electromagnetic properties of particles differ from the properties of surrounding them medium. Scattering can also occurs on fluctuations of density of the medium or variations of the refractive index of the medium.

The theory of electromagnetic wave propagation in scattering (and absorption) media has been developed by many authors; it includes the solutions of Maxwell equations with particles derived for the particles of different shape, size, and permittivity (van de Hulst 1981; Bohren and Huffman 1983; Shifrin 1988; Deirmendjian 1969; Kerker 1969; Barber and Hill 1990; Ishimaru 1991).

Briefly, the main statements and results of the theory are the following:

- Mathematical description of scattering amplitude and cross section
- Single and multiple scattering approaches
- Light scattering by spherical (non)uniform particles of different size
- Light scattering by particles of complex (nonspherical) geometry
- Light propagation through random discrete media of distributed particles
- Light propagation through continuous turbulent media
- Definition of effective properties of multi-phase dispersed media.

Several fundamental theoretical problems related to ocean optics are outlined below.

The first problem is the scattering of an electromagnetic plane wave by a homogeneous sphere of any size parameter. The theory is known as the theory Mie (1908); it is based on exact solution of Maxwell equations and the vector wave equation in spherical coordinates. According to the Mie theory, scattering efficiency factor Q_s, extinction efficiency factor Q_e, and absorption efficiency factor Q_a of a spherical particle are defined as

$$Q_s(a) = \frac{2}{x^2}\sum_{n=1}^{\infty}(2n+1)\left(|a_n|^2 + |b_n|^2\right), \tag{3.33}$$

$$Q_e(a) = \frac{2}{x^2}\sum_{n=1}^{\infty}(2n+1)\mathrm{Re}(a_n + b_n), \tag{3.34}$$

$$Q_a(a) = Q_e - Q_s, \tag{3.35}$$

where $x = ka = 2\pi a/\lambda$ is the diffraction (or size) parameter ($0.1 < x < 100$), a is the radius of spherical particle, and λ is the electromagnetic wavelength. The Mie scattering complex coefficients are given by (van de Hulst 1981)

$$a_n = \frac{\psi_n(x)\psi_n'(mx) - m\psi_n(mx)\psi_n'(x)}{\zeta_n(x)\psi_n'(mx) - m\psi_n(mx)\zeta_n'(x)}, \tag{3.36}$$

$$b_n = \frac{m\psi_n(x)\psi_n'(mx) - \psi_n(mx)\psi_n'(x)}{m\zeta_n(x)\psi_n'(mx) - \psi_n(mx)\zeta_n'(x)}, \tag{3.37}$$

where $\psi_n(x) = \sqrt{\pi x/2}\cdot J_{n+1/2}(x)$ and $\zeta_n(x) = \sqrt{\pi x/2}\cdot H_{n+1/2}^{(1)}(x)$ expressed in terms of the Bessel function and the Hankel function of the first kind; $m = \sqrt{\varepsilon}$ is the complex refractive index of a particle material ($\mu = 1$). The corresponding spectral cross sections of a spherical particle are given $\sigma_{s,e,a} = \pi a^2 Q_{s,e,a}$. The spectral volume coefficients for a system of particles of different size are given by

$$< \sigma_{s,e,a} > = \pi \int_0^\infty Q_{s,e,a}(a) a^2 f(a)\, da, \qquad (3.38)$$

where f(a) is the size distribution probability density function $\left(\int_0^\infty f(a)\, da = 1 \right)$. Volume coefficients are involved in the radiative transfer equation (RTE).

The results in (3.36) and (3.37) can be generalized on the case where a spherical particle is suspended in a medium with refractive index $\neq 1$. This conversion can be done by replacing $m \rightarrow m_1/m_2$, $\lambda \rightarrow \lambda/m_2$, $x = 2\pi a/\lambda \rightarrow 2\pi a m_2/\lambda$ where m_1 is the refractive index (real or complex) of a particle material, m_2 is the refractive index (real) of surrounding media. Although Mie theory is a rigorous and applicable for all size of spheres as well materials, there may occur problems with computing Mie coefficients when both refractive indexes m_1 and m_2 are complex or when m_1 is real but m_2 is complex. Environmental examples are *bioparticles* or *gaseous bubbles* in seawater that are of special interest for ocean optics, microwave radiometry, and acoustics.

Numerous calculations and investigations have shown that there are different regimes of the scattering depending on regions (values) of the diffraction parameter x and complex refractive index m of the scatterer. Correspondently, the following approximations are

- Rayleigh approximation: $x \ll 1$ and m is arbitrary (applies to scattering by small particles)
- Rayleigh–Gans approximation: $2x|m - 1| \ll 1$ and $|m - 1| \ll 1$ (applies to scattering by nonspherical particles)
- Perelman approximation (also called S-approximation): x arbitrary but $m \rightarrow 1$ (x region is wider than by Rayleigh and Rayleigh–Gans approximations)
- Resonant Mie scattering: $1 \leq x \leq 20$, ($a \sim \lambda$), m is complex; both parameters x and m are arbitrary
- van de Hulst approximation (also called the anomalous diffraction approximation): $x \rightarrow \infty$ and $|m - 1| \ll 1$ (applied to scattering by large soft particles)
- Geometrical optics approximation: $x \geq 20$ and m is real (applies to scattering by large particles).

The volume coefficients $< \sigma_{s,e,a} >$ vary depending on the particle size distribution function f(a) so that the boundaries of x regions will vary as well. Figure 3.3 shows numerical example of efficiency factors for atmospheric aerosol where all x regions are presented. More detailed data are available in (Kokhanovsky 2008).

The second problem considered here is light propagation in random inhomogeneous (turbulent) scattering media. The problem is associated with fluctuations of the refractive index that is important factor in atmospheric optics. A solution can

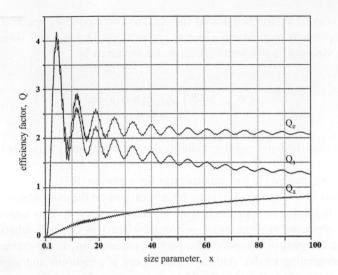

FIGURE 3.3 Aerosol optics. Dependences of Mie efficiency factors on the size parameter $x = ka$ at complex refractive index of particles $m = 1.45 - 0.005i$ and wavelength $\lambda = 650\,\text{nm}$.

be obtained using the wave Equations (3.10) and (3.11). The scalar wave equation is written as

$$\nabla^2 U(\vec{r},\omega) + k_0^2 U(\vec{r},\omega) = -4\pi F(r,\omega)U(\vec{r},\omega), \qquad (3.39)$$

where

$$F(\vec{r},\omega) = \frac{1}{4\pi} k_0^2 \left[n^2(\vec{r},\omega) - 1 \right], \qquad (3.40)$$

where $U(\vec{r},\omega)$ denotes the components of the electromagnetic field, $F(\vec{r},\omega)$ is called the scattering potential of the medium; $n(\vec{r},\omega)$ is the refractive index and $k_0 = \omega/c$. The rigorous solution for *potential scattering* is given by (Born and Wolf 1999)

$$U(\vec{r},\omega) = e^{ik_0 \vec{s}_0 \vec{r}} + \int_V F(\vec{r}',\omega)U(\vec{r}',\omega) \frac{e^{ik_0|\vec{r}-\vec{r}'|}}{|\vec{r}-\vec{r}'|} d^3 r', \qquad (3.41)$$

$$G(\vec{r}-\vec{r}',\omega) = \frac{e^{ik_0|\vec{r}-\vec{r}'|}}{|\vec{r}-\vec{r}'|}, \qquad (3.42)$$

where the first term corresponds to incident field propagating in the direction specified by a real unit vector $\vec{s}_0$ and the second term describes the behaviour of *the total field* $U(\vec{r},\omega)$ within the scattering volume V; $G(\vec{r}-\vec{r}',\omega)$ is the free-space Green's function. Note that there are several approximation solutions of integral Equations (3.41) and (3.42); the most known and practically convenient for optics are the first-order Born approximation and the Rytov approximation (1937). In both approximations, the total field is expressed as the sum $U(\vec{r},\omega) = U^{(i)}(\vec{r},\omega) + U^{(s)}(\vec{r},\omega)$ of the incident field (index "i") and the scattering field (index "s").

In the first-order Born approximation, the total field under the integral (3.35) is replaced by the incident field, i.e., $U(\vec{r},\omega) \approx U^{(i)}(\vec{r},\omega)$ and an analytical solution can be found for the far field $(k_0 r \to \infty)$. In the Rytov approximation, it is assumed that the refractive index is close to unity $n(\vec{r},\omega) = 1 + \Delta n(\vec{r},\omega)$, $\Delta n \ll 1$, and $\Delta n(\vec{r}) = (1/2)\delta\beta(\vec{r})$, so that $n^2(\vec{r}) - 1 \approx \delta\beta(\vec{r})$ and the scattering potential (3.40) becomes $F(\vec{r},\omega) = \dfrac{1}{4\pi}k_0^2\delta\beta(\vec{r})$, where $\beta(\vec{r})$ is some function. The total field sets as $U(\vec{r},\omega) \approx e^{\psi(\vec{r},\omega)}$ and the solution of (3.41) is expressed in the form of the exponential series sum $U(\vec{r},\omega) = e^{[\psi_0(\vec{r},\omega)+\delta\psi_1(\vec{r},\omega)+\delta^2\psi_2(\vec{r},\omega)+...]}$ (known as the *Rytov expansion*), where δ is the perturbation parameter of fluctuations.

The Born approximation is commonly used for the solution of backscattering problem whereas the Rytov approximation is used mostly for the solution of forward-scattering problem. Moreover, the Rytov solution may be considered superior to the first-order multiple scattering theory (Ishimaru 1991). There are numerous theoretical and experimental studies of light scattering including laser (lidar) applications; the reader can find many relevant references from literature sources.

The third problem worthy our attention is definition of effective electromagnetic parameters of a random scattering media. It is related to the multiple scattering problems. Here we are concerned with effective wave number, K_{eff}, and the effective complex refraction index, n_{eff}, that can be used for modeling wave propagation through the scattering volume of spherical particles. Typical environmental examples are fog, clouds, or dense sea salt aerosol. In this case, a stochastic scattering environment is considered as a continuous macroscopic turbulent medium with the average properties of randomly fluctuating fields. This is a classical topic, with a large literature and we refer the reader to well-known books (van de Hulst 1981; Ishimaru 1991; Tsang and Kong 2001; Tsang et al. 2000).

A concept of effective parameters is based on optical coherent theory of multiple-particle scattering that operates with the Foldy–Twersky's integral equation

$$< \psi^a > = \varphi_i^a + \int u_s^a < \psi^s > \rho(\vec{r}_s)d\vec{r}_s, \qquad (3.43)$$

where $\psi^a = < \psi^a > + \psi_f^a$ is a total scalar field in a random medium, $< \psi^a >$ the average (or coherent) field, and ψ_f^a the fluctuation (incoherent) field, φ_i^a the incident field at the absence of any particles, $< \psi^s >$ the average scattered field in the medium, u_s^a the scattering characteristics of the particle, $\rho = N/V$ the number density of particles per unit volume. The theory describes multiple scattering as the contribution of incident field and fields scattered from N particles (Ishimaru 1991).

In particular, the effective wave number and refractive index of inhomogeneous scattering media of *spherical particles* is defined through the solution of Equation (3.43) and the Mie theory. The general relationship embodies using so-called forward-scattering theorem or "optical theorem" which is given by formula $\sigma_{tot} = (4\pi/k_0^2)\text{Im}\{S(0)\}$, where $\sigma_{tot} = \sigma_s + \sigma_a$ is a total cross section coefficient and $\text{Im}\{S(0)\}$ the imaginary part of the forward-scattering amplitude $S(\theta = 0)$

with an angle of zero. For a spherical particle $S(0) = (1/2) \sum_{n=1}^{\infty} (2n+1)(a_n + b_n)$ and a_n, b_n are the Mie complex coefficients defined by (3.36) and (3.37) (van de Hulst 1981).

In the simplified case of *independent discrete spherical particles of the same size* suspended in free space, formula for effective complex wave number K_{eff} of a scattering medium is given by Foldy (1945)

$$K_{eff}^2 = k_0^2 + \frac{4\pi\rho}{k_0} S(0) \text{ or } K_{eff} \approx k_0 + \frac{2\pi\rho}{k_0} S(0) \, (\rho \ll 1). \tag{3.44}$$

Correspondingly, effective complex refractive index of a scattering medium is

$$n_{eff}^2 = \frac{K_{eff}^2}{k_0^2} = 1 + \left(\frac{4\pi\rho}{k_0^3} \right) \cdot S(0) \text{ or } n_{eff} \approx \frac{K_{eff}}{k_0} = 1 + \left(\frac{2\pi\rho}{k_0^2} \right) \cdot S(0) \, (\rho \ll 1). \tag{3.45}$$

A concept of effective electromagnetic parameters is used in many areas of physics including microwave remote sensing (e.g., Sihvola 1999; Tsang and Kong 2001; Raizer 2017), electrodynamics of composite media and semiconductors, ellipsometry, micro- and nanooptics. This subject can be entitled "The effective medium theory" (e.g., Choy 2016). The usefulness of the theory is obvious for hydrological and atmospheric optics involving radiance measurements and applications. For instance, the presence in seawater of particulate constituents, organic or nonorganic compounds significantly change optical properties of the upper ocean that is detectable using lidar or spectroscopic methods; possible variations of optical signals can be explained by light scattering and absorption in seawater-type liquid composite macro-media.

3.3   RADIOMETRY, PHOTOMETRY, AND QUANTITIES

Optical remote sensing involves the measurements of electromagnetic radiation being reflected or emitted by the object at selected wavelengths. There are two basic methods to measure energy of light—radiometric and photometric. The first (called radiometry) measures radiation energy at selected wavelength bands from entire ultraviolet, visible, and infrared ranges ($100-10^6$ nm) of the electromagnetic spectrum. The second (called photometry) measures radiation energy of visible light only (400–700 nm) to which the human eye is sensitive and perceives light naturally. Terminology has historical roots and the only difference between radiometry and photometry is in their units of measurements. Correspondingly, optical instrumentations and applications differ as well. For example, modern infrared radiometer is usually a digital electro-optical device whereas a regular photometer is designed as analog device.

In optical remote sensing, radiometric and photometric standard units, symbols, and quantities are used (Slater 1980; Egan 2004; Träger 2012). They are the following:
In radiometry—

Radiant energy (Joule—J = Ws): electromagnetic energy that travels in traverse waves.

Radiant energy density (J/m³): the radiant flux per unit area emitted or incident per unit area.

Radiant flux (Watt—W): the amount of radiant energy emitted, transmitted, or received per unit time.

Radiant flux density (W/m²): radiant flux per unit area.

Irradiance (W/m²): radiant flux density incident on a surface.

Radiant spectral flux density (W/m²·µm): radiant flux density per unit of wavelength interval.

Radiant intensity (W/sr): flux emanating from a surface per unit solid angle.

Radiance (W/m²·sr): radiant flux density emanating from a surface per unit solid angle.

Spectral radiance (W/m²·sr·µm): radiance per unit wavelength interval.

Radiant emittance (W/m²): radiant flux density emitted by a surface.

In photometry—

Luminous energy (Joule—J = Ws): photometrically weighted radiant energy (energy of light).

Luminous energy density (J/m³): photometrically weighted radiant flux density.

Luminous flux (lumen—lm): radiant flux evaluated in terms of a standardized visual response.

Luminous intensity (candela—cd): luminous flux per unit solid angle in a certain direction.

Luminous exitance (lm/m²): total luminous flux given out per unit area.

Luminance (cd/m²): photometrically weighted radiance (brightness perception).

Illuminance (lux—lx): luminous flux incident on a surface, per unit area.

Figure 3.4 illustrates radiometric concept.

Spectral radiance, irradiance, and reflection are fundamental quantities (AOPs) of interest in optical oceanography: they completely characterize positional, temporal, directional, and spectral properties of light field (Jerlov 1968, 1976; Kirk 2011; Mobley 1994, 2001; Dickey et al. 2006). These quantitative are used in bio-geo-optical measurements, color, water body, and water quality optical studies. Remote sensing concept, terminology, and experimental methods were developed in works (Boileau and Gordon 1966; Spitzer and Arief 1983; Slater 1980; Egan 2004).

FIGURE 3.4   Illustrating radiometric concept.

3.4 THE RADIATIVE TRANSFER EQUATION

RTE provides the theoretical basis in ocean optics and remote sensing. The propagation of light in ocean is governed by the integral-differential equation of radiative transfer, which operates with radiant energy and absorbing-scattering parameters of the medium.

Mathematical details of radiative transfer theory are considered in books (Chandrasekhar 1960; Sobolev 1963; Preisendorfer 1965; Ishimaru 1991; Apresyan and Kravtsov 1996). Radiative transfer in environmental systems including ocean and atmosphere are discussed in the book (Thomas and Stamnes 1999; Marshak and Davis 2005; Kuznetsov et al. 2012; Stamnes and Stamnes 2015; Stamnes et al. 2017). The theory also developed and adapted for imagery of scattering media (Zege et al. 1991) and thermal radiation analysis of dispersed media (Dombrovsky and Baillis 2010).

In optical oceanography, horizontally homogeneous plane-parallel stationary medium is considered and RTE is given by

$$\mu \frac{dL(z,\lambda,\theta,\varphi)}{dz} = -c(\lambda)L(z,\lambda,\theta,\varphi) + L^*(z,\lambda,\theta,\varphi), \tag{3.46}$$

$$L^*(z,\lambda,\theta,\varphi) = \int_{4\pi} L(z,\lambda,\theta',\varphi')\beta(z,\lambda,\theta',\varphi')d\Omega' + S(z,\lambda,\theta,\varphi), \tag{3.47}$$

$$S(z,\lambda,\theta,\varphi) = [1-\alpha]B(T) + S^*(z,\lambda,\theta',\varphi') + \frac{\alpha}{4\pi}\int_{4\pi} p(\Omega',\Omega)L(z,\lambda,\theta',\varphi')d\Omega', \tag{3.48}$$

where $L(z,\lambda,\theta,\varphi)$ is spectral radiance at depth z, $L^*(z,\lambda,\theta,\varphi)$ is the path function, (θ,φ) and (θ',φ') are incident and scattering directions, $c(\lambda) = \alpha(\lambda) + b(\lambda)$ is the beam attenuation coefficient, $\alpha(\lambda)$ is the absorption coefficient and $b(\lambda)$ is the scattering coefficient, $\beta(z,\lambda,\theta,\varphi)$ is the volume scattering function (which is scattered intensity per unit incident irradiance per unit volume of water), and λ is the wavelength. The source term $S(z,\lambda,\theta,\varphi)$ describes an *internal* light source and includes thermal emission term $[1 - \alpha]B(T)$, first-order scattering term $S^*(z,\lambda,\theta',\varphi')$, and multiple scattering term $\frac{\alpha}{4\pi}\int_{4\pi} P(\Omega',\Omega)L(z,\lambda,\theta',\varphi')d\Omega'$, $B(T)$ is the Planck function, $P(\Omega,\Omega')$ is the phase function, and $\Omega' = \sin\theta'd\theta'd\varphi'$ is the differential solid angle. The RTE (3.46)–(3.48) describes the connection of inherent optical properties (IOP) of natural water and radiance.

More simple representation of the RTE can be written in terms of *total upwelling radiance,* reflected from the ocean-atmosphere system and received by a sensor

$$L_u = L_a + T_pL_r + T_pL_w + \text{complications}, \tag{3.49}$$

where L_a is the atmospheric path radiance, L_r is the reflected from the ocean surface radiance, L_w is irradiance from the ocean surface, and T_p is the direct transmittance

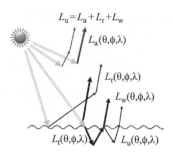

FIGURE 3.5 Atmospheric correction. L_u is the total upwelling radiance above the sea surface, L_a is the sun's unscattered beam, L_a is atmospheric path radiance, L_r is surface-reflected radiance, L_w is water-leaving radiance, and L_t is radiance transmitted through the surface. Thick arrows denote single-scattering contributions and thin arrows denote multiple scattering contributions.

of atmosphere given by $T_p = \exp(-\tau/\mu)$, where τ is the total optical depth dependent on atmospheric scattering and absorption and $\mu = \cos\theta$ is cosine of the angle of view. Equation (3.49) describes the atmospheric path radiance received by the sensor (Figure 3.5) and known as atmospheric correction. Radiative transfer model in form (3.49) is widely used in optical oceanography providing convenient algorithms for interpretation of space-based multispectral imagery of the ocean. However, RTE related to atmospheric optics and propagation of polarized light through the ocean-atmosphere system requires more accurate analysis (Section 3.7).

3.5 OPTICAL PROPERTIES OF WATER

The optical properties of water play important roles in monitoring of coastal aquatic ecosystems, bio-optical oceanography, marine photochemistry, mixed-layer dynamics, laser bathymetry, remote sensing of biological productivity, sediment load, or pollutants. Detailed spectral measurements of optical properties of water have been done by Smith and Baker (1981), Pope and Fry (1997); radiance and emission of sea surface were investigated by Saunders (1968), Wu and Smith (1997). These and other works including the contributions from Russian scientists (e.g., Monin 1983; Shifrin 1988; Kopelevich et al. 2007, 2017) provided a basis for further experimental and theoretical studies in the field of ocean optics. In particular, a branch known as bio-optics has become a principal focus for satellite oceanography and ecological control.

Bio-optics is a powerful semi-empirical approach for estimating marine properties and primary production in seawater. Numerous studies have been made in the field of light propagation in the ocean. The reader can refer to several books (Shifrin 1988; Dera 1992; Mobley 1994, 2010; Arst 2003; Kirk 2011; Woźniak and Stramski 2004; Wozniak and Dera 2007; Watson and Zielinski 2013) and reviews (Dickey et al. 2006, 2011) which provide comprehensive information on this subject. Books (Bukata et al. 1995; Pozdnyakov and Grassl 2003; Miller et al. 2005; Mishra et al. 2017; Pozdnyakov et al. 2017) discuss remote sensing studies of aquatic waters and optical characteristics of marine biochemical processes. In this section, we consider several most important parameters of bio-optics that influence on the quality and hydrodynamic content of ocean optical imagery.

According to common terminology, the bulk or large-scale, optical properties of water are divided into two classes: inherent (IOPs) and apparent properties (AOPs). Recall from Chapter 1 that IOPs are those properties that depend only upon the medium and therefore are independent of the ambient light field; AOPs are those properties that depend both on the medium (the IOPs) and on the light field in which they are measured. The following properties of water are usually considered:

- IOP's—the absorption coefficient, the volume scattering function, the index of refraction, the beam attenuation coefficient and the single-scattering albedo
- AOP's—radiance, reflectance, irradiance, and diffuse attenuation coefficient.

Figure 3.6 developed by Mobley (2010) summarizes the theoretical framework in optical oceanography and remote sensing. The various IOPs and AOPs are fundamentals of radiative transfer in ocean water; saying more about these parameters, we will focus on the index of refraction, absorption, scattering, albedo, and reflectance.

3.5.1 THE INDEX OF REFRACTION

The index of refraction of water and aqueous solutions has been studied in visible and infrared range by many authors; these data are available (e.g., in Ray 1972; Hale and Querry 1973; Akhadov 1980; Ghosh 1998). The *complex index of refraction* $m(\omega) = n(\omega) + \iota\kappa(\omega)$ can be determined from Kramers–Kronig relation (1926–1927) connecting the real and imaginary parts:

$$n(\omega) - 1 = \left(\frac{2}{\pi}\right) P \int_0^\infty \frac{\omega'\kappa(\omega')}{\omega'^2 - \omega^2} d\omega', \tag{3.50}$$

$$\kappa(\omega) = -\left(\frac{2\omega}{\pi}\right) P \int_0^\infty \frac{n(\omega') - 1}{\omega'^2 - \omega^2} d\omega', \tag{3.51}$$

where P stands for Cauchy principle value that indicates that only the principle part of the integral should be taken. The real part of the index of refraction $n(\omega)$ is responsible for dispersive optical phenomena (called "the real index of refraction"); the imaginary part $\kappa(\omega)$ is related to the phenomena of light absorption (called the "extinction coefficient"). These values represent bulk electromagnetic properties. A detailed description of the Kramers–Kronig relation is in the book (Lucarini et al. 2005). There is a simple relation between the extinction coefficient $\kappa(\omega)$ and the absorption coefficient $\alpha(\omega)$ of the medium, i.e., $\kappa(\omega) = c\alpha(\omega)/2\omega$ or $\alpha(\lambda) = 4\pi\kappa(\lambda)/\lambda$ ($\omega = 2\pi c/\lambda$, c is the speed of light in vacuum). The real index of refraction for seawater $n(\lambda,T,S,p)$ is a function of wavelength, λ temperature, T, salinity, S, and

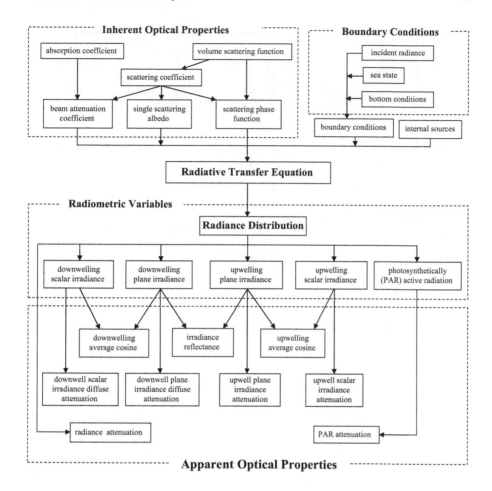

FIGURE 3.6 Organization chart for optical oceanography. (Adapted and modified from Mobley 2010.)

pressure p; detailed data are given by Austin and Halikas (1976). Figure 3.7 shows spectra of optical constants $n(\lambda)$ and $\kappa(\lambda)$ for pure water (Mobley 2010). There is the narrow "window" in $\kappa(\lambda)$ that yields the corresponding window in the spectral absorption coefficient $\alpha(\lambda)$.

The index of refraction of natural water is influenced by the presence of suspended particulate matter. The particles found in natural waters often have a bimodal index of refraction distribution. Living phytoplankton typically have low indices of refraction, in the range of n = 1.01–1.09 relative to the index of refraction of pure seawater. Detritus and inorganic particles generally have high indices of refraction, in the range of n = 1.15–1.20 relative to seawater (Jerlov 1976). Typical values are 1.05 for phytoplankton and 1.16 for inorganic particles. Standard value of the real index of refraction for pure seawater at visible range λ = 300–700 nm is n = 1.34.

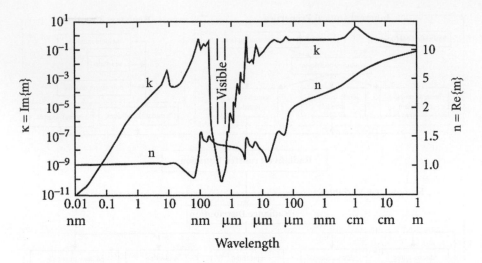

FIGURE 3.7   Spectra of optical constants for pure water. (Adapted from Mobley 2010.)

3.5.2 ABSORPTION

The spectral absorption coefficient $\alpha(\lambda)$ is the most important optical characteristics of natural water. A large number of experimental and model studies have been made to define the values and spectral dependencies of $\alpha(\lambda)$. According to the common specification (Kirk 2011), the total spectral absorption coefficient can be defined as a sum of several contributing absorption coefficients

$$\alpha(\lambda) = \alpha_w(\lambda) + \alpha_{NAP}(\lambda) + \alpha_{ph}(\lambda) + \alpha_{CDOM}(\lambda) + \alpha_y(\lambda) + \alpha_{TSM}(\lambda), \quad (3.52)$$

that correspond to pure water $\alpha_w(\lambda)$, non-algal particles (NAP) $\alpha_{NAP}(\lambda)$, phytoplankton $\alpha_{ph}(\lambda)$, colored (or chromophoric) dissolved organic matter (CDOM, known also as yellow substance) $\alpha_{CDOM}(\lambda)$, and total suspended matter $\alpha_{TSM}(\lambda)$. Each component is a subject of matter for many oceanographic and bio-optical studies; their review is beyond the scope of our book. Mobley (2010) refers to bio-optical models for the total spectral absorption coefficient of water

$$\alpha(\lambda) = \left[\alpha_w(\lambda) + 0.06 a_c^{*'}(\lambda) C^{0.65}\right]\left[1 + 0.2\exp\left(-0.014(\lambda - 440)\right)\right], \quad (3.53)$$

and/or

$$\alpha(\lambda) = \alpha_w(\lambda) + C\left[a_c^0(\lambda) + 0.1\exp\left(-0.015(\lambda - 440)\right)\right]. \quad (3.54)$$

Here $\alpha_w(\lambda)$ is the absorption coefficient of pure water (m^{-1}), $a_c^{*'}(\lambda)$ is a nondimensional, statistically derived chlorophyll-specific absorption coefficient, C is the chlorophyll concentration (mg/m^3), $a_c^0(\lambda)$ is the chlorophyll-specific absorption coefficient for phytoplankton (m^2/mg), and λ is the wavelength (nm). Examples of spectral dependencies α(λ) are shown in Figure 3.8 (Mobley 2010). Note that light absorption models designed for specific regions, seasons, and seawater conditions provides a better understanding of the variability of the absorption spectra.

3.5.3 SCATTERING

Shuleikin (1968) in 1923 was the first who initialed the importance of light scattering; he has developed complete explanation of the color of the sea. After this pioneering work, numerous experimental studies have been done to explore light scattering in water; these results are presented and discussed in books (Jerlov 1968, 1976; Mobley 1994; Stover 1995; Mishchenko et al. 2000; Jonasz and Fournier 2007). Basically, there are two main types of light scattering within the aquatic medium—molecular scattering and particle scattering.

The first type is light scattering by a particle-free pure fluid, which is described by the Einstein-Smoluchowski theory of scattering (e.g., Landau and Lifshitz 1987). According to this theory, scattering is due to microscopic fluctuations of the real index of refraction (or equivalently, the dielectric constant, $\varepsilon = n^2$), causing by localized fluctuations of density and concentration. Fabelinskii (1968), who has developed a common theory of molecular scattering of light in liquids and gases, wrote: "Light scattering appears only when the light flux encounters some optical inhomogeneities on its way...The fluctuations of physical quantities that lead to the appearance of an optical inhomogeneity in the investigated medium are significant for light scattering. These include density and temperature fluctuations, and the density fluctuations depend on the pressure and entropy fluctuations." Although the Rayleigh molecular theory of scattering does not apply to pure liquid, the predicted angular distribution of scattering is similar that given by the Rayleigh theory for gases, i.e., identical in the forward and backward directions.

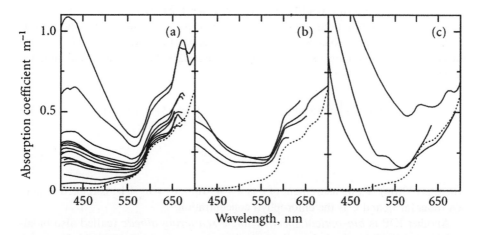

FIGURE 3.8 Spectral dependencies of light absorption for different concentrations of dissolved substances, phytoplankton, and detritus. (Adapted from Mobley 2010.) (a) for waters dominated by phytoplankton, (b) waters with a high concentration of nonpigmented particles, and (c) waters rich in yellow matter.

In oceanographic literature (e.g., Mobley 1994, 2010; Kirk 2011) the volume scattering function (VSF) at 90° $\beta(90°,\lambda,T,S)$ is used; it is expressed as the sum

$$\beta(90°,\lambda,T,S) = \beta_d(90°\lambda,T) + \beta_c(90°,\lambda,S) \qquad (3.55)$$

of scattering by density fluctuations $\beta_d(90°\lambda,T)$ and that by concentration fluctuations $\beta_c(90°\lambda,S)$. These functions are measured and defined empirically (Mobley 2010). The corresponding spectral scattering coefficient is $b(\lambda) = 2\pi \int_0^\pi \beta(\theta,\lambda)\sin\theta\,d\theta$. The net result for the total spectral scattering coefficient for either pure water or pure sea water is given by (Mobley 2010)

$$b_w(\lambda) = 16.06\left(\frac{\lambda_0}{\lambda}\right)^{4.32} \beta_w\left(90°,\lambda_0\right), \left[m^{-1}\right] \qquad (3.56)$$

which has dependence of $\lambda^{-4.32}$ rather than λ^{-4} for Rayleigh scattering.

The second type of scattering is volume scattering by ensemble of solid particles suspended in water media that can be described by Rayleigh or Mie theories. These formulas for spherical particles are given by (3.33)–(3.38). The electromagnetic theory of scattering is also developed for nonspherical particles and non-uniform particles. The corresponding scattering electromagnetic models are considered in books (Mishchenko et al. 2000; Kokhanovsky 2004a; Jonasz and Fournier 2007). Particles in water and bio-models are discussed in works (Morel 2001, 2009; Gregory 2006; Sánchez and Piera 2016). It is assumed that ocean is a very dilute suspension of random scatterers and only incoherent scattering is considered.

The spectral beam attenuation coefficient $c(\lambda)$ is just the sum of the spectral absorption and scattering coefficients: $c(\lambda) = \alpha(\lambda) + b(\lambda)$. Since both $\alpha(\lambda)$ and $b(\lambda)$ are highly variable functions of physical parameters and concentration of the constituents of natural waters so is $c(\lambda)$. Bio-optical models for $c(\lambda)$ are not simple because the complicated dependences of scattering by water particles. An empirical model is given by (Mobley 2010)

$$c(\lambda) = c_w(\lambda) + \left[c\left(490\,nm\right) - c_w\left(490\,nm\right)\right]\left[1.563 - 1.149\times10^{-3}\lambda\right], \quad (3.57)$$

$$c(490\,nm) = 0.39C^{0.57}, \qquad (3.58)$$

where $c(490\,nm)$ is a measurement of c at $\lambda = 490\,nm$, $c_w = \alpha_w + b_w$ is related to pure sea water [m^{-1}], and C is the chlorophyll concentration.

Another IOP is *bio-optical spectral single-scattering albedo* (called also bioalbedo) $\omega_0 = b/c$. This albedo is a dimensionless parameter that provides information *solely* on the relative amount of scattering and absorption occurring *within* a natural water mass. High values of ω_0 indicate that scattering dominates over absorption in water column and low values of ω_0 indicate that absorption dominates over scattering. Both values of ω_0 could refer to entire range of water type from very clear to very turbid (Bukata et al. 1995). Theoretically, bioalbedo can be calculated using the Rayleigh or Mie formulas, modified for a polydispersed system of solid spherical particles. However, rigorous quantification of bioalbedohas remained elusive because of great variety of "bio-" components and particle characteristics in marine environment. More important and better studied case is light scattering by gas bubbles in water (Marston et al. 1982; Langley and Marston 1984; Zhang et al. 1998; Jonasz and Fournier 2007). These effects influence on the spectral reflectance and the *ocean surface albedo.*

3.5.4 ALBEDO

Albedo is Latin, meaning whiteness. The ocean surface albedo (OSA) is the ratio of the upwelling to downwelling solar irradiance (flux) at the air-sea interface. OSA has various applications in satellite remote sensing and in climate modeling. Spectral OSA or broadband OSA is one of the important parameters in surface radiation budget. The OSA determines the amount of solar energy penetrating into the ocean or reflected back into space.

Over recent decades, few experiments were conducted to estimate the values of OSA (e.g., Payne 1972; Paltridge and Platt 1976; Katsaros et al. 1985; Preisendorfer and Mobley 1986; Feng et al. 2016; Sinnett and Feddersen 2016). Several parameterization schemes and models of the OSA have been proposed as well (e.g., Jin et al. 2004, 2011; Enomoto 2007; Qu et al. 2015; Séférian et al. 2018). As a whole, the OSA is a function of solar zenith angle (SZA), sea surface state, and bio-optical properties (i.e., suspended matter in water).

More specifically, the albedo is defined as the ratio between the reflected energy and the incident energy over a unit area and, therefore, there is some similarity between the albedo and the reflectance. In fact, spectral monochromatic albedo at specified wavelength is identical to the reflectance. The broadband or shortwave albedo can be defined from the BRDF, see Equation (3.65), that is hemispherical measure.

By definition, the spectral monochromatic albedo $\alpha_0(\theta_i,\lambda)$ (known also as narrowband albedo) and the broadband albedo (or simply albedo) $\alpha_0(\theta_i,\Delta\lambda)$ are given by (Liang 2001)

$$\alpha_0(\theta_i,\lambda) = \frac{F_u(\theta_i,\lambda)}{F_d(\theta_i,\lambda)} \quad \alpha_0(\theta_i,\Delta\lambda) = \frac{\int_{\lambda_1}^{\lambda_2} F_u(\theta_i,\lambda)d\lambda}{\int_{\lambda_1}^{\lambda_2} F_d(\theta_i,\lambda)d\lambda}, \tag{3.59}$$

where $F_u(\theta_i,\lambda)$ is the upwelling flux, $F_d(\theta_i,\lambda)$ is the downward flux, θ_i is the SZA, and $\Delta\lambda = \lambda_2 - \lambda_1$ denotes the broad range of wavelengths. The albedo gives the fraction of sunlight reflected by the surface. The fraction absorbed by the surface is $(1-\alpha_0)F_d$. Measured fluxes from satellites are usually available in visible, near-infrared, and shortwave infrared ranges. According to Payne (1972), a daily average of $\alpha_0 \approx 0.06$.

Theoretically, spectral (direct) albedo can be estimated from the Fresnel reflection coefficients R_h and R_v for perpendicular (h) and parallel (v) polarizations and the Snell's law; for unpolarized light and mirror reflection the relationship is following:

$$\alpha_0(\theta_i,\lambda) = \frac{1}{2}(R_h + R_v) = \frac{1}{2}\left[\frac{\sin^2(\theta_i - \theta_r)}{\sin^2(\theta_i + \theta_r)} + \frac{\tan^2(\theta_i - \theta_r)}{\tan^2(\theta_i + \theta_r)}\right], \quad (3.60)$$

$$n_i \sin\theta_i = n_r \sin\theta_r, \quad (3.61)$$

where θ_i and θ_r are angles of incidence and refraction, respectively, n_i is the index of refraction of air ($n_i = 1.00029$ for all wavelengths) and $n_r = n_w(\lambda)$ is the real index of refraction of seawater which varies with wavelength (for pure water $n_w = 1.34$ at visible range). Formula (3.62) is derived from the Fresnel equations (Stratton 1941; Born and Wolf 1999). However, experimental and Fresnel-based theoretical OSA always will differ from one other by the value because of the impact of real world environmental—first of all, diffusive reflectance, variable transparency of seawater (due to suspended matter), and the influence of sea surface irregularities (surface roughness and foam/whitecap). To provide a better assessment, a number of semi-empirical parameterizations of OSA are proposed and tested (e.g., Jin et al. 2004, 2011). For example, wind-speed parameterization of OSA can be constructed by the following additive scheme

$$\alpha_0(\theta,\lambda,U) = W(U)\alpha_f(\lambda) + [1 - W(U)]\alpha_{f-f}(\theta,\lambda), \quad (3.62)$$

$$\alpha_{f-f}(\theta,\lambda) = \alpha_{sp}(\theta,\lambda) + \alpha_d(\theta,\lambda), \quad (3.63)$$

where the following three components of albedo are involved: (1) direct (or glint) component of albedo $\alpha_{sp}(\theta,\lambda)$ which depends on surface slope statistics, (2) diffuse component of albedo $\alpha_d(\theta,\lambda)$, and (3) foam or whitecap albedo, $\alpha_f(\theta,\lambda)$. Here $\alpha_{f-f}(\theta,\lambda)$ is albedo of foam-free water medium, $W(U) = aU^b$ is the foam/whitecap coverage fraction with a variety of empirical coefficient, e.g., $a = 2.95 \times 10^{-6}$ and $b = 3.52$ (Monahan and O'Muircheartaigh 1980), and U is standard wind speed (m/s). According to observations (Whitlock et al. 1982; Frouin et al. 1996; Moore et al. 1998, 2000; Nicolas et al. 2001) and model calculations (Koepke 1984; Gordon and Jacobs 1977; Stabeno and Monahan 1986; Kokhanovsky 2004b), foam albedo varies in a wide range of $\alpha_f = 0.55$–0.95 at $\lambda = 400$–$700\,$nm. Thus, the resulting OSA depends on SZA, θ, wavelength λ, and wind speed U.

Although composition albedo model (3.62) and (3.63) seems to be simple enough, there are many uncertainties in the assessment of OSA (and its individual components as well) due to a variety of input parameters and unpredictability of environment. Meanwhile, the OSA, mostly multispectral OSA, is a valuable indicator of correlations between hydrodynamic and bioproduction processes that offers an additional quantitative resource in thematic analysis and application of ocean optical data.

3.5.5 Radiance and Reflectance

As mentioned in Section 3.2.5, radiance and reflectance are the most important quantities in optical remote sensing (correspondingly, *radiance image* and *reflectance image* are distinguished). Radiance is the variable directly measured by remote sensing instruments. Radiance is dimensional quantity; it has unit of (W/m²·sr). Reflectance of a body is defined as the ration of reflected radiant power to incident radiant power; it is nondimensional quantity. There are additional reflectance quantities which are

Remote-sensing reflectance, $R_{rs}(\lambda, \theta, \varphi)$, is defined as the ratio of water-leaving upwelling radiance L_u to downwelling irradiance L_d penetrating into the water

$$R_{rs}(\lambda, \theta, \varphi) = \frac{L_u(\lambda, \theta, \varphi)}{L_d(\lambda)} = \frac{\text{upwelling radiance}}{\text{downwelling radiance}}, [sr^{-1}] \qquad (3.64)$$

Remote-sensing reflectance has become more valuable quantity in multispectral satellite-based observations of ocean environment (e.g., using the Moderate Resolution Imaging Spectroradiometer, MODIS/Aqua). In particular, $R_{rs}(\lambda, \theta, \varphi)$ is an artificial unit which is used in order to obtain ocean colour data. It is derived from top-of-atmosphere radiance/reflectance measurements through atmospheric correction and instrument calibration.

The Bidirectional Reflectance Distribution Function (BRDF), $f_{bd}(\theta_i, \varphi_i, \theta_r, \varphi_r)$, is a ratio of outgoing reflected radiance from the surface L_r through solid angle in direction $\Omega(\theta_r, \varphi_r)$ to incident irradiance L_i from illumination direction, $\Omega(\theta_i, \varphi_i)$. BRDF is

$$f_{bd}(\theta_i, \varphi_i, \theta_r, \varphi_r) = \frac{dL_r(\theta_r, \varphi_r)}{dL_i(\theta_i, \varphi_i)} = \frac{dL_r(\theta_r, \varphi_r)}{L_i(\theta_i, \varphi_i) \cos\theta_i d\Omega_i(\theta_i, \varphi_i)}, [sr^{-1}] \qquad (3.65)$$

$$f_{bd}(\theta_i, \varphi_i, \theta_r, \varphi_r) = f_{bd}(\theta_r, \varphi_r, \theta_i, \varphi_i). \qquad (3.66)$$

The BRDF is fundamental theoretical concept that describes the relationship between the geometry of the source radiance and the sensor viewing geometry (Hapke 2012). In sun-sensor configuration, $\Omega_i(\theta_i, \varphi_i)$ is position of sun, $\Omega_r(\theta_r, \varphi_r)$ is position of specular (mirror) reflectance direction (Figure 3.9). The BRDF specifies the behaviour of surface scattering as a function of illumination and view angles in a particular wavelength band; it is widely used in radiometry, spectroscopy, remote sensing, and computer vision. Several following types of reflectance (and BRDF) are known:

- *Lambertian reflectance (also called diffuse reflectance)* is defined by the Lambert law. The measured reflected radiance is independent of the viewing direction. The BRDF for a Lambertian surface is a constant $f_{bd} = \rho_d/\pi$ (ρ_d is the directional-hemispherical reflectance of a Lambert surface). Surface radiance is $L_r = (\rho_d/\pi)I_0\cos\theta_i$ (I_0 is source intensity). Surface appears equally bright from all directions.

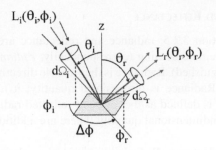

FIGURE 3.9 The BRDF concept.

- *Specular (mirror) reflectance*

$$f_{bd}(\theta_i, \varphi_i, \theta_0, \varphi_0) = \rho_s \frac{\delta(\theta_i - \theta_0)\delta(\varphi_i + \pi - \varphi_0)}{\cos\theta_i}. \quad (3.67)$$

Surface radiance is $L_r = \rho_s I_0 \delta(\theta_i - \theta_0)\delta(\varphi_0 + \pi - \varphi_0)$, ρ_s is the specular reflectance defined through the Fresnel coefficients, and (θ_0, φ_0) is viewing direction. In this case, $\rho_s \equiv \alpha_0$.
- Combinations of Lambertian and specular reflectance

$$f_{bd}(\theta_i, \varphi_i, \theta_r, \varphi_r) = \frac{\rho_{bd}\eta}{\pi} + \rho_s(1-\eta)\frac{\delta(\theta_i - \theta_0)\delta(\varphi_i + \pi - \varphi_0)}{\cos\theta_i}, \quad (3.68)$$

where the constant η is the weight coefficient, controlling the proportion between specular and diffuse reflectance terms.

In common case, reflectance $\rho(\lambda)$ and BRDF are related as

$$\rho(\lambda) = \frac{\int_{2\pi_r} L_r(\Omega_r, \lambda)\cos\theta_r d\Omega_r}{\int_{2\pi_i} L_i(\Omega_i, \lambda)\cos\theta_i \, d\Omega_i} = \frac{\int_{2\pi_r}\int_{2\pi_i} f(\Omega_i, \Omega_r)L_i(\Omega_i, \lambda)\cos\theta_i \cos_r \, d\Omega_i \, d\Omega_r}{\int_{2\pi_i} L_i(\Omega_i, \lambda)\cos\theta_i \, d\Omega_i}.$$

$$(3.69)$$

For isotropic incident radiance, the general equation for the irradiance reflectance is

$$\rho(\lambda) = \frac{\int_{2\pi_r} L_r(\Omega_r, \lambda)\cos\theta_r d\Omega_r}{\int_{2\pi_i} L_i(\Omega_i, \lambda)\cos\theta_i \, d\Omega_i} = \frac{\int_{2\pi_r}\int_{2\pi_i} f(\Omega_i, \Omega_r)L_i(\Omega_i, \lambda)\cos\theta_i \cos_r \, d\Omega_i \, d\Omega_r}{\int_{2\pi_i} L_i(\Omega_i, \lambda)\cos\theta_i \, d\Omega_i}$$

$$= \frac{1}{\pi}\int_{2\pi_r}\int_{2\pi_i} f(\theta, \varphi, \theta, \varphi)\cos\theta_r \cos\theta_r \, d\Omega_i d\Omega_r.$$

$$(3.70)$$

The quantity $\rho(\lambda)$ corresponds to the spherical albedo (or albedo).

Reflectance of the ocean environment is a complicated function of many variables; it can be modeled in the same way as albedo, Equations (3.62) and (3.63), i.e., as a weighted sum

$$\rho(\lambda, U) = W(U)\rho_f(\lambda) + [1 - W(U)][\rho_s(\lambda, U) + \rho_d(\lambda)], \qquad (3.71)$$

where $\rho_f(\lambda)$, $\rho_s(\lambda, U)$, and $\rho_d(\lambda)$ are contributions from foam, specular (glint), and diffuse reflectance, respectively. Algorithms for computing $\rho(\lambda, U)$ are reported in works (Sayer et al. 2010; Séférian et al. 2018). Formulation (3.71) provides parameterization of $\rho(\lambda, U)$ by wind speed U through empirical relationship $W(U) = aU^b$ and the Cox and Munk wind-dependent wave slope distribution model $P[Z'_x(U), Z'_y(U)]$ (in local system of coordinates). Model calculations indicate that the reflectance of the sea varies with geometry, wind speed, wavelength, and water properties. Uncertainties in the definition of the reflectance still remain due to the vagueness of IOP's (Lee et al. 2010) and also imperfect specification of the optical model. Figure 3.10 shows an example. Reflectance spectra also depend on a chlorophyll concentration. In present spectral variations in the ocean reflection especially at NIR (700–900 nm) are estimated from multispectral remote sensed data collected, e.g., by WiFS and Aqua/MODIS.

According to studies (Saunders 1968; Sidran 1981; Morel and Gentili 1991, 1993, 1996; Mobley 1999; Morel et al. 2002; Caillault et al. 2007), reflectance of water body or seawater has values $\rho(\lambda) = 0.01$–0.05 at $\lambda = 400$–900 nm depending on wavelength and SZA. Model data and measurements (Wu and Smith 1997) shows that emissivity of a rough sea surface $\kappa(\lambda) = 1 - \rho(\lambda)$ varies in the range of $\kappa(\lambda) = 0.85$–0.95 at atmospheric window $\lambda = 0.8$–$13\,\mu m$. Reflectance of sea foam and whitecap has been studied in works (Whitlock et al. 1982; Koepke 1984; Frouin et al. 1996; Moore et al. 2000; Nicolas et al. 2001; Kokhanovsky 2004b; Devetzoglou et al. 2014; Ma et al. 2015; Schwenger and Repasi 2017). These data shows the ranges of $\rho_f(\lambda) = 0.7$–1.0 (in visible) and $\rho_f(\lambda) = 0.1$–0.7 (in near-infrared). Foam reflectance is close to the

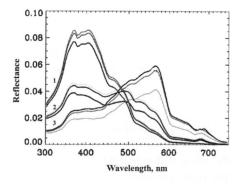

FIGURE 3.10 Reflectance spectra of sea surface at different chlorophyll concentrations: 1–0.1, 2–1.0 and 3–10 mg/m³. The curves differ by sky conditions. (Adapted from Mobley 2010.)

Lambert law; it weakly depends on SZA. Wave breaking is also an important factor influencing on sea reflectance (Stramski and Tegowski 2001).

3.6 POLARIZATION

Polarization is a fundamental property of electromagnetic waves that is described by physical laws of optics. According to historical data, the polarized behaviour of light was reported first in 1669 by *Erasmus Bartholinus* who observed the double refraction of light in calcite crystals. Since then, the nature of polarized light has become a subject of numerous scientific studied (e.g., Hulburt 1934; Können 1985). Over the last several decades, polarized radiation has been an important scientific topic in many fields of applied optics. These works included the theory of polarized light, light propagation in non-uniform scattering/absorption media, developments of optical polarimetry, ellipsometry, spectroscopy, and imagery (Brosseau 1998; Goldstein 2011; Collett and Schaefer 2012; Zhao et al. 2016).

In classical optics, polarization is defined as the orientation of electric field (E-field) of an electromagnetic wave (Born and Wolf 1999; Hecht 2017). Polarization of light in general is described by an ellipse. Two special cases of elliptical polarization are linear polarization and circular polarization. The case when the electric field always lies in the same plane is called linear polarization; the case when the electric field rotates while propagating is called circular polarization. In our brief description, we refer to books (Schott 2009; Collett and Schaefer 2012).

3.6.1 The Polarization Ellipse

According to Fresnel's theory, components $E_x(z,t)$ and $E_y(z,t)$ describe sinusoidal oscillations of E-field in the x–z and y–z planes, respectively,

$$E_x(z,t) = E_{0x} \cos(\tau + \varphi_x), \tag{3.72}$$

$$E_y(z,t) = E_{0y} \cos(\tau + \varphi_y), \tag{3.73}$$

with the maximum amplitudes E_{0x} and E_{0y} of the E-field at any time t; $\tau = \omega t - kz$ refers to the *propagator* and φ_x and φ_y are the phases. By themselves, these equations are not particularly revealing. However, eliminating the time-space propagator $\omega t - kz$ between the two equations leads to the equation of an ellipse

$$\frac{E_x^2}{E_{0x}^2} + \frac{E_y^2}{E_{0y}^2} + \frac{-2E_xE_y}{E_{0x}E_{0y}}\cos\varphi = \sin^2\varphi, \tag{3.74}$$

where $\varphi = \varphi_y - \varphi_x$ is the phase difference between the two components. Equation (3.74) describes an ellipse rotated through an angle of orientation ψ

$$\tan 2\psi = \frac{2E_{0x}E_{0y}\cos\varphi}{E_{0x}^2 - E_{0y}^2}, \tag{3.75}$$

that represents the pattern traced by the E-field in time at a fixed plane $z = z_0$. Due to phase differences, in the consistent waves, the orientation of the polarization ellipse may rotate over time and the amplitude may vary.

A special case is linear polarization, defined as

$$E_x(z,t) = E_{0x} \cos(\tau + \varphi_x), E_y(z,t) = 0 \text{ perpendicular (horizontal) polarization,}$$

(3.76)

$$E_x(z,t) = 0, E_y(z,t) = E_{0y} \cos(\tau + \varphi_y) \text{ parallel (vertical) polarization.} \quad (3.77)$$

More generally, if $\varphi = 0$ or π, Equation (3.74) reduces to relationship

$$E_y = \pm \frac{E_{0y}}{E_{0x}} E_x, \quad (3.78)$$

related to linearly polarized radiation at orientation defined by the slope $\pm \left(\dfrac{E_{0y}}{E_{0x}} \right)$.

When maximum amplitudes $E_{0y} = E_{0x}$ the slope is 1, representing linear polarization oriented along the $\pm 45°$ axis ($\varphi = 0$ is $+45°$ and $\varphi = \pi$ is $-45°$). Linearly polarized radiation is the most commonly used in remote sensing.

In the case of $\varphi = \pi/2$ or $\varphi = (3/2)\pi$, Equation (3.74) reduces to

$$\frac{E_x^2}{E_{0x}^2} + \frac{E_y^2}{E_{0y}^2} = 1, \quad (3.79)$$

which is the standard (unrotated) version of an ellipse.

Additionally, if $\varphi = \pi/2$ or $\varphi = (3/2)\pi$ and $E_{0y} = E_{0x} = E_0$, then

$$\frac{E_x^2}{E_0^2} + \frac{E_y^2}{E_0^2} = 1, \quad (3.80)$$

which is the equation of a circle yielding circular polarization (right handed if $\varphi = \pi/2$ or left handed if $\varphi = (3/2)\pi$).

The concept of polarization ellipse provides some insight into the nature of polarized radiation; however, it has practical limitations (Goldstein 2011). First, the theoretical description presented is only applicable to fully polarized radiation which is very uncommon in nature. Second, the polarization ellipse describes the electromagnetic field which is difficult (or impossible) to measure directly by optical remote sensor. Therefore, in remote sensing, polarized radiation is described in terms of the Stokes parameters.

3.6.2 THE STOKES POLARIZATION PARAMETERS

Stokes (1852) showed that at the time average $\langle E_i(z,t)E_j(z,t) \rangle = \lim\limits_{T_0 \to \infty} \int\limits_0^{T_0} E_i(z,t)E_j(z,t)\,dt$, $\{i,j = x,y\}$, T_0 is total averaging time, the equation for polarization ellipse (3.74) can be expressed in the form

$$\left(E_{0x}^2 + E_{0y}^2\right)^2 = \left(E_{0x}^2 - E_{0y}^2\right)^2 + \left(2E_{0x}E_{0y}\cos\varphi\right)^2 + \left(2E_{0x}E_{0y}\sin\varphi\right)^2. \quad (3.81)$$

Equation (3.81) is written as $S_0^2 = S_1^2 + S_2^2 + S_3^2$ where each terms is defined as

$$S_0 = E_{0x}^2 + E_{0y}^2, \quad (3.82a)$$

$$S_1 = E_{0x}^2 - E_{0y}^2, \quad (3.82b)$$

$$S_2 = 2E_{x0}E_{y0}\cos\varphi, \quad (3.82c)$$

$$S_3 = 2E_{x0}E_{y0}\sin\varphi, \quad (3.82d)$$

where S_0, S_1, S_2, and S_3 are referred to as the *Stokes polarization parameters* for a plane wave. The S_0 term describes the total energy in a beam. The S_1 term describes the amount of linear horizontal or vertical polarization. The S_2 term describes the amount of linear $\pm 45°$ polarization and the S_3 term describes the amount of right- or left-handed circular polarization.

The Stokes parameters can be arranged as a column matrix, which is referred to as the Stokes vector for elliptically polarized light:

$$S = \begin{pmatrix} S_0 \\ S_1 \\ S_2 \\ S_3 \end{pmatrix} = \begin{pmatrix} E_{0x}^2 + E_{0y}^2 \\ E_{0x}^2 - E_{0y}^2 \\ 2E_{0x}E_{0y}\cos\varphi \\ 2E_{0x}E_{0y}\sin\varphi \end{pmatrix} = \begin{pmatrix} S_0 \\ S_0\cos(2\chi)\cos(2\psi) \\ S_0\cos(2\chi)\sin(2\psi) \\ S_0\sin(2\chi) \end{pmatrix} = \begin{pmatrix} I \\ Q \\ U \\ V \end{pmatrix},$$

$$\quad (3.83)$$

$$\tan(2\psi) = \left(\frac{S_2}{S_1}\right), 0 \le \psi \le \pi, \quad (3.84)$$

$$\sin(2\chi) = \left(\frac{S_3}{S_0}\right), -\pi/4 \le \chi \le \pi/4, \quad (3.85)$$

where ψ and χ are angles of polarization and ellipticity associated with the *Poincaré sphere*. If $\chi = \pm\pi/4$, we have circular polarization and if $\chi = 0$, we have linear polarization.

The Stokes parameters are real observable quantities expressed in terms of optical intensities or radiometric energies. The Stokes parameters are also commonly denoted as a 4×1 vector $S = (I,Q,U,V)^T$, T indicates the transpose, and $I = S_0$ is the intensity according to (3.83)–(3.85). The Stokes vector describes the state of polarization (SOP) of totally or partially polarized light. Here we have several important cases:

- Linear polarization $Q \ne 0$, $U \ne 0$, $V \ne 0$
- Circular polarization $Q = 0$, $U = 0$, $V \ne 0$
- Fully polarized light $I^2 = Q^2 + U^2 + V^2$

- Partially polarized light $I^2 > Q^2 + U^2 + V^2$
- Unpolarized light $Q = U = V = 0$.

Important parameters are the degree of polarization (DOP) and the degree of linear polarization (DoLP)

$$\text{DOP} = \left(\sqrt{S_1^2 + S_2^2 + S_3^2}\right) \Big/ S_0, \qquad (3.86)$$

$$\text{DoLP} = \left(\sqrt{S_1^2 + S_2^2}\right) \Big/ S_0, \qquad (3.87)$$

which represent states of polarization.

3.6.3 THE MUELLER MATRIX

The Mueller matrix $\mathbf{M}$ (introduced by Hans Mueller during the early 1940s) totally characterizes the interaction of polarized light with an object in the absence of nonlinear effects. The Mueller matrix is defined as the matrix which transforms an incident Stokes vector $\mathbf{S}$ into the exiting (reflected, transmitted, or scattered) Stokes vector $\mathbf{S}'$,

$$\mathbf{S}' = \begin{bmatrix} s_0' \\ s_1' \\ s_2' \\ s_3' \end{bmatrix} = [\mathbf{MS}] = \begin{bmatrix} m_{00} & m_{01} & m_{02} & m_{03} \\ m_{10} & m_{11} & m_{12} & m_{13} \\ m_{20} & m_{21} & m_{22} & m_{23} \\ m_{30} & m_{31} & m_{32} & m_{33} \end{bmatrix} \begin{bmatrix} s_0 \\ s_1 \\ s_2 \\ s_3 \end{bmatrix}. \qquad (3.88)$$

The Mueller matrix is a 4×4 matrix of real values and represents the polarization-altering properties of an object. The Mueller matrix is a function of the frequency of light, scattering geometry, and the direction of light propagation. The Mueller matrix is an appropriate formalism for characterizing polarization measurements; its elements contain all of the polarization properties. The Mueller matrix determines how the SOP is transformed as light interacts with an object or an optical element or system.

For example, the Mueller matrix of rotated polarizing elements is $\mathbf{M}' = \mathbf{R}(i)$ $\mathbf{MR}(-i)$, where $\mathbf{R}(i)$ is a rotation matrix and i is an rotation angle (positive in a counterclockwise sense looking toward the source). In this case, the new Stokes vector $\mathbf{S}'$ is defined as

$$\mathbf{S}' = \begin{bmatrix} I' \\ Q' \\ U' \\ V' \end{bmatrix} = [\mathbf{M}'\mathbf{S}] = \begin{bmatrix} 1 & 0 & 0 & 0 \\ 0 & \cos(2i) & \sin(2i) & 0 \\ 0 & -\sin(2i) & \cos(2i) & 0 \\ 0 & 0 & 0 & 1 \end{bmatrix} \begin{bmatrix} I \\ Q \\ U \\ V \end{bmatrix}, \qquad (3.89)$$

where (I,Q,U,V) and (I',Q',U',V') refer to the incident light and the reflected, scattered, or transmitted light, correspondingly.

For a nondepolarizing optical element, the *Jones matrix* $\mathbf{E}' = \mathbf{JE}$ is introduced. The Jones matrix consists of 2×1 Jones vectors to describe the field components and 2×2 Jones matrices to describe polarizing components. Since 2×2 matrix is simpler than the Mueller matrix, the Jones formulation is limited to treating only completely polarized light; it cannot describe unpolarized or partially polarized light. The Jones matrix is used for treating interference phenomena or in cases of field amplitude superposition.

Detailed description of Mueller/Jones matrix formalism, calculus and applications can be found in books (Azzam and Bashara 1987; Gil Perez and Ossikovski 2016; Zhang et al. 2017). Mueller matrixes, polarized light field, reflectance and transmittance for ocean water are discussed in works (Voss and Fry 1984; Kattawar and Adams 1989; Kattawar 2013; Zappa et al. 2012; Freda et al. 2015; Mobley 2015; Zhou et al. 2017).

3.6.4 OPTICAL POLARIMETRY AND REMOTE SENSING

Optical polarimetry is relatively new direction in remote sensing of environment (Kokhanovsky 2003; Egan 2004; Shell 2005; Schott 2009). Over the last several decades, a number of optical polarimetric systems has been developed and deployed on aircraft and satellite platforms. The most known for us remote sensing applications (excluding SAR polarimetry) are the following: (1) land/island polarimetric imagery with Space Shuttle missions, (2) atmospheric aerosol and clouds measurements with POLDER and similar instrumentation, (3) airborne imagery of ocean waves (Baxter 2012; Bowles et al. 2015), (4) optical and infrared polarimetry of the sea surface (Tyo et al. 2006a,b), (5) observation of submarine illumination, (6) bio-optical oceanography including hydrosol measurements, and (7) astrophysics (Mishchenko et al. 2010). Polarimetric imagery is also successfully used for target detection and object's recognition. These and other applications are reviewed by Snik et al. (2014).

Field experiments show that polarimetric optical imagery can provide some enhancement of visible low-contrast oceanic scenes due to the difference between polarized und unpolarized upward irradiance reflectance. Light measured by polarimeter is less sensitive to variations of sunlight conditions but more sensitive to sea surface state (due to the effects of depolarization from microfacets). In principle, polarized light creates a better *texturization* of ocean optical images that is important for spectral analysis and selection of relevant wave components. Polarimertic observations can also improve identification and measurements of wave breaking and foam/whitecap fields perfectly registered in optical images at high wind. This provides more accurate statistical analysis of foam/whitecap area fractions. However, space-based fully polarimetric optical/infrared observations of the ocean involving Stokes–Mueller matrix polarimetry technique is still a challenging task.

3.7 PROPAGATION IN ATMOSPHERE

3.7.1 ELEMENTS OF ATMOSPHERIC OPTICS

A branch of optics studying the behaviour of solar radiance in the Earth's atmosphere is called atmospheric optics (Rozenberg 1966; McCartney 1976). This includes both understanding naturally occurring optical effects and interactions between electromagnetic radiation (sunlight) and atmospheric constituents. Vertical structure of the Earth atmosphere is shown schematically in Figure 3.11. According to standard nomenclature, the vertical profile is divided into four distinct layers known as troposphere, stratosphere, mesosphere, and thermosphere (and also fifth layer known as exosphere). Optical wave propagation and radiation in cloudiness atmosphere is defined by the following major factors:

- Profiles of temperature and pressure
- Profiles of oxygen, water vapor, and cloud liquid water concentrations
- Fluctuation of the index of refraction
- Molecular absorption by atmospheric gases (O_2, O_3, N_2, CO_2)
- Scattering and absorption by atmospheric particles, micro-dispersed aerosols, thin clouds, and other fine-scale atmospheric phenomena.

Figure 1.2 (in Section 1.2) illustrates atmospheric windows for ultraviolet, visible, and infrared ranges of electromagnetic spectrum. Atmospheric windows are used for optical remote sensing of the Earth's surfaces; however, in the case of ocean observations, it is difficult to predict exactly optimal *transmissive* imaging frequency because of the influence of nonuniform near-surface atmospheric boundary layer that can "block" the atmospheric electromagnetic transparency. Therefore, high-resolution multispectral optical/infrared imagery can detect different "false" texture elements that are not related to surface waves.

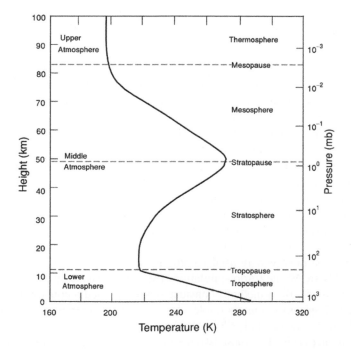

FIGURE 3.11   Vertical structure of the Earth atmosphere. (Based on Liou 2002.)

Propagation of optical radiation in the ocean-atmosphere system can be described by a vector RTE in a plane-parallel scattering medium given by

$$\mu \frac{dL(\tau,\mu,\phi)}{d\tau} = L(\tau,\mu,\phi) - \Im(\tau,\mu,\phi), \tag{3.90}$$

$$\Im(\tau,\mu,\phi) = \frac{\omega_0(\tau)}{4\pi} \int_0^{2\pi} \int_{-1}^{1} P(\tau,\mu,\phi,\mu',\phi')L(\tau,\mu',\phi')d\mu'd\phi', \tag{3.91}$$

$$P(\tau,\mu,\phi,\mu',\phi') = R(\pi - i_2)M(\tau,\Theta)R(-i_1). \tag{3.92}$$

In (3.90) a beam light of arbitrary polarization is represented by the Stokes vector $L = (I,Q,U,V)^T$, where the four elements (I,Q,U,V) are the Stokes parameters, τ is the optical depth, $\mu = \cos\theta$, θ is the zenith angle ($\theta < 90°$ for upward and $\theta > 90°$ for downward), ϕ is the azimuthal angle, ω_0 is the single-scattering albedo and $\Im$ is the source matrix. In (3.92) P is the phase matrix. In (3.93) R is the rotation matrix, i_1 and i_2 are the rotation angles and M is the scattering Mueller matrix for turbid medium with the scattering angle $\Theta = \Theta (\mu,\phi,\mu',\phi')$.

The integral RTE can be re-written in form (Chandrasekhar 1960; Liou 2002)

$$L(\tau,\mu < 0,\phi) = L(\tau_\ell,\mu,\phi)e^{-(\tau_\ell-\tau)/\mu} - \int_{\tau_\ell}^{\tau} e^{-(\tau'-\tau)/\mu}\Im(\tau',\mu,\phi)d\tau'/\mu, \tag{3.93}$$

$$L(\tau,\mu > 0,\phi) = L(\tau_u,\mu,\phi)e^{-(\tau_u-\tau)/\mu} + \int_{\tau}^{\tau_u} e^{-(\tau'-\tau)/\mu}\Im(\tau',\mu,\phi)d\tau'/\mu, \tag{3.94}$$

where τ_ℓ and τ_u are the lower and upper limits of the medium under consideration ($\tau_\ell < \tau < \tau_u$) which are defined in accordance with boundary conditions. In the last three decades, special attention has been paid to find effective methods for solving the RTE. The simplest method is known as the two-stream approximation (e.g., Ishimaru 1991; Thomas and Stamnes 1999) in which radiation is propagating in only two discrete directions. It was considered first by *Arthur Schuster* in 1905. In present modified versions of this method with multiple scattering and absorption are widely used in radiative transfer flux calculations in climate models, atmospheric optics, and microwave remote sensing of environment.

As a whole, the RTE is a complex mathematical framework having various ways of analysis and solutions. Mathematical methods and schemes have been developed in early works (e.g., Marchuk et al. 1980; Elepov 1980; Spinrad et al. 1994; Bukata et al. 1995). The most widely used techniques to solve vector RTE for polarized radiation fields are Monte Carlo simulations (MCS) and successive order of scattering (SOS) methods. Both methods consider scattering media containing two types of scatterers: Rayleigh scatterers (molecules) and Mie scatterers (aerosols). Recent numerical analysis (Zhai et al. 2009; Mukherjee et al. 2017) shows that differences between radiances solved by these two methods with different mixture schemes are of the order of 0.1%.

Meanwhile, MCS have advantages because describe statistical character of light propagation (photon trajectories) in a discrete random medium that exactly corresponds to stochastic nature of a coupled ocean-atmosphere system. Selection of initial conditions for MCS includes solar radiance distribution, angles of incident, geometry of sensor, and structural characteristics of environment. MCS can solve time-dependent and 3D problems with arbitrary geometry. MCS provides numerical simulations of realistic radiation process that is necessary for the analysis of remotely sensed data and the retrieval of ocean and atmosphere parameters from satellite data.

Different aspects of atmospheric optics including atmospheric correction, developments of radiation models and numerical schemes for remote sensing are discussed in books (Kondratyev 1969; Marchuk et al. 1980; Kondratyev et al. 1992; Goody and Yung 1995; Thomas and Stamnes 1999; Liou 2002; Melnikova and Vasilyev 2005; Thomas 2006; Timofeyev and Vasil'ev 2008; Stamnes et al. 2017). Below we consider some results specifically related to optical/infrared imagery.

3.7.2 ATMOSPHERIC TURBULENCE

Atmospheric turbulence is a major problem in optical communication and imaging. Atmospheric turbulence is caused by many different phenomena—convective currents (thermal turbulence), wind shear and obstruction of wind flow (mountain turbulence), jet streams, wakes, and the other environmental disturbances (Wyngaard 2010).

Turbulent atmosphere is a random stochastic medium in which fluctuations of the refractive index along the propagation path cause wave front distortions in the optical waves. The most well-known effect of optical turbulence is the twinkling of stars that produce irregular change in brightness of the image. In addition, atmospheric turbulence disrupts the coherence of optical (laser) beam; this limits the capabilities of an optical device to provide perfect imagery of an object. Atmospheric turbulence impacts on detection capacity as well.

Electromagnetic wave propagation through turbulent media has been studied by many authors; classical books (Chernov 1960; Tatarski 1961; Uscinski 1977; Zuev 1982; Ishimaru 1991; Sasiela 2007) provide comprehensive data. More specifically, optical turbulence, signal transmission, and propagation of laser beam in atmosphere are considered in books (Panofsky and Dutton 1984; Andrews et al. 2001; Andrews and Phillips 2005; Korotkova 2014).

A theory of optical turbulence is based on the following statements.

Turbulent motions in atmosphere—variations of wind velocity and evolution of eddies, cause random fluctuations of the index of refraction of air (commonly referred to as optical turbulence). The index of refraction is a function of temperature and pressure which is approximated as

$$n \cong 1 + 77.6 \times 10^{-6} \left(1 + \frac{0.00752}{\lambda^2} \right) \frac{P}{T}, \tag{3.95}$$

where λ is the optical wavelength in micrometers, P is the pressure in millibars, and T is the temperature in Kelvin. The effect of pressure is usually rather insignificant.

At sea level and $\lambda = 0.5\,\mu m$, $\Delta n \approx 10^{-6}\,\Delta T$. Changes in optical signal due to scattering and absorption by molecules and aerosols are not considered here.

The structure function of the refractive index is

$$D_n(r_1,r_2) = <\left[n(r_1) - n(r_2)\right]^2 >,\qquad(3.96)$$

where $n(r_1)$ and $n(r_2)$ are the values of the refractive index at positions r_1 and r_2. It was shown (e.g., Ishimaru 1991) that under the assumption of horizontal homogeneity and local isotropy, the fluctuations follow to the Kolmogorov statistics and the structure function has the form

$$D_n(r) = C_n^2 r^{2/3},\ \ell_0 << r << L_0,\qquad(3.97)$$

Where ℓ_0 are L_0 the inner and outer scales of turbulence and C_n^2 is so-called *the refractive index structure parameter* (or structure constant). It is a measure of the magnitude of the fluctuations of the refractive index which is given in units of the $-2/3$ power of meters. Values of C_n^2 have typical range from $10^{-17}\,m^{-2/3}$ or less at weak turbulence regime and up to $10^{-13}\,m^{-2/3}$ at strong turbulence regime.

The corresponding power spectrum in three-dimension is

$$\Phi_n(K) = 00033 C_n^2 K^{-11/3},\ 2\pi/L_0 << K << 2\pi/\ell_0,\qquad(3.98)$$

where K is the spatial frequency (wavenumber). Equation (3.98) is well-known Kolmogorov spectrum. There are other models of the spectrum of refractive-index fluctuations; the most known are Tatarskii spectrum and von Kármán spectrum (Tatarski 1961; Ishimaru 1991). These models are more appropriate for various calculations and applications.

Optical turbulence causes a variety of deleterious effects relevant to laser (lidar) measurements and remote sensing imagery of nonstationary objects. Briefly, the following optical distortions due to atmospheric turbulence can occur (Andrews et al. 2001):

- Beam spreading—beam divergence of a plane wave due to diffraction
- Beam wander—angular deviation of the centroid of a beam
- Image dancing—fluctuation in the beam at the focal plane of the receiver
- Beam scintillation—irradiance fluctuations within the beam cross section
- Loss of spatial coherence—limits the effective aperture diameter in an imaging system
- Astronomical seeing—blurring and twinkling of astronomical objects (stars)
- Distortion of incoming optical wavefront.

The propagation of unpolarized optical wave through a random stationary turbulent medium is governed by the reduced ($\Delta n/n << 1$) wave equation (Section 3.2.2) which is given by

$$\nabla^2 U(\vec{r}) + k_0^2 n^2(\vec{r}) U(\vec{r}) = 0,\qquad(3.99)$$

where $U(\vec{r})$ is the optical field as a function of space coordinates $\vec{r} = \{x, y, z\}$, $n(\vec{r}) = n_0 + \Delta n(\vec{r})$ is a random index of refraction, n_0 and $\Delta n(\vec{r})$ are mean and fluctuation parts, and $k_0 = 2\pi/\lambda$ is the electromagnetic wavenumber. Equation (3.99) are solved using different methods and wave models depending on configuration of $n(\vec{r})$. In particular, various wave propagation problems can be analyzed analytically using classical Born approximation and/or Rytov approximation (e.g., Born and Wolf 1999).

3.7.3 IMAGING THROUGH TURBULENCE

The problem of remote sensing imaging through the turbulent atmosphere is similar to the problem of light-beam propagation through the atmosphere. The distortion effect of atmospheric turbulence on optical imaging systems and image quality has been recognized since the 1950s. This phenomenon is well known from astronomy and observations of celestial objects through optical telescopes (Roggemann et al. 1996; Lukin and Fortes 2002; Goodman 2015; McKechnie 2016). In particular, strong atmospheric turbulence results to overall reduction of spatial resolution of optical images. Turbulence degrades the images by inducing geometric distortion, defocus, warping, and blurring. This degradation process arises under the influence of many atmospheric factors mentioned above, but the most important impact on remote sensing imaging produces fluctuations of the index of refraction originated with turbulent air motion.

There are several techniques developed over the years to eliminate these atmospheric effects. The most known are (1) short-exposure imagery referred to a *speckle imaging*, (2) adaptive optics method, and (3) hybrid imaging techniques that use elements of adaptive optics and image reconstruction techniques (Roggemann et al. 1996).

Let's consider imaging process through atmosphere. Image formation is described by a linear convolution

$$I(x, y, t) = \iint O(x', y', t) P(x' - x, y' - y, t) \, dx' dy' + \eta = O(x, y, t) \otimes P(x, y, t) + \eta,$$

(3.100)

where $I(x, y, t)$ is the observed atmospheric turbulence-degraded optical image, $O(x, y, t)$ is the true image, $P(x, y, t)$ is the Point Spread Function (PSF), η denotes additive noise, and symbol "$\otimes$" denotes 2D convolution (additive noise is not considered here). The problem of recovering the true image $O(x, y, t)$ from the given observed degraded image $I(x, y, t)$ is called *image restoration* in signal processing literature. In general, restoration methods require complete knowledge of the spatiotemporal PSF which involves both instrument and medium parameters (called also statistical or turbulent PSF).

The fundamental quantities for characterizing the performance of an optical imaging system in the presence of atmospheric-turbulence effects are (1) the structure constant, C_n^2, (2) the turbulent PSF, (3) the optical transfer function (OTF) which is the Fourier transform of the normalized PSF, and (4) the atmospheric turbulence modulation transfer function (MTF) which is a measure of resolution in frequency domain limited by the Nyquist frequency. All these quantities are defined using parameters of an optical imaging system and statistical models of turbulence (e.g., Boreman 2001; Blaunstein and Kopeika 2018). Under certain conditions, the total "optical system + turbulent atmosphere + aerosol" MTF_{total} can be written as the product of three components:

$$MTF_{total}(\nu) = MTF_{opt}(\nu) \times MTF_{trb}(\nu) \times MTF_{ars}(\nu), \qquad (3.101)$$

$$MTF_{opt}(\nu) = \frac{2}{\pi}\left[\cos^{-1}(\nu/\nu_c) - (\nu/\nu_c)\sqrt{1-(\nu/\nu_c)^2}\right], \nu/\nu_c < 1, \quad (\text{otherwise it is } 0),$$
$$(3.102)$$

$$MTF_{trb}(\nu) = \exp\left\{-3.44(\lambda f\nu/r_0)^{5/3}\left[1 - a(\lambda f\nu/D)^{1/3}\right]\right\}, \qquad (3.103)$$

$$MTF_{asl}(\nu) \approx \exp(-\tau_{sct}) + \left[1 - \exp(-\tau_{sct})\right]\exp\left[-(\nu/\nu_c)^2\right]. \qquad (3.104)$$

Components $MTF_{opt}(\nu)$, $MTF_{trb}(\nu)$, and $MTF_{asl}(\nu)$ refer to a perfect diffraction-limited optical imaging system (index "opt"), turbulent atmosphere (index "trb") and atmospheric aerosol (index "asl"). In Equation (3.102) the spatial frequency is ν (in cycles/mm or line/mm), optical cut-off frequency is $\nu_c = 1/(\lambda f\#)$, and the number (f#) describes the ratio between the focal length f of the objective and the diameter D of the entrance pupil, f# = f/D. In Equation (3.104) τ_{sct} is the opacity related to scattering and absorption by aerosol and/or dust particles (e.g., Kokhanovsky 2008).

Equation (3.103) describes long exposure (a = 0) and short exposure (a = 1) cases. $MTF_{atm}(\nu)$ is defined by Fried's parameter (Fried 1966)

$$r_0 = \left[0.423k_0^2 \sec\varsigma \int_0^{h_{max}} C_n^2(z)dz\right]^{-3/5}, \qquad (3.105)$$

where β is the angle between the point direction and normal to the ground (i.e., zenith angle), $C_n^2(z)$ is vertical profile of the structure constant (called also atmospheric turbulence strength), h_{max} is maximal altitude, and $k_0 = 2\pi/\lambda$. Fried's parameter or atmospheric coherent length is defined as the diameter of the circular pupil for which the diffraction-limited image and the seeing limited image have the same angular resolution. At visible range $r_0 \propto \lambda^{6/5}$, its typical value is 10–20 cm at good seeing conditions.

MTF is one of the best tools available to quantify the overall imaging performance of an observation system in terms of resolution and contrast. MTF can be defined from Equation (3.100). In Fourier transform domain (noise term is omitted)

$$I = O(x,y,t) \otimes G(x,y,t) \to \tilde{F}(I) = \tilde{F}(O) \otimes \tilde{F}(PSF), \qquad (3.106)$$

$$\tilde{F}(I) = \tilde{F}(O) \otimes OTF \to MTF = |OTF|. \qquad (3.107)$$

The angular resolution of an optical system is severely limited by atmospheric turbulence (aerosol impact is not considered). Instead diffraction-limited angular resolution $\alpha_{opt} \sim \lambda/D$, it will be $\alpha_{atm} \sim \lambda/r_0$. The reduction is by factor D/r_0 which, e.g., for IKONOS imagery is $D/r_0 \sim 0.7/(0.1 \div 0.2) \sim 3 \div 7$ (note that for regular optical telescope $D/r_0 \sim 25$). For example, in the case of an atmosphere/sensor imaging system with the ratio of $D/r_0 = 5$, the speckle (glitters) patterns will be spread over an area approximately five times larger than the diffraction-limited PSF. The reason of data degradation is optical turbulence which exists even at excellent seeing conditions. It means, in particular, that ocean surface features registered in the optical (IKONOS) images always will comprise some geometrical distortions (blurring effects) due to turbulence-induced changes in pixel resolution. In some situations, this can limit capabilities of thematic data analysis and the detection capacity.

3.8   SUMMARY

In this chapter, the basic aspects of optical oceanography and ocean optics are covered.

This cross-disciplinary science includes the following topics:

- electromagnetic wave propagation
- the nature of light, radiometric terms, and radiative transfer theory
- IOPs and AOPs of water
- the identity and characteristics of marine absorbers and scatterers
- bio-optical models of primary production
- ocean colour
- underwater visibility
- remote sensing
- atmospheric corrections for remote sensing.

Light propagation and transmission in ocean and atmosphere are described by fundamental electromagnetic wave theory based on the Maxwell's equations. In advanced concept, ocean and atmosphere environments both can be represented as non-uniform media with stochastic properties. Optical signatures are associated with the impact of internal inhomogeneities resulting to random fluctuations of the refractive index. In ocean water, those are variations of density, temperature, salinity as well the influence of marine bio-particles and sediments. In the atmosphere, air motions, wind, aerosol scattering and absorption are the main causes of optical turbulence. Interaction of light with the ocean, sunlight reflection, propagation, and scattering are key issues to the understanding remote sensing capabilities especially needed to detection of hydrodynamic phenomena.

A majority of optical remote sensing covered in this chapter, is based on the RTE that has long heritage and is very popular tool for data analysis among remote sensing community. Rigorous mathematical solutions involving scattering problems frequently use 1D models for computing light propagation in linear deterministic media. Alternatively, numerical schemes based on MCS enable to predict propagation of polarized light in 3D turbulent medium that is more realistic option in context with geophysical analysis of satellite remotely sensed data. It may require some efforts in creating hybrid dynamic radiative transfer models of the polarized light field for high-resolution imagery of the ocean.

The atmosphere plays an important role in light propagation and formation of optical images. The atmospheric influences are based on changes of the refractive index, molecular absorption, scattering by aerosol and dust particles. Atmospheric turbulence reduces the azimuth resolution and the image resolution. Modern observation models involve combined PSFs and MTFs dependent on sensor's parameters and statistical properties of turbulent atmosphere.

While we attempted to cover key topics of ocean optics, there are many important data and results that did not receive our attention they deserved. Readers are encouraged to examine the references cited in this chapter.

REFERENCES

Akhadov, Y. Y. 1980. *Dielectric Properties of Binary Solutions: A Data Handbook.* Pergamon, Oxford, UK.

Andrews, L. C. and Phillips, R. L. 2005. *Laser Beam Propagation through Random Media,* 2nd edition. SPIE Press, Bellingham, WA.

Andrews, L. C., Phillips, R. L., and Hopen, C. Y. 2001. *Laser Beam Scintillation with Applications.* SPIE Press, Bellingham, WA.

Apresyan, L. A. and Kravtsov, Y. A. 1996. *Radiation Transfer: Statistical and Wave Aspects (translated from Russian by M. G. Edelev).* Gordon and Breach Publishers, Amsterdam, The Netherlands.

Arst, H. 2003. *Optical Properties and Remote Sensing of Multicomponental Water Bodies.* Springer, Praxis Publishing, Chichester, UK.

Austin, R. W. and Halikas, G. 1976. *The Index of Refraction of Seawater. Technical report.* SIO Ref. No.76-1. Scripps Institute of Oceanography, La Jolla, CA. Available on the Internet http://misclab.umeoce.maine.edu/education/VisibilityLab/reports/SIO_76-1.pdf.

Azzam, R. M. A. and Bashara, N. M. 1987. *Ellipsometry and Polarized Light.* North Holland Personal Library, Elsevier, Amsterdam, The Netherlands.

Barber, P. W. and Hill, S. C. 1990. *Light Scattering by Particles: Computational Methods.* World Scientific Publishing, Singapore.

Baxter, R. 2012. *Ocean wave slope and height retrieval using airborne polarimetric remote sensing.* Dissertation. Georgetown University, DC. Available on the Internet https://repository.library.georgetown.edu/handle/10822/557750.

Blaunstein, N. and Kopeika, N. (Eds.). 2018. *Optical Waves and Laser Beams in the Irregular Atmosphere.* CRC Press, Boca Raton, FL.

Bohren, C. F. and Huffman, D. R. 1983. *Absorption and Scattering of Light by Small Particles.* John Wiley & Sons, New York.

Boileau, A. R. and Gordon, J. J. 1966. Atmospheric properties and reflectances of ocean water and other surfaces for a low sun. *Applied Optics,* 5(5):803–813. doi:10.1364/ao.5.000803.

Boreman, G. D. 2001. *Modulation Transfer Function in Optical and Electro-optical Systems.* SPIE Press, Bellingham, WA.

Born, M. and Wolf, E. 1999. *Principles of Optics: Electromagnetic Theory of Propagation, Interference and Diffraction of Light,* 7th edition. Cambridge University Press, Cambridge, UK.

Bowles, J. H., Korwan, D. R., Montes, M. J., Gray, D. J., Gillis, D. B., Lamela, G. M., and Miller, W. D. 2015. Airborne system for multispectral, multiangle polarimetric imaging. *Applied Optics*, 54(31):F256–F267.

Brosseau, C. 1998. *Fundamentals of Polarized Light: A Statistical Optics Approach*. John Wiley & Sons, New York.

Bukata, R. P., Jerome, J. H., Kondratyev, K. Ya, and Pozdnyakov, D. V. 1995. *Optical Properties and Remote Sensing of Inland and Coastal Waters*. CRC Press, Boca Raton, FL.

Caillault, K., Fauqueux, S., Bourlier, C., Simoneau, P., and Labarre, L. 2007. Multiresolution optical characteristics of rough sea surface in the infrared. *Applied Optics*, 46(22):5471–5481.

Chandrasekhar, S. 1960. *Radiative Transfer*. Dover Publications, New York.

Chernov, L. A. 1960. *Wave Propagation in a Random Medium (translated from Russian by R. A. Silverman)*. McGraw-Hill, New York. (Dover Publications, Mineola, New York, 1988).

Choy, T. C. 2016. *Effective Medium Theory: Principles and Applications (International Series of Monographs on Physics 165)*, 2nd edition. Oxford University Press, Oxford, UK.

Collett, E. and Schaefer, B. 2012. *Polarized Light for Scientists and Engineers*. The Polawave Group, Long Branch, NJ.

Deirmendjian, D. 1969. *Electromagnetic Scattering on Spherical Polydispersions*. American Elsevier Publishing Company, New York.

Dera, J. 1992. *Marine Physics*. Elsevier Science Publishing, New York.

Devetzoglou, M. A. and Evans, J. R. G. 2014. Diffuse reflectance of foams. *Journal of Marine Research*, 72(1):19–29.

Dickey, T., Lewis, M., and Chang, G. 2006. Optical oceanography: Recent advances and future directions using global remote sensing and in situ observations. *Reviews of Geophysics*, 44(1) RG1001 doi:10.1029/2003RG000148.

Dickey, T. D., Kattawar, G. W., and Voss, K. J. 2011. Shedding new light on light in the ocean. *Physics Today*, 64(4):44–49. doi:10.1063/1.3580492.

Dombrovsky, L. A. and Baillis, D. 2010. *Thermal Radiation in Disperse Systems: An Engineering Approach*. Begell House Publishers Inc., Redding, CT.

Egan, W. G. 2004. *Optical Remote Sensing: Science and Technology*. Marcel Dekker, New York.

Elepov, B. S. 1980. *The Monte Carlo Methods in Atmospheric Optics*. Springer, Berlin, Germany.

Enomoto, T. 2007. Ocean surface albedo in AFES. *JAMSTEC Report of Research and Development*, 6:21–30. Available on the Internet http://www.godac.jamstec.go.jp/catalog/data/doc_catalog/media/JAM_RandD06_02.pdf.

Fabelinskii, I. L. 1968. *Molecular Scattering of Light (translated from Russian by R. T. Beyer)*. Plenum Press, New York.

Feng, Y., Liu, Q., Qu, Y., and Liang, S. 2016. Estimation of the ocean water albedo from remote sensing and meteorological reanalysis data. *IEEE Transactions on Geoscience and Remote Sensing*, 54(2):850–868.

Foldy, L. L. 1945. The multiple scattering of waves. I. General theory of isotropic scattering by randomly distributed scatterers. *Physical Review*, 67(3–4):107–119. doi:10.1103/physrev.67.107.

Freda, W., Piskozub, J., and Toczek, H. 2015. Polarization imaging over sea surface–A method for measurements of Stokes components angular distribution. *Journal of the European Optical Society – Rapid publications*, 10:15060-1–15060-7. doi:10.2971/jeos.2015.15060.

Fried, D. L. 1966. Optical resolution through a randomly inhomogeneous medium for very long and very short exposures. *Journal of the Optical Society of America*, 56(10):1372–1379. doi:10.1364/JOSA.56.001372.

Frouin, R., Schwindling, M., and Deschamps, P.-Y. 1996. Spectral reflectance of sea foam in the visible and near-infrared: In situ measurements and remote sensing implications. *Journal of Geophysical Research*, 101(C6):14361–14371.

Ghosh, G. 1998. *Handbook of Optical Constants of Solids: Handbook of Thermo-Optic Coefficients of Optical Materials with Applications*. Academic Press, San Diego, CA.

Gil Perez, J. J. and Ossikovski, R. 2016. *Polarized Light and the Mueller Matrix Approach*. CRC Press, Boca Raton, FL.

Goldstein, D. H. 2011. *Polarized Light*, 3rd edition. CRC Press, Boca Raton, FL.

Goodman, J. W. 2015. *Statistical Optics*, 2nd edition. John Wiley & Sons, Hoboken, NJ.

Goody, R. M. and Yung, Y. L. 1995. *Atmospheric Radiation: Theoretical Basis*, 2nd edition. Oxford University Press, New York.

Gordon, H. R. and Jacobs, M. M. 1977. Albedo of the ocean–atmosphere system: Influence of sea foam. *Applied Optics*, 16(8):2257–2260.

Gregory, J. 2006. *Particles in Water: Properties and Processes*. CRC Press, Boca Raton, FL.

Hale, G. M. and Querry, M. R. 1973. Optical constants of water in the 200-nm–200-μm wavelength region. *Applied Optics*, 12(3):555–563.

Hapke, B. 2012. *Theory of Reflectance and Emittance Spectroscopy*, 2nd edition. Cambridge University Press, Cambridge, UK.

Hecht, E. 2017. *Optics*, 5th edition. Pearson Education Limited, Harlow, UK.

Hulburt, E. O. 1934. The polarization of light at sea. *Journal of the Optical Society of America*, 24(2):35–42.

Ishimaru, A. 1991. *Electromagnetic Wave Propagation, Radiation, and Scattering*. Englewood Cliffs, Prentice Hall, NJ.

Jerlov, N. G. 1968. *Optical Oceanography*. Elsevier, Amsterdam, The Netherlands.

Jerlov, N. G. 1976. *Marine Optics*. Elsevier, Amsterdam, The Netherlands.

Jin, Z., Charlock, T. P., Smith, Jr., W. L., and Rutledge, K. 2004. A parameterization of ocean surface albedo. *Geophysical Research Letters*, 31(22) L22301. doi:10.1029/2004gl021180.

Jin, Z., Qiao, Y.,Wang, Y., Fang, Y., and Yi, W. 2011. A new parameterization of spectral and broadband ocean surface albedo. *Optical Express*, 19(27):26429–26443.

Jonasz, M. and Fournier, G. 2007. *Light Scattering by Particles in Water: Theoretical and Experimental Foundations*. Elsevier, London, UK.

Katsaros, K. B., McMurdie, L. A., Lind, R. J., and DeVault, J. E. 1985. Albedo of a water surface, spectral variation, effects of atmospheric transmittance, sun angle and wind speed. *Journal of Geophysical Research*, 90(C4):7313–7321.

Kattawar, G. W. 2013. Genesis and evolution of polarization of light in the ocean. *Applied Optics*, 52(5):940–948.

Kattawar, G. W. and Adams, C. N. 1989. Stokes vector calculations of the submarine light field in an atmosphere-ocean with scattering according to a Rayleigh Phase Matrix: Effect of interface refractive index on radiance and polarization. *Limnology and Oceanography*, 34(8):1453–1472.

Kerker, M. 1969. *The Scattering of Light and Other Electromagnetic Radiation*. Academic Press, New York.

Kirk, J. T. O. 2011. *Light and Photosynthesis in Aquatic Ecosystems*, 3rd edition. Cambridge University Press, Cambridge, UK.

Koepke, P. 1984. Effective reflectance of oceanic whitecaps. *Applied Optics*, 23(11):1816–1824.

Kokhanovsky, A. A. 2003. *Polarization Optics of Random Media*. Springer, Praxis Publishing, Chichester, UK.

Kokhanovsky, A. A. (Ed.). 2004a. *Light Scattering Media Optics. Problems and Solutions*, 3rd edition. Springer, Praxis Publishing, Chichester, UK.

Kokhanovsky, A. A. 2004b. Spectral reflectance of whitecaps. *Journal of Geophysical Research*, 109(C5) C05021. doi:10.1029/2003JC002177.

Kokhanovsky, A. A. 2008. *Aerosol Optics: Light Absorption and Scattering by Particles in the Atmosphere*. Springer, Praxis Publishing, Chichester, UK.

Kondratyev, K. Y. 1969. *Radiation in the Atmosphere (International Geophysics Series, Volume 12)*. Academic Press, New York.

Kondratyev, K. Y., Kozoderov, V. V., and Smokty, O. I. 1992. *Remote Sensing of the Earth from Space: Atmospheric Correction*. Springer, Berlin, Germany.

Können, G. P. 1985. *Polarized Light in Nature*. Cambridge University Press, Cambridge, UK.

Kopelevich, O. V., Sheberstov, S. V., Burenkov, V. I., Vazyulya, S. V., and Likhacheva, M. V. 2007. Assessment of underwater irradiance and absorption of solar radiation at water column from satellite data. In Proceedings of SPIE 6615, Current Research on Remote Sensing, Laser Probing, and Imagery in Natural Waters (Eds. I. M. Levin, G. D. Gilbert, V. I. Haltrin, and C C. Tree), Vol. 6615, pp. 661507-1–661507-11. doi:10.1117/12.740441.

Kopelevich, O. V. 2017. Use of light in the exploration and research of the seas and oceans. *Light & Engineering*, 25(2):4–17 (in Russian). Available on the Internet https://www.researchgate.net/publication/322339891_Use_of_light_in_the_exploration_and_research_of_the_seas_and_oceans.

Korotkova, O. 2014. *Random Light Beams: Theory and Applications*. CRC Press, Boca Raton, FL.

Kravtsov, Y. A. 2005. *Geometrical Optics in Engineering Physics*. Alpha Science International Ltd., Harrow, UK.

Kuznetsov, A., Melnikova, I., Pozdnyakov, D., Seroukhova, O., and Vasilyev, A. 2012. *Remote Sensing of the Environment and Radiation Transfer: An Introductory Survey*. Springer, Berlin, Germany.

Landau, L. D. and Lifshitz, E. M. 1987. *Fluid Mechanics (Course of Theoretical Physics, Volume 6)*, 2nd edition. Elsevier–Butterworth-Heinemann, Burlington, MA.

Langley, D. S. and Marston, P. L. 1984. Critical-angle scattering of laser light from bubbles in water: Measurements, models, and application to sizing of bubbles. *Applied Optics*, 23(7):1044–1054.

Lee, Z. P., Arnone, R. A., Hu, C.-M., Werdell, P.-J., and Lubac, B. 2010. Uncertainties of optical parameters and their propagations in an analytical ocean color inversion algorithm. *Applied Optics*, 49(3):369–381.

Liang, S. 2001. Narrowband to broadband conversions of land surface albedo I: Algorithms. *Remote Sensing of Environment*, 76(2):213–238. doi:10.1016/S0034-4257(00)00205-4.

Liou, K.-N. 2002. *An Introduction to Atmospheric Radiation*, 2nd edition. Academic Press, San Diego, CA.

Lucarini, V., Saarinen, J. J., Peiponen, K.-E., and Vartiaine, E. M. 2005. *Kramers-Kronig Relations in Optical Materials Research*. Springer, Berlin, Germany.

Lukin, V. P. and Fortes, B. V. 2002. *Adaptive Beaming and Imaging in the Turbulent Atmosphere*. SPIE Press, Bellingham, WA.

Ma, L. X., Wang, F. Q., Wang, C. A., Wang, C. C., and Tan, J. Y. 2015. Investigation of the spectral reflectance and bidirectional reflectance distribution function of sea foam layer by the Monte Carlo method. *Applied Optics*, 54(33):9863–9874.

Marchuk, G. I., Mikhailov, G. A., Nazareliev, M. A., Darbinjan, R. A., Kargin, B. A., and Elepov, B. S. 1980. *The Monte Carlo Methods in Atmospheric Optics*. Springer, Berlin, Germany.

Marshak, A. and Davis, A. B. (Eds.). 2005. *3D Radiative Transfer in Cloudy Atmospheres*. Springer, Berlin, Germany.

Marston, P. L., Langley, D. S. and Kingsbury, D. L. 1982. Light scattering by bubbles in liquids: Mie theory, physical-optics approximations, and experiments. *Applied Scientific Research*, 38(1):373–383. doi:10.1007/BF00385967.

McCartney, E. J. 1976. *Optics of the Atmosphere: Scattering by Molecules and Particles*. John Wiley & Sons, New York.

McKechnie, T. S. 2016. *General Theory of Light Propagation and Imaging through the Atmosphere*. Springer, Switzerland.

Melnikova, I. N. and Vasilyev, A. V. 2005. *Short-Wave Solar Radiation in the Earth's Atmosphere: Calculation, Observation, Interpretation*. Springer, Berlin, Germany.

Mie, G. 1908. Beiträge zur Optik trüber Medien, speziell kolloidaler Metallösungen. *Annalen Der Physik*, 330(3):377–445. doi:10.1002/andp.19083300302.

Miller, R. L., Del Castillo, C. E., and McKee, B. A. (Eds.). 2005. *Remote Sensing of Coastal Aquatic Environments: Technologies, Techniques and Applications*. Springer, Dordrecht, The Netherlands.

Mishchenko, M. I., Hovenier, J. W., and Travis, L. D. (Eds.). 2000. *Light Scattering by Nonspherical Particles: Theory, Measurements, and Applications*. Academic Press, San Diego, CA.

Mishchenko, M. I., Rosenbush, V. K., Kiselev, N. N., Lupishko, D. F., Tishkovets, V. P., Kaydash, V. G., Belskaya, I. N., Efimov, Y. S., and Shakhovskoy, N. M. 2010. *Polarimetric Remote Sensing of Solar System Objects*. Akademperiodyka, Kyiv, Ukraine.

Mishra, D. R., Ogashawara, I., and Gitelson, A. A. 2017. *Bio-optical Modeling and Remote Sensing of Inland Waters*. Elsevier, Amsterdam, The Netherlands.

Mobley, C. D. 1994. *Light and Water: Radiative Transfer in Natural Waters*. Academic Press, San Diego, CA.

Mobley, C. D. 1999. Estimation of the remote-sensing reflectance from above-surface measurements. *Applied Optics*, 38(36):7442–7455. doi:10.1364/ao.38.007442.

Mobley, C. D. 2001. Radiative transfer in the ocean. In *Encyclopedia of Ocean Sciences* (Eds. J. H. Steele, S. Thorpe, and K. Turekian), Elsevier, New York, pp. 2321–2330.

Mobley, C. D. 2010. The optical properties of water. In *Handbook of Optics* (Eds. M. Bass), 3rd edition. McGraw-Hill, New York, Vol. 4, pp. 1.3–1.53.

Mobley, C. D. 2015. Polarized reflectance and transmittance properties of windblown sea surfaces. *Applied Optics*, 54(15):4828–4849. doi:10.1364/AO.54.004828.

Monahan, E. C. and O'Muircheartaigh, I. G. 1980. Optimal power-law description of oceanic whitecap coverage dependence on wind speed. *Journal of Physical Oceanography*, 10(2):2094–2099.

Monin, A. S. (Ed.). 1983. *Optika okeana (Ocean Optics), volumes 1 and 2*. Nauka, Moscow (in Russian).

Moore, K. D., Voss, K. J., and Gordon, H. R. 1998. Spectral reflectance of whitecaps: Instrumentation, calibration and performance in coastal waters. *Journal of Atmospheric and Oceanic Technology*, 15(2):496–509.

Moore, K. D., Voss, K. J., and Gordon, H. R. 2000. Spectral reflectance of whitecaps: Their contribution to water-leaving radiance. *Journal of Geophysical Research*, 105(C3):6493–6499.

Morel, A. 2001. Bio-optical properties of oceanic waters: A reappraisal. *Journal of Geophysical Research*, 106(C4):7163–7180.

Morel, A. 2009. Are the empirical relationships describing the bio-optical properties of case 1 waters consistent and internally compatible? *Journal of Geophysical Research*, 114(C1) C01016. doi:10.1029/2008JC004803.

Morel, A., Antoine, D., and Gentili, B. 2002. Bidirectional reflectance of oceanic waters: Accounting for Raman emission and varying particle scattering phase function. *Applied Optics*, 41(30):6289–6306. doi:10.1364/AO.41.006289.

Mukherjee, L., Zhai, P.-W., Hu, Y., and Winker, D. M. 2017. Equivalence of internal and external mixture schemes of single scattering properties in vector radiative transfer. *Applied Optics*, 56(14):4105–4112. doi:10.1364/AO.56.004105.

Nicolas, J.-M., Deschamps, P.-Y., and Frouin, R. 2001. Spectral reflectance of oceanic whitecaps in the visible and near infrared: Aircraft measurements over open ocean. *Geophysical Research Letters*, 28(23):4445–4448.

Paltridge, G. W. and Platt, C. M. R. 1976. *Radiative Processes in Meteorology and Climatology*. Elsevier Scientific Publishing Company, Amsterdam-Oxford-New York.

Panofsky, H. A. and Dutton, J. A. 1984. *Atmospheric Turbulence: Models and Methods for Engineering Applications*. John Wiley & Sons, New York.

Payne, R. E. 1972. Albedo of the sea surface. *Journal of Atmospheric Science*, 29(5):959–970.

Pope, R. M. and Fry, E. S. 1997. Absorption spectrum (380–700 nm) of pure water. II. Integrating cavity measurements. *Applied Optics*, 36(33):8710–8722.

Pozdnyakov, D. and Grassl, H. 2003. *Color of Inland and Coastal Waters: A Methodology for Its Interpretation*. Springer, Praxis Publishing, Chichester, UK.

Pozdnyakov, D. V., Pettersson, L. H., and Korosov, A. A. 2017. *Exploring the Marine Ecology from Space: Experience from Russian-Norwegian cooperation (Springer Remote Sensing/Photogrammetry)*. Springer, Switzerland.

Preisendorfer, R. W. 1965. *Radiative Transfer on Discrete Spaces*. Pergamon Press, New York.

Preisendorfer, R. W. and Mobley, C. D. 1986. Albedos and glitter patterns of a wind-roughened sea surface. *Journal of Physical Oceanography*, 16(6):1293–1316.

Qu, Y., Liang, S., Liu, Q., He, T., Liu, S., and Li, X. 2015. Mapping surface broadband albedo from satellite observations: A review of literatures on algorithms and products. *Remote Sensing*, 7(1):990–1020.

Raizer, V. 2017. *Advances in Passive Microwave Remote Sensing of Oceans*. CRC Press, Boca Raton, FL.

Ray, P. S. 1972. Broadband complex refractive indices of ice and water. *Applied Optics*, 11(8):1836–1844.

Roggemann, M. C., Welsh, B. M., and Hunt, B. R. 1996. *Imaging through Turbulence*. CRC Press, Boca Raton, FL.

Rozenberg, G. V. 1966. *Twilight: A Study in Atmospheric Optics (translated from Russian by R. B. Rodman)*. Springer, New York.

Sánchez, A.-M. and Piera, J. 2016. Methods to retrieve the complex refractive index of aquatic suspended particles: Going beyond simple shapes. *Biogeosciences*, 13:4081–4098.

Sasiela, R. J. 2007. *Electromagnetic Wave Propagation in Turbulence: Evaluation and Application of Mellin Transforms*, 2nd edition. SPIE Press, Bellingham, WA.

Saunders, P. 1968. Radiance of sea and sky in the infrared window 800–1200 cm^{-1}. *Journal of the Optical Society of America*, 58(5):645–652.

Sayer, A. M., Thomas, G. E., and Grainger, R. G. 2010. A sea surface reflectance model for (A)ATSR, and application to aerosol retrievals. *Atmospheric Measurement Techniques*, 3:813–838. doi:10.5194/amt-3-813-2010.

Schott, J. R. 2009. *Fundamentals of Polarimetric Remote Sensing*. SPIE Press, Bellingham, WA.

Schwenger, F. and Repasi, E. 2017. Simulation of oceanic whitecaps and their reflectance characteristics in the short wavelength infrared. *Applied Optics*, 56(6):1662–1673.

Séférian, R., Baek, S., Boucher, O., Dufresne, J.-L., Decharme, B., Saint-Martin, D., and Roehrig, R. 2018. An interactive ocean surface albedo scheme (OSAv1.0): Formulation and evaluation in ARPEGE-Climat (V6.1) and LMDZ (V5A). *Geoscientific Model Development*, 11:321–338.

Shell, J. R. 2005. *Polarimetric remote sensing in the visible to near infrared*. Dissertation. Rochester Institute of Technology. Rochester, New York. Available on the Internet http://dirsig.org/docs/shell.pdf.

Shifrin, K. S. 1988. *Physical Optics of Ocean Water (translated from Russian by D. Oliver)*. AIP Translation Series American Institute of Physics, Woodbury, NY.

Shuleikin, V. V. 1968. *Fizika Moria (Physics of the Sea)*, 4th edition revised and expanded. Izdatel'stvo Akademii Nauk S.S.S.R., Moscow, USSR (in Russian).

Sidran, M. 1981. Broadband reflectance and emissivity of specular and rough water surfaces. *Applied Optics*, 20(18):3176–3183.

Sihvola, A. 1999. *Electromagnetic Mixing Formulas and Applications (IEE Electromagnetic Waves Series 47).* The Institution of Electrical Engineers, London, UK.

Sinnett, G. and Feddersen, F. 2016. Observations and parameterizations of surfzone albedo. *Methods in Oceanography,* 17:319–334. doi:10.1016/j.mio.2016.07.001.

Slater, P. N. 1980. *Remote Sensing: Optics and Optical Systems.* Addison-Wesley Publishing Company, Reading, MA.

Smith, R. C. and Baker, K. S. 1981. Optical properties of the clearest natural waters (200–800 nm). *Applied Optics,* 20(2):177–184.

Snik, F., Craven-Jones, J., Escuti, M., Fineschi, S., Harrington, D., De Martino, A., Mawet, D., Riedi, J., and Tyo, J. S. 2014. An overview of polarimetric sensing techniques and technology with applications to different research fields. In *Proceedings of SPIE 9099, Polarization: Measurement, Analysis, and Remote Sensing XI* (Eds. B. Chenault and D. H. Goldste), Baltimore, MD, Vol. 90990B, pp. 90990B-1–90990B-20. doi:10.1117/12.2053245.

Sobolev, V. V. 1963. *A Treatise on Radiative Transfer (translated from Russian by S. I. Gaposchkin).* Van Nostrand, Princeton, NJ.

Spinrad, R. W., Carder, K. L., and Perry, M. J. (Eds.). 1994. *Ocean Optics.* Oxford University Press, New York.

Spitzer, D. and Arief, D. 1983. Relationship between the sky radiance reflected at the sea surface and the downwelling irradiance. *Applied Optics,* 22(3):378–379.

Stabeno, P. J. and Monahan, E. C. 1986. The influence of whitecaps on the albedo of the sea surface. In *Oceanic Whitecaps and Their Role in Air-Sea Exchange Processes* (Eds. E. C. Monahan and G. M. Niocaill). D. Reidel Publishing Company, Dordrecht, Holland, pp. 261–266.

Stamnes, K. and Stamnes, J. J. 2015. *Radiative Transfer in Coupled Environmental Systems: An Introduction to Forward and Inverse Modeling.* John Wiley & Sons, Weinheim, Germany.

Stamnes, K., Thomas, G. E., and Stamnes, J. J. 2017. *Radiative Transfer in the Atmosphere and Ocean,* 2nd edition. Cambridge University Press, Cambridge, UK.

Stokes, G. G. 1852. On the change of refrangibility of light. *Philosophical Transactions of the Royal Society of London,* 142:463–562. doi:10.1098/rstl.1852.0022.

Stover, J. C. 1995. *Optical Scattering: Measurement and Analysis,* 2nd edition. SPIE Press, Bellingham, WA.

Stramski, D. and Tegowski, J. 2001. Effects of intermittent entrainment of air bubbles by breaking wind waves on ocean reflectance and underwater light field. *Journal of Geophysical Research,* 106(C12):31345–31360.

Stratton, J. A. 1941. *Electromagnetic Theory.* McGraw-Hill Book Co., New York.

Tatarski, V. I. 1961. *Wave Propagation in a Turbulent Medium (translated from Russian by R. A. Silverman).* McGraw-Hill, New York (Reprint, Dover Publication, 2016).

Thomas, M. E. 2006. *Optical Propagation in Linear Media: Atmospheric Gases and Particles, Solid-State Components, and Water.* Oxford University Press, Oxford, New York.

Thomas, G. E. and Stamnes, K. 1999. *Radiative Transfer in the Atmosphere and Ocean.* Cambridge University Press, Cambridge, UK.

Timofeyev, Y. M. and Vasil'ev, A. V. 2008. *Theoretical Fundamentals of Atmospheric Optics.* Cambridge International Science Publishing, Cambridge, UK.

Träger, F. (Ed.). 2012. *Springer Handbook of Lasers and Optics,* 2nd edition. Springer, Heidelberg, Germany.

Tsang, L. and Kong, J. A. 2001. *Scattering of Electromagnetic Waves: Advanced Topics.* John Wiley & Sons, New York.

Tsang, L., Kong, J. A., and Ding, K.-H. 2000. *Electromagnetic Waves: Theories and Applications.* John Wiley & Sons, New York.

Tyo, J. S., Goldstein, D. H., Chenault, D. B., and Shaw, J. A. 2006a. Polarization in remote sensing — Introduction. *Applied Optics*, 45(22):5451–5452.

Tyo, J. S., Goldstein, D. H., Chenault, D.B., and Shaw, J. A. 2006b. Review of passive imaging polarimetry for remote sensing applications. *Applied Optics*, 45(22):5453–5469.

Uscinski, B. J. 1977. *The Elements of Wave Propagation in Turbulent Media*. McGraw-Hill, New York.

van de Hulst, H. C. 1981. *Light Scattering by Small Particles*. Dover Publications. Corrected edition, Mineola, NY.

Voss, K. J. and Fry, E. S. 1984. Measurement of the Mueller matrix for ocean water. *Applied Optics*, 23(23):4427–4439. doi:10.1364/AO.23.004427.

Walker, R. E. 1994. *Marine Light Field Statistics*. John Wiley & Sons, New York.

Watson, J. and Zielinski, O. (Eds.). 2013. *Subsea Optics and Imaging*. Woodhead Publishing, Cambridge, UK.

Whitlock, C. H., Bartlett, D. S., and Gurganus, E. A. 1982. Sea foam reflectance and influence on optimum wavelength for remote sensing of ocean aerosols. *Geophysical Research Letters*, 9(6):719–722.

Wozniak, B. and Dera, J. 2007. *Light Absorption in Sea Water*. Springer, New York.

Woźniak, S. B. and Stramski, D. 2004. Modeling the optical properties of mineral particles suspended in seawater and their influence on ocean reflectance and chlorophyll estimation from remote sensing algorithms. *Applied Optics*, 43(17):3489–3503.

Wu, X. and Smith, W. L. 1997. Emissivity of rough sea surface for 8–13 µm: Modeling and verification. *Applied Optics*, 36(12):2609–2619.

Wyngaard, J. C. 2010. *Turbulence in the Atmosphere*. Cambridge University Press, Cambridge, UK.

Zappa, C. J., Banner, M. L., Schultz, H., Gemmrich, J. R., Morison, R. P., LeBel, D. A., and Dickey, T. 2012. An overview of sea state conditions and air-sea fluxes during RaDyO. *Journal of Geophysical Research*, 117(C7) C00H19. doi:10.1029/2011JC007336.

Zege, E. P., Ivanov, A. P., and Katsev, I. L. 1991. *Image Transfer through a Scattering Medium*. Springer, New York.

Zhai, P.-W., Hu, Y., Trepte, C. R., and Lucker, P. L. 2009. A vector radiative transfer model for coupled atmosphere and ocean systems based on successive order of scattering method. *Optics Express*, 17(4):2057–2079. doi:10.1364/OE.17.002057.

Zhang, X., Lewis, M., and Johnson, B. 1998. Influence of bubbles on scattering of light in the ocean. *Applied Optics*, 37(27):6525–6536.

Zhang, Y., Giardino, C., and Li, L. (Eds.). 2017. *Water Optics and Water Colour Remote Sensing*. MDPI, Basel, Switzerland.

Zhao, Y., Yi, C., Kong, S. G., Pan, Q., and Cheng, Y. 2016. *Multi-band Polarization Imaging and Applications*. National Defense Industry Press Beijing–Springer, Berlin, Germany.

Zhou, G., Wang, J., Xu, W., Zhang, K., and Ma, Z. 2017. Polarization patterns of transmitted celestial light under wavy water surfaces. *Remote Sensing*, 9(4) 324. doi:10.3390/rs9040324.

Zuev, V. E. 1982. *Laser Beams in the Atmosphere (translated from Russian by S. Wood)*. Consultants Bureau, New York.

4 Optical Remote Sensing Technologies

4.1 INTRODUCTION

Optical remote sensing is a powerful operational tool for monitoring of the Earth's environment. In particular, multi- and hyperspectral imagery provides accurate and valuable information essential for understanding and managing marine ecosystems. Since optical sensor measures the amount of electromagnetic energy reflected or emitted by the object, the *detection efficiency* (i.e., sensitivity or ability to detect radiation) becomes the major concern of the observation technology. The information quality of imaging data is highly dependent on the design and overall performance of optical system. Ocean hydrodynamic studies require smart sensor solutions—the use of state-of-the-art sensor technologies in combination with complete integration of data acquisition and processing capabilities.

In visible and infrared ranges of the electromagnetic spectrum, ocean environment represents relatively low-contrast extended source with purely dynamic stochastic response. The radiation incident upon water is absorbed, scattered, and reflected in different proportions, depending on the wavelength and surface conditions. The longest visible wavelength and the near-infrared (NIR) radiation is absorbed more than the visible wavelengths; therefore, ocean images look blue or blue-green due to stronger reflectance at these shorter wavelengths and darker at red or NIR wavelengths of light.

The effects of water flow and turbulence at the ocean-atmosphere interface are quite important and are generally involved into the formation of reflected radiation. In this context, utilization of optical information requires the implementing advanced acquisition systems capable to register highly variable optical signals simultaneously at many spectral bands with the same temporal resolution. Optical information is usually available in the form of digital images destined for normal human perception of visible light. Such a product allows the researcher to perform thematic analysis using well-developed methods of digital image processing. Further quantitative and qualitative assessment requires only the knowledge of spectral characteristics of optical data. Incidentally, this is a principal difference between optical and microwave imaging data, where utilization of last ones is multi-stage and sometimes ambiguous procedure.

Technological and scientific principles of ocean remote sensing can be formulated as the following. A basic requirement is the sensor's capability to register multiscale signatures, associated with ocean dynamics with highest spatial and temporal resolutions. In general, ocean remote sensing must incorporate multisensor systems and interdisciplinary scientific approach in order to provide detailed analysis and specification of hydrodynamic processes and/or events being detected. Multiple observing

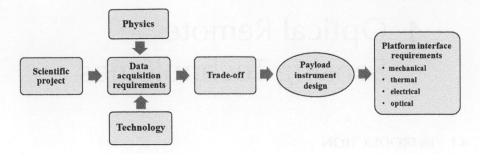

FIGURE 4.1 Payload design process.

platforms can greatly improve observation process, utilization, and interpretation of remotely sensed data. Implementation and assessment of ocean data are also important processes which require the knowledge of ocean physics and electromagnetic wave propagation theory. Several correction factors—the contribution of specular reflection in the form of sun glint, atmospheric scattering and absorption, and the bi-directionality of both the downwelling and the upwelling radiance distributions, should be taken into account in order to provide adequate quantitative analyses of ocean optical data. Cloud cover and atmospheric fog often reduce information capability of optical sensors, limiting analysis of ocean data. Most importantly, remotely sensed data are limited to the upper optical depth of the ocean. Satellite temporal and spatial coverage is severely restricted by orbital parameters and viewing geometries. The sensor–platform combination determines the characteristics of the resulting imaging data. In complex situations, aircraft and/or satellite information must be complemented with *in situ* measurements to calibrate remote sensors as well to characterize optical properties of the water column. This is important especially at bio-optical variability that also may effect on surface hydrodynamics. Today optical observations can be performed using commercially available instrumentation. However, the success depends on the understanding of the mission goals and tasks and capabilities of *cognitive* analysis of satellite data. Figure 4.1 illustrates the payload design process which can be accommodated for remote sensing missions.

4.2 OPTICAL PAYLOAD CLASSIFICATION

Optical payload—the carrying capacity of an aircraft or launch vehicle, is defined by mission requirements. The payload includes all instrumentation that provides data acquisition from airspace. As mentioned in Chapter 1, optical sensors are divided on two categories: passive and active and, correspondingly, scanning and non-scanning. Active optical sensors such as lidar, laser or laser altimeter, reflectometer, interferometer, etc., emit their own electromagnetic energy to illuminate the object and then detect and measure the radiation that is reflected or scattered from this object. Passive sensors such as aerial photo camera, video (TV) camera, multi- or hyperspectral imager, etc., detect and measure the naturally reflected solar radiation from the object (exception is IR scanner which can measure reflected and emitted thermal radiation).

Modern electro-optical imaging systems are classified into several following types depending on the number of operated spectral bands: panchromatic or polarimetric (1 band), multispectral (3–10 bands), superspectral (10–50 bands), and hyperspectral (50–300 bands). Technical specification of each sensor is defined in terms of temporal, spatial, spectral, and radiometric resolution. Optical payloads can be roughly classified into a number of instrument types based on their design technology and operational functions. Optical sensors which are potentially capable to provide hydro-physical information are considered below.

4.2.1 AERIAL PHOTOGRAPHY

Photography is a tradition and primarily method of recording the image of an object. Historically, photograph is considered as a raw documented material in many civilian and military missions including remote sensing of environment as well. Photography is obtained using an analog photo camera, operated with optical lenses and films. Modern photographic cameras enable to cover all flying altitudes usable with commerce and military aircrafts with speeds varying between 40 and 600 km/h, usually have large exposure latitude ranges, film resolution ratings between 40 and 125 line pairs/mm and diffuse rms granularities between 9 and 40. The availability of forward/image-motion compensation of aerial camera platforms increases the quality of aerial photography that is an important factor for reliable detection of moving objects (e.g., ships or oil spills) and specification of their geometrical characteristics.

Aerial and space photo cameras use either panchromatic, PAN (black and white) or natural color, or color infrared. PAN imagery captures all visible light within a wavelength range between 0.4 and 0.7 μm that create natural looking image acceptable for the human eye. Black and white infrared film has a sensitivity range from 0.4 to 0.9 μm. The infrared photography utilizes a black filter to absorb all visible spectrum (<0.7 μm). Color film is also classified into two broad categories, normal color and color infrared. Normal color film, similar to PAN film, is sensitive to the 0.4–0.7 μm range of the spectrum. Color infrared has its sensitivity extended to 0.9 μm. The separation of the film sensitivity by spectral channels is quite useful in identification of the Earth solid surfaces; however, in the case of open ocean surface (except coral and bio-productive coast regions) such a detailed spectral differentiation is not so important. Advantages of PAN and normal color films are (1) more natural for human eye than infrared, (2) more details can be seen than infrared, (3) resolution is better than infrared, and (4) higher penetration efficiency into water. On the other hand, infrared emphasizes turbid water, moist area, and aerated water.

PAN sensor produces images with a much better spatial resolution than multi- or hyperspectral sensor onboard on the same platform. For example, the QuickBird satellite imagery has pixel resolution of 0.6 m × 0.6 m for PAN and 2.4 m × 2.4 m for multispectral bands. Analogically, the IKONOS imager provides 1 and 4 m pixel resolution for PAN and multispectral channels, respectively.

During the years, aerial photo cameras have been employed in many applications providing topographic mapping, maritime patrol, and reconnaissance missions.

The most important application of aerial photography in ocean remote sensing is detailed imagery of multiscale surface waves, wakes, oil spills, and other complex hydrodynamic patterns. In the past, aerial photography provided detailed studies of wave breaking fields and foam and whitecap coverage geometry at variable surface conditions (Chapter 7). Increasingly, however, aerial and space analog photographs are being acquired using more refined and sophisticated digital optical sensors. Due to the versatility of the new digital technology the term "camera" is no longer used to describe capabilities of optical imaging system. This is entirely because of the fact that the system represent "multispectral" unit which not only produces PAN and color images but also captures other spectral information such as infrared or ultraviolet.

4.2.2 Imaging Spectroscopy

Spectroscopy is a commonly used technique for analyzing the composition of materials. Remote sensing imaging spectroscopy (known also as hyperspectral imagery, Section 4.2.4) is used for identification and classification of the Earth objects and atmospheric aerosol. Imaging spectroscopy is defined as acquisition of images in hundreds of contiguous narrow (<10 nm) spectral bands and inherently registered in the spectral range from the near ultraviolet to the shortwave infrared. Imaging spectroscopy enables to collect the near-laboratory-quality spectra. Analysis of spectroscopic data is a real challenge because they are highly correlated. Sophisticated statistical methods need to be applied to such data in order to specify the components of the material being studied.

Since the late 1980s, imaging spectroscopy has been successfully applied in many geophysical disciplines including geology, mineralogy, vegetation, atmospheric science, hydrology, and oceanography. In particular, methods of optical spectroscopy are potentially capable of providing detection, analysis and identification of sea surface organic and nonorganic compounds, biochemical compositions of marine sediments, oil spill products (gasoline, benzene and petroleum) as well atomic radioactive elements and dust. These studies are important to coastal hydrodynamics and marine ecology. Sea surface spectroscopy can be conducted from ships or aircraft.

4.2.3 Multispectral Imagery

Multispectral (MS) imagery is a method of measuring radiation energy simultaneously at several discrete spectral bands or selected wavelengths of the electromagnetic spectrum. MS imagery provides detection and recognition of surface objects with greater detail and accuracy. Unlike PAN, MS data allow distinguishing the areas of interest from background scene by specific multispectral signatures and their variations. Methods of MS imagery are also applied for automatic target detection.

MS images can be obtained using as analog photo camera as well digital sensors. Earth MS photography has been important objective in early space missions—Apollo 9, Skylab, and Souyz. MS photo cameras operated with two or four individual spectral bands. Photo films were recorded using the narrowest pass band color (RGB) filters (known as *pseudo* bands or false color).

A conventional type of MS imager is an optical/infrared/ultraviolet scanner with a number of spectral channels varied from 2 to 24. The MS scanner was used for the first time in the Landsat program. The Landsat-1 and -2 had three spectral bands; the other system was a four-band MS scanner, and Landsat-7 had seven bands. The reflectance data obtained by these instruments were first converted to electronic signals then to digital format and telemetered to the receiving ground station.

Standard specification of an electro-optical satellite-based MS system (e.g., IKONOS) includes one grayscale PAN image (0.4–0.7 µm) and four-band MS images that correspond to RGB color space. MS data usually represent red (635–700 nm), green (490–560 nm), blue (450–490 nm) and infrared (0.8–10 µm) or longer, classified as near-infrared (NIR), middle infrared (MIR), and far infrared (FIR or thermal).

4.2.4 Hyperspectral Imagery

Hyperspectral (HS) imagery also known as imaging spectroscopy is a remote sensing technology for measurements of the radiance at hundreds contiguous inherently registered spectral images. Collected HS data produce a 3D data structure (as a result of spatial and spectral sampling), referred as a "hyperspectral data cube" or "hypercube" (Figure 4.2). Each element (pixel) of the image contains complete spatial information across entire electromagnetic spectrum of radiance. The spectrum is a plot of radiance or reflectance vs. electromagnetic wavelength. This plot is used to detect and identify spectral signatures and characterize environmental features within the scene known also as "fingerprints."

The two general methods to obtain HS data are pushbroom scanning and snapshot imaging. Pushbroom scanners read images overtime; snapshot imaging operates with digital cameras and generate images instantly. HS data, hypercubes, can be obtained from different platforms using scanning imaging sensors. Airborne sensors and satellites are typically used for remote sensing of the Earth surfaces to provide

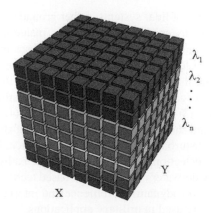

FIGURE 4.2 (See color insert.) Hyperspectral data cube.

classification of the area of interest. A typical data processing and interpretation involve principle component transforms, methods of feature extractions, fusion, classification, segmentation, and statistical estimation (Chapter 6). The results can be presented in the format of classification maps or material signature maps.

Nowadays, HS imagery is widely used in many geophysical and remote sensing studies. The most important applications are soil moisture, vegetation, and urban classifications. Applications to environment include estimating atmosphere parameters, gases, carbon dioxide, water vapor, clouds, and aerosols. In hydrology HS sensors provide monitoring of coastal waters, water quality parameters, chlorophyll, planktonic species, colored dissolved organic matter, suspended sediments, turbidity, mapping, and characterizing biomass, submerged aquatic vegetation, and coral reefs. Moreover, HS imaging techniques are used successfully for mapping and classification of snow and ice cover properties.

The use of HS imagery for ocean hydrodynamic studies is limited. The obvious applications are detection and monitoring of sea surface pollutants such as oil spills, organic and nonorganic compounds, and radioactive sediments. HS techniques especially with high-resolution capability enable to provide study of coastal hydrodynamics and mapping of sea floor in shallow-water areas. We believe that HS techniques can provide detection of turbulent gas-liquid two-phase flows (bubbly and jet flows).

4.2.5   LIGHT DETECTION AND RANGING

Lidar (also called LIDAR, LiDAR, LADAR, or lidar—laser radar) is a technology which uses laser light source. Lidar can provide various types of information, e.g., (1) topographical lidar and laser altimeter—range to target, (2) differential scattering-absorption lidar—object's properties and (3) Doppler lidar—velocity of target. The primary wavelengths of laser radiation include the ultraviolet (180–400 nm), visible (0.4–0.7 μm), and infrared (between 700 nm and 1 mm). Different lidar techniques exploit different types of receiving signals depending on scattering process; most common are Rayleigh scattering, Mie scattering, Raman scattering, and fluorescence.

Nowadays, various types of lidar systems, either ground based, mobile, airborne, or satellite are used for many environmental studies including meteorology, atmospheric physics, and oceanography. In particular, lidar scanning technology is an accurate and effective tool for creating 3D topographical maps and highly accurate surveys of surface structural features. Applications of lidars in oceanography can be listed as the following: (1) ocean colour observations , (2) estimating phytoplankton fluorescence and biomass, (3) seawater quality monitoring, (4) coastal and nearshore monitoring (coral reefs and shallow waters), (5) detection of ship wakes, (6) measuring small-scale surface disturbances (roughness, turbulence, bubbles), (7) oil slick detection, (8) fish school detection, and (9) shallow-water bathymetry. Lidar systems provide 3D imaging of underwater objects. Laser-based technique is also capable for detection of underwater hydrodynamic turbulence and internal waves. Nowadays, a number of lidar systems are used in military applications.

4.2.6 LASER RAMAN SPECTROSCOPY

Raman spectroscopy is based on the Raman effect (1928), which is the inelastic scattering (known as Raman scattering) of photons by molecules or molecular aggregates. Laser Raman spectroscopy (LRS) provides measurements of scattered photon energy with shifted frequency and polarization due to the interaction of monochromatic laser light with a sample or system being studied. The shift in scattered energy gives vital information about the inherent vibrational modes in the system. If emitted light photon has the same wavelength as the absorbing photon, the phenomenon is known as Rayleigh scattering. If the frequency of the scattered photon is lowered, then the phenomenon is called Stokes scattering. If the energy of the photon is shifted upward, it is called anti-Stokes scattering. The Raman shift is an intrinsic property of a molecule that makes the LRS a powerful experimental tool for diagnostics and monitoring of the system molecular structure and/or molecular composite. LRS uses wavelengths ranging from ultraviolet through visible to NIR. Remote sensing application of LRS in ocean studies are relatively new and includes the determination of (sub)surface seawater temperature, investigations of sea surface monolayer, detection of biomolecular interactions, and oil spill properties.

Methods of LRS can be developed and applied for researching molecular hydrodynamics. In particular, LRS offers unique analytical capabilities for studying turbulent flows involved molecular processes in the upper ocean layer. Transient flow conditions with variable molecular structures can occur, e.g., under the influence of a wave, generating *acoustic* streaming from a deep ocean source. Microwave sensors are not sensitive to molecular dynamics at all; therefore, LRS could be considered as a potential and complementary technique for probing (sub)surface hydrodynamic processes at the molecular level.

4.2.7 FOURIER TRANSFORM SPECTROSCOPY

Fourier transform spectroscopy (FTS) and/or Fourier transform infrared spectroscopy (FTIS) is optical/IR technique providing measurements of the electromagnetic radiation in time-domain or space-domain using interference of light. A basic principle of FTS/FTIS is applying Fourier transform to convert raw data (called an "interferogram") into the actual spectrum. A typical FTS uses a Michelson interferometer. The interferogram contains the source's frequency information modulated in a time domain as a function of the moving mirror's displacement. The interferogram is Fourier-transformed to obtain the spectrum. Basically, the FTIR spectrometer is devoted to detect and study the concentration and spatial distribution of atmospheric trace gases and of natural ozone in the wavelength range 2–16 μm.

Nowadays, space-based FTS/FTIS instruments looking downwards provide new insights on the Earth's atmosphere, climate, and meteorological processes. These instruments also are targeting galaxies, stars, planets, and many smaller objects within the universe. All these techniques—visible and ultraviolet FTS, FTIR, ultraviolet FT Raman spectroscopy, and the NIR FT resonance spectroscopy, have the potential to provide specific studies of ocean environment.

4.2.8 Spectrometers and Radiometers

Spectrometers, known also as spectrographs, spectroscopes, or spectrophotometers, measure a portion of the light spectrum to provide identification of materials. Optical spectrometers have been deployed as payloads in space missions for many years. Spectrometers split light spectrum on a large number of spectral lines and measure their wavelengths and intensities (and sometimes polarization as well). Spectrometers operate over a very wide range of wavelengths, from gamma rays and X-rays to the FIR. Spectrophotometer measures the spectrum in absolute units. The majority of spectrophotometers are used in regions near the visible spectrum. Spaceborne spectrometers are also used for mapping of atmosphere parameters, gases, and pollutants. The Infrared Spectrograph (NASA) is one of Spitzer's three science instruments. It provides both high- and low-resolution spectroscopy at mid-infrared wavelengths (5–40 µm).

Radiometers measure the radiant energy or radiant flux in ultraviolet, visible, and/or IR ranges of the electromagnetic spectrum. To measure the radiation at selected wavelengths in visible range, different calibrated optical filters are used. An ultraviolet radiometer measures either radiance or irradiance from a source of ultraviolet radiation. An IR radiometer measures radiance to determine the temperature of objects.

4.2.9 Thermal Infrared Imagery

Thermal infrared (TIR) refers to the region of electromagnetic spectrum between 3.0 and 20 µm. The main difference between TIR and NIR is that TIR is dominated by emitted radiation and NIR is dominated by reflected radiation similar to light. TIR techniques measure the radiant temperature which is defined by emissivity and kinetic temperature of an object according to Planck law (blackbody radiation), Wein's displacement law, Stefan–Boltzman law, and Kirchhoff's radiation law. Most TIR remote sensing applications use two atmospheric wavelength windows in the 3–5 µm range and 8–15 µm range. TIR imaging has been growing fast and playing an important role in various fields of geosciences and remote sensing. Nowadays, TIR sensors are integrated with the MS and/or HS imaging systems; TIR is standard payload in many satellite missions dedicated to remote sensing of environment. Here is a short list:

> TIROS (Television Infrared Observation Satellite, 1960); GOES (Geostationary Operational Environmental Satellite, NOAA); Landsat 4, 5, 7; HCMM (Heat Capacity Mapping Mission); CZCS (Coastal Zone Color Scanner on Nimbus 7); AVHRR (Advanced Very High Resolution Radiometer, NOAA); TIMS (Thermal Infrared Multispectral Scanner, Airborne); ATLAS (Airborne Terrestrial Application Sensor); MODIS (Moderate-Resolution Imaging Spectroradiometer); ASTER (Advanced Spaceborne Thermal Emission and Reflection Radiometer); INSAT-3D (IMA/ISRO, India), FY-4M (CMA/NSMC, China); HIMAWARI-8 (JMA, Japan); GOES-R (NOAA/NASA); AIRS (Advanced Infrared Sounder); HIRDLS (High Resolution Dynamic Limb Sounder, NASA); METEOSAT (EUMETSAT, Europe); TES

(Tropospheric Emission Spectrometer); NPOESS (National Polar-orbiting Operational Environmental Satellite System, joint NOAA/NASA/DoD).

TIR application areas are agriculture, soil moisture, forest fire, volcanology, meteorology, seismology, hydrology, oceanography, atmosphere science and meteorology and intelligence/military. In particular, a number of satellite TIR sensors provide monitoring and mapping of bulk-skin sea surface temperature across the globe.

An important advantage of TIR is the capability to detect physical temperature with high accuracy and identify thermal features of objects that cannot be done by conventional visible-range optical sensors. Disadvantages of TIR remote sensing are difficulty of the instrument intercalibration in general and difficulty in data interpretation due to uncertainties of the object's emissivity.

An important TIR application in ocean hydrodynamics is study of (1) thermal fluxes across the ocean-atmosphere interface and (2) behaviour of fluid flow over complex body shape (this area is known as thermo-fluid-dynamics). This scientific research can be considered in context with pretentious concept of "thermal wake." Thermal wake is a specific surface *thermo-hydrodynamic* pattern having temperature gradient different from the temperature of surround environment. Natural thermal wake (mostly "cold" wake) can occur due to hurricane or typhoon impacts that have been observed and described in oceanographic literature. However, remote sensing capabilities for detecting this phenomenon in open ocean remain unclear at the moment.

4.2.10 OTHER SENSORS

There are several other optical techniques which are available for ocean explorations.

- *Scattering (or light-scattering or scattering-based) SCA sensors* measure light scattered from particles; the measure is defined in terms of volume scattering function (VSF). In marine optics, the VSF provides total information—concentration, size, composition, and shape of the particles. SCA sensors are also used for measuring turbidity and suspended solids concentrations of seawater.
- *Surface plasmon resonance (SPR)* is a physical process that can occur when plane-polarized light hits a thin film under total internal reflection conditions. Potentially, SPR spectroscopy can be applied for detection and identification of surface monomolecular films (monolayers) and organic/nonorganic compounds at the air-water interface.
- *Gamma-ray spectrometer (GRS)* is a device to measure the gamma-ray spectrum of an object. GRS is used for monitoring of environmental radiation. Nowadays, GRS is widely used in planetary missions. Potentially, the GRS can be applied for detection of nuclear radiation from reactor-power ships or submarines as well for monitoring of radioactive pollutant in the ocean.

- *Spectral Fluorometer (SF)*, (known also as spectrofluorometer, fluorometers, bioluminescence sensors) is a device that measures the fluorescence or light emitted by different fluorescing objects. SF instruments are widely used in oceanography to detect chlorophyll *a*, phytoplankton, and bio-organisms in water.
- *Flow cytometry (FC)* is laser-base or impedance-based technology measuring light scattering and fluorescence properties of individual particles. FC is used to study marine phytoplankton and cell structures.

Optical sensors and techniques, described in this section are being used to provide the study of ocean environment over the years but with constant improving the quality and information content of the collected data. Nowadays, technological advances and capabilities of airspace optical sensors allow the researches to increase the variety of ocean observations and greatly expand our experiences in operational coast/ocean monitoring. This is not to say that optical methods are optimal in all cases. However, optical remote sensing, in particular, satellite multispectral imagery of high resolution is a promising avenue of novel ocean study with a growing potential for sophisticated applications and services.

4.3   SATELLITE IMAGING CAPABILITY

The use of optical payload in the satellites offers many advantages in ocean observations. First of all, this provides wide synoptic coverage at fine spatial resolutions and production of various datasets that are distinguished by information capacity from all other (active/passive microwave) sensors. The capability of an optical sensor is defined by satellite characteristics and thereby, different tasks require different payload combinations. However, due to limitations in satellite payload technology, many satellites carry multipurpose optical instruments capable for collection of ultimate datasets which then are utilized in a variety of oceanographic products and services. In this section, we present a gallery of remarkable optical images of the oceans in order to demonstrate capabilities of high-resolution satellite imagery for observing striking ocean features.

4.3.1   WAVE PHENOMENA

Surface wave phenomena are revealed in the optical images very clear. Among them wave refraction on the coast (Figure 4.3), shallow-water waves (Figure 4.4), and packets of nonlinear internal waves (Figures 4.5 and 4.6) are the most impressive examples. All these phenomena are characterized by complex geometry of periodic-like wave patterns of different scales and propagating directions, as well variable configurations of the wavefront. The observed reflection of solar radiation causes by the interactions between surface waves and the coastline and/or seafloor. Such optical data are very much appreciated for spatial analysis and hydrodynamic treatment and modeling.

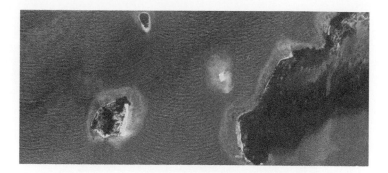

FIGURE 4.3 **(See color insert.)** Satellite image. Wave refraction. (Image from Digital Globe.)

FIGURE 4.4 Satellite image. Shallow-water waves. Dorre Island, Western Australia. (Image from Copernicus Sentinel-2.)

FIGURE 4.5 **(See color insert.)** Satellite image. Internal waves. The island of Trinidad in the southeastern Caribbean Sea. (Image from NASA.)

FIGURE 4.6 Satellite image. Nonlinear internal solitary waves. The Lombok Strait, Indonesia (Aqua/MODIS). (Image from NASA.)

4.3.2 CORAL REEFS

Coral reefs produce in the images the myriad of beautiful colors which is result of multiple optical reflections from the sea bed bio-communities. In the images, shown in Figures 4.7 and 4.8 underwater reefs form a hazy blue-green halo around the "whole" or island. Color blends generating a larger scale collage of features are identified perfectly. Such optical effects are associated with mixed optical properties of the water column above reef-building corals. Complex composition and structure of coral reef habitats may also form a hazy blue-green halo around the island. The contribution of multiple scattering and absorption of light is highly possible.

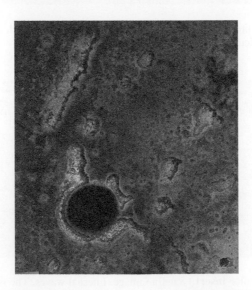

FIGURE 4.7 **(See color insert.)** Satellite image. Corals around Belize Great Blue Hole. (Image from Digital Globe.)

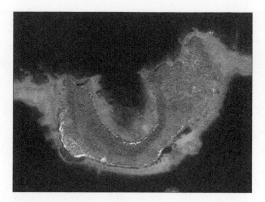

FIGURE 4.8 **(See color insert.)** Satellite image. Coral reefs around the island. Matangi Island, South Pacific. (Image from Digital Globe.)

4.3.3 BIOPRODUCTIVITY

Sea bioproductivity creates color water patterns (Figures 4.9 and 4.10). The image shows swirling blooms of phytoplankton coloring the surface waters blue and green. Marine scientists believe that microscopic photosynthetic organisms, the chlorophyll, and other pigments create phytoplankton blooms spreading across the ocean on hundreds and sometimes thousands of kilometers. Ocean colour is an important indicator of phytoplankton populations that can be monitored globally using satellite high-resolution hyperspectral imagers.

4.3.4 ALGAL BLOOM

Algal bloom forms in freshwater or marine water systems. Figures 4.11 and 4.12 illustrate algae blooms in the Baltic Sea observed from space. The researchers found

FIGURE 4.9 **(See color insert.)** Satellite image. Bioproductivity. Black Sea (Aqua MODIS). (Image from NASA.)

FIGURE 4.10 **(See color insert.)** Satellite image. Phytoplankton bloom. Northeast of the Falkland Islands (Terra MODIS). (Image from NASA.)

FIGURE 4.11 **(See color insert.)** Satellite image. Algal bloom. Central Baltic Sea. (Image from Copernicus Sentinel-2A.)

that the algae is concentrated in locations where the vertical and horizontal water movements in the Baltic Sea generate the best nutrient and light conditions for algal growth, which are then drawn out by the water circulation. The bloom has helped to create a "dead zone," which is an area of oxygen poor waters that can prove deadly to marine life.

4.3.5 MARINE POLLUTION

Marine pollution or oil spills typically appear in the optical images as distinct areas or spots (Figures 4.13 and 4.14). The varying brightness in these areas reflects

FIGURE 4.12 **(See color insert.)** Satellite image. Algal bloom at coastal waters (Landsat 8). (Image from NASA.)

FIGURE 4.13 **(See color insert.)** Satellite image. Oil pollutions. The Gulf of Mexico (Terra MODIS). (Image from NASA.)

FIGURE 4.14 **(See color insert.)** Satellite image. Drilling slicks. The Gulf of Mexico. (Image from Digital Globe.)

different thicknesses of oil films. The wispy patterns of the oil spill are formed due to the transport of the oil by waves and currents. Oil displays a moderately larger reflectance than water. Thin oil layers reflect light over a wide spectral range—as far as the blue. Thick oil layers appear to be the same color as bulk oil, typically brown or black. They usually represent water-in-oil emulsions containing up to 70% of water. Therefore, an infrared sensor often is unable to detect them.

4.4 SUMMARY

Satellite image capability depends as much on payload design, measurement strategy, and architecture as well as on specific sensor technology. It is well known fact that there is the classic trade-off between fewer highly capacity multipurpose instruments vs. a large number of specialized "small" sensors dedicated for specific measurements. In this chapter, we considered different type of instruments which are potentially capable to provide remote sensing studies of ocean environment. They include aerial, multispectral and hyperspectral cameras, infrared radiometers, spectrometers, light detectors (lidars), and some others specific sensors. Sensor design should follow the scientific goals, methodology, and measurement requirements which are directly connected with fundamental physics, the state of technology, and cost. Therefore, it is still difficult to define optimal payload design for detection purposes; however, we may suppose that multisensor and multipurpose satellite constellation system will be more preferable than single-type sensor. Selected and presented satellite images just confirm this statement and illustrate great optical capabilities to observed complex oceanic phenomena which perhaps, influence on the surface hydrodynamics in some way or another.

BIBLIOGRAPHY

Astarita, T. and Carlomagno, G. M. 2013. *Infrared Thermography for Thermo-Fluid-Dynamics.* Springer, Berlin, Germany.

Beer, R. 1992. *Remote Sensing by Fourier Transform Spectrometry.* John Wiley & Sons, New York.

Bell, R. 1972. *Introductory Fourier Transform Spectroscopy.* Academic Press, New York.

Borengasser, M., Hungate, W. S., and Watkins, R. 2008. *Hyperspectral Remote Sensing: Principles and Applications.* CRC Press, Boca Raton, FL.

Budzier, H. and Gerlach, G. 2011. *Thermal Infrared Sensors: Theory, Optimisation and Practice.* John Wiley & Sons, Chichester, UK.

Bunkin, A. and Voliak, K. 2001. *Laser Remote Sensing of the Ocean: Methods and Applications.* John Wiley & Sons, New York.

Chang, C.-I. 2003. *Hyperspectral Imaging: Techniques for Spectral Detection and Classification.* Springer, New York.

Churnside, J. H. 2014. Review of profiling oceanographic lidar. *Optical Engineering,* 53(5):051405-1–051405-13.

Dong, P. and Chen, Q. 2018. *LiDAR Remote Sensing and Applications.* CRC Press, Boca Raton, FL.

Eismann, M. 2012. *Hyperspectral Remote Sensing.* SPIE Press, Bellingham, WA.

Emery, W. and Camps, A. 2017. *Introduction to Satellite Remote Sensing: Atmosphere, Ocean, Land and Cryosphere Applications.* Elsevier, Amsterdam, The Netherlands.

Fujii, T. and Fukuchi, T. 2005. *Laser Remote Sensing.* CRC Press, Boca Raton, FL.

Hou, W. W. 2013. *Ocean Sensing and Monitoring: Optics and Other Methods.* SPIE Press, Bellingham, WA.

Karp, S. and Stotts, L. S. 2013. *Fundamentals of Electro-Optic Systems Design: Communications, Lidar, and Imaging.* Cambridge University Press, Cambridge, UK.

Kramer, H. J. 2002. *Observation of the Earth and Its Environment: Survey of Missions and Sensors,* 4th edition. Springer, Berlin, Germany.

Kuenzer, C. and Dech, S. (Eds.). 2013. *Thermal Infrared Remote Sensing: Sensors, Methods, Applications.* Springer, New York.

Landgrebe, D. A. 2003. *Signal Theory Methods in Multispectral Remote Sensing.* John Wiley & Sons, Hoboken, NJ.

Larkin, P. 2018. *Infrared and Raman Spectroscopy: Principles and Spectral Interpretation,* 2nd edition. Elsevier, Amsterdam, The Netherlands.

Liang, S. (Ed.). 2017. *Comprehensive Remote Sensing.* Elsevier, Amsterdam, The Netherlands.

Manolakis, D., Lockwood, R., and Cooley, T. 2016. *Hyperspectral Imaging Remote Sensing: Physics, Sensors, and Algorithms.* Cambridge University Press, Cambridge, UK.

Measures, R. M. 1984. *Laser Remote Sensing: Fundamentals and Applications.* John Wiley & Sons, New York.

Moore, C., Barnard, A., Fietzek, P., Lewis, M. R., Sosik, H. M., White, S., and Zielinski, O. 2009. Optical tools for ocean monitoring and research. *Ocean Science,* 5:661–684.

Pavia, D. L., Lampman, G. M., Kriz, G. S., and Vyvyan, J. A. 2013. *Introduction to Spectroscopy,* 5th edition. Cengage Learning, Stamford, CT.

Pu, R. 2017. *Hyperspectral Remote Sensing: Fundamentals and Practices (Remote Sensing Applications Series).* CRC Press, Boca Raton, FL.

Qian, S.-E. (Ed.). 2016. *Optical Payloads for Space Missions.* John Wiley & Sons, Chichester, UK.

Rees, W. G. 2013. *Physical Principles of Remote Sensing,* 3rd edition. Cambridge University Press, Cambridge, UK.

Robinson, I. S. 2010. *Discovering the Ocean from Space: The Unique Applications of Satellite Oceanography.* Springer, New York.

Robles-Kelly, A. and Huynh, C. P. 2013. *Imaging Spectroscopy for Scene Analysis*. Springer, London, UK.

Saptari, V. 2004. *Fourier Transform Spectroscopy Instrumentation Engineering*. SPIE Press, Bellingham, WA.

Smith, B. C. 2011. *Fundamentals of Fourier Transform Infrared Spectroscopy*, 2nd edition. CRC Press, Boca Raton, FL.

Smith, E. and Dent, G. 2005. *Modern Raman Spectroscopy: A Practical Approach*. John Wiley & Sons, Chichester, UK.

Tang, H. and Li, Z.-L. 2014. *Quantitative Remote Sensing in Thermal Infrared: Theory and Applications*. Springer, Berlin, Germany.

Ünsalan, C. and Boyer, K. L. 2011. *Multispectral Satellite Image Understanding: From Land Classification to Building and Road Detection*. Springer, London, UK.

Vandenabeele, P. 2013. *Practical Raman Spectroscopy: An Introduction*. John Wiley & Sons, Chichester, UK.

Zibordi, G., Donlon, C. J., and Parr, A. C. (Eds.). 2014. *Optical Radiometry for Ocean Climate Measurements, Volume 47 (Experimental Methods in the Physical Sciences)*. Academic Press – Elsevier, Amsterdam, The Netherlands.

5 Satellite Optical Imagery

5.1 INTRODUCTION

High resolution (HR) and/or very high resolution (VHR) airspace optical remote sensing technology provides detailed images of the Earth's surface. Common regular sensors are nadir-viewing instruments with a horizontal spatial resolution in the range from 10 to 100 m and swath widths of order 100 km. In the past decades, HR and VHR sensors have emerged with spatial resolution in the range from 1 to 5 m and less than 1 m, correspondingly.

Most HR sensors provide panchromatic, PAN (a single waveband) and multispectral, MS (multiple waveband) imagery, with a number of synchronized spectral channels in the visible and IR ranges of the electromagnetic spectrum. This increases the information content and satellite-imagery capability. In order to reduce atmospheric effects and to increase image quality, the operating wavelengths are selected to coincide with atmospheric windows.

The use of optical sensors can be limited by weather conditions, since they are unable to provide imagery through thick cloud, rain or fog, and therefore the observations are typically restricted to fair weather and daytime-only operation. Some sensors have pointing capability which enables imagery of specified areas to be acquired more frequently on a regular basis. Many countries have and/or are planning HR optical imaging missions for various civilian and military purposes. Future trends include a greater number of spectral channels with improving payload design and spectral and spatial resolutions.

HR optical imagers are the most common instruments in remote sensing of environment with a wide range of applications. They include geological and topographic mapping, cartography, oceanography, meteorology, land cover and urban monitoring, agriculture, environmental control, post-disaster assessment, military airspace surveillance and reconnaissance, and developments of 3D Digital Elevation Model (DEM). Many efforts have been made to provide operational monitoring and managing *coastal and marine socio-ecological systems* using optical sensors. Henceforth, in this chapter, HR optical imagery is considered as advanced remote sensing technique capable to accomplish detection and recognition of ocean hydrodynamic features. In the following sections, we describe and elaborate some technological and methodological aspects of the subject matter.

5.2 BRIEF HISTORY

The history of the optical satellites starts from classified military systems that performed photography of Earth's surface from the 1960s to 1970s. This type of machine is known as reconnaissance satellite. For instance, U.S. CORONA satellites operated in the time period 1960–1972 and U.S.S.R. Zenit 2–8 series satellites (all military satellites were known as Kosmos series) operated from 1961 to 1994.

The Corona satellites used special 70 mm film with a 24-inch (610 mm) focal length camera, manufactured by Eastman Kodak. Zenits used the SA-20 camera with a focal length of 1 m and the SA-10 camera with a focal length of 0.2 m. Unlike Corona program, Zenit program provided the return capsule carried both the film and the cameras; it was an advantage at that time.

Satellite imagery became publically available from the 1970s. The first civilian Earth observation satellite, Landsat, was launched in 1972; since then, capabilities of both optical satellites and data processing have been increased dramatically. The number of satellites for remote sensing increased rapidly through the 1980s and 1990s. Among them, Landsat-4 and -5 missions with improved spatial resolutions up to 30 m included shortwave IR bands that greatly improved geological applications. In 1986, remarkable SPOT satellite remote sensing program with a highly valued capability and 10-m resolution was launched; nowadays, some satellites from the SPOT-1, -2, -3, -4, and -5 constellation offer imagery across the globe every day. SPOT payload comprises HR optical imager operated either in PAN mode or MS mode (the green, red, and NIR bands of the electromagnetic spectrum). On SPOT-5, the HR stereoscopic imaging instrument provides acquisition of stereopairs.

The real revolution in satellite optical imagery arrived in the late 1990s with developments of digital technology and computer science. IKONOS is the first commercial Earth observation satellite. It collects 1-m resolution PAN and four MS (blue, green, red, and NIR) 4-m resolution data simultaneously. IKONOS-2 was launched on September 24, 1999 and was called as one of the most significant achievement in the history of the space age. IKONOS space images began to sell on January 1, 2000.

QuickBird, is other commercial Earth observation satellite, launched on October 18, 2001. It is the first constellation mission of three satellites with highest resolution capability. QuickBird collects PAN imagery at 60–70 cm resolution and MS imagery at 2.4–2.8 m resolution. At this resolution, urban buildings, roads, airfields, and other infrastructure objects are visible in the images in great detail. Additionally, image processing toolboxes including GIS software packages were developed especially for digital analysis of QuickBird data. These data are also used for mapping applications such as Google Earth and Google Maps.

WorldView-1 was launched on September 18, 2007. This satellite collects more than 1,000,000 square kilometers of imagery per day. WorldView-1 provides 50-cm resolution PAN imagery with the swath width of 17.6 km at nadir. Next modification, WorldView-2, provides PAN and eight MS band (instead of the typical four MS bands) imagery. Five RapidEye satellite sensors, launched on August 29, 2008 at Baikonur Cosmodrome in Kazakhstan, acquire five MS bands but no PAN imagery.

Nowadays, most HR and VHR satellite images are commercially available. Some new optical satellites are now with the DigitalGlobe cancelation (Quickbird, WorldView-1, -2, and GeoEye). The cancelation includes also WorldView-3 and -4, Airbus' Pléiades-1A,-1B, and the Korean Multi-Purpose Satellites (KOMPSAT-3,-3A) with the 70–30 cm resolution range. Recently Russian Roscosmos has reported that satellites AIST-2D, Resurs-P and Canopus-B have also HR imaging capabilities. These and others similar remote sensing instruments provide image acquisition at fine resolutions in digital format offering effective and efficient data analysis, modeling, and application products. Furthermore, it is expected to expand a number of advanced remote sensing instruments, e.g., to integrate the improved Chinese SuperView satellites and Airbus' Pléiades satellites.

Newest generation of optical surveillance satellites provides superior resolution; however, actual ground resolution is limited by atmospheric turbulence (this effect is known as degradation of astronomical seeing). Military and security agencies plan the successor with resolution of 25 cm that is realistic number but it will require enhanced digital processing. Further improvements are concerned not only spatial resolution but also obtaining real-time information and the ability to track missiles and small vehicles. However, there is no so far specialized intelligent Earth observing satellite system dedicated to detecting "hidden" ocean targets and/or induced hydrodynamic phenomena (except surface ship wakes and/or oil spills which are perfectly visible in optical images).

Table 5.1 presents a list of the satellite optical sensors with <5 m spatial resolution. Detailed descriptions and specifications of satellite systems are available in the books (Slater 1980; Kramer 2002; Angelo 2006; Maini and Agrawal 2007; Olsen 2007; Tan 2014; Joseph 2015; Baghdadi and Zribi 2016). Textbook (Dowman et al. 2012) gives the reader the comprehensive knowledge about characteristics and use of HR and VHR satellite optical data.

TABLE 5.1
List of the Satellite Optical Sensors with <5 m Spatial Resolution

Satellite—Sensor	Launch Date	Number of Bands and Spectral Bands (nm)	Spatial Resolution (m)	Swath Width (km)	Revisit Interval (day)
Digital Globe WorldView-1	September 18, 2007	PAN	0.5	17.7	1.7
Digital Globe WorldView-2	October 8, 2009	8(400–1040)–1 PAN (450–800)	1.85–0.46	16.4	1.1
NOAA WorldView-3	August 13, 2014	8(400–1040)–1 PAN (450–800)–8 SWIR (1195–2365)	1.24–3.7–0.31	13.1	1–4.5
Digital Globe QuickBird	October 18, 2001	4(430–918)–1 PAN (450–900)	2.62–0.65	18	2.5
GeoEye Geoeye-1	September 6, 2010	4(450–920)–1 PAN (450–800)	1.65–0.41	15.2	<3
GeoEye IKONOS	September 24, 1999	4(445–853)–1 PAN (526–929)	3.2–0.82	11.3	~3
SPOT-5 HGR	May 4, 2002	3(500–850)–1 PAN (480–710)–1 SWIR (1580–1750)	2.5 and 5–10–20	60	2–3
CATROSAT	May 5, 2005	PAN (500–850)	2.5	30	5
ALOS AVNIR-2	January 24, 2006	4(420–890)–1 PAN (520–770)	2.5–10	70	2

5.3   SATELLITE CHARACTERISTICS

Satellites have many unique characteristics which make them particularly useful for remote sensing of the Earth. Among them, orbit, swath, configuration design (type), resolution, revisit time are the most important characteristics for optical observations. These characteristics define sensor's capability and information content of HR and/or VHR optical data; in particular, the choice of satellite's swath and spatial and radiometric resolution is critical for providing detection and monitoring of ocean hydrodynamic features. To understand a possible concept better, we will give an overview of general satellite characteristics using standard specifications (NGA 2009).

5.3.1   TYPES OF OPTICAL SENSORS

The term "Optical Sensor" usually refers to a digital data acquisition device, which comprises both geometric and measurement parts. According to Jensen (2014), optical imaging systems are divided into six categories, which are as follows: "(a) traditional aerial photography, (b) multispectral imaging using a scanning mirror and discrete detectors, (c) multispectral imaging with linear arrays (often referred to as 'pushbroom' technology), (d) imaging with a scanning mirror and linear arrays (often referred to as 'whiskbroom' technology), (e) imaging spectrometry using linear and area arrays, and (f) digital frame camera aerial photography based on area arrays." Here is a brief description of the most important configurations (Figure 5.1).

Framing sensor (traditional panoramic camera) is an oldest type of equipment, which provided photography of the Earth using one or more spectral bands.

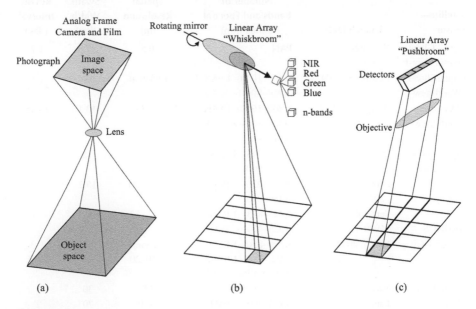

FIGURE 5.1   Types of satellite observations: (a) frame, (b) whiskbroom, and (c) pushbroom. (Modified and updated from Jensen 2014.)

Framing cameras build up strips of images by acquiring successive images in the along-track direction. In the last, framing cameras used a film to record photographic images. As pixel array detectors become available, framing cameras become more common for airspace imagery. Digital framing camera detects and records all the picture element (pixel) data for a single image (frame) at an instant of time at one spectral band. Framing cameras were widely used in aerial photography and the reconnaissance satellites, e.g., U.S. Corona series and Keyhole (KH) satellite systems KH-7 and KH-9 as well as in Apollo and SKYLAB satellites in the mid-1970s. The first use of CCD-based digital framing camera was in Russian RESURS-DK which had 4000 mm focal length, 0.5 m lens aperture diameter, and 1–2 m PAN/MS ground resolution. RESURS-DK operated from 2006 to 2016. Nowadays, digital frame cameras are mostly used in space explorations and planetary missions (e.g., each NASA Mars Exploration Rover carries 10 framing cameras). Remote sensing applications with digital framing cameras require the use of large and heavy optical payload to provide HR and/or VHR MS imagery of the Earth's surface. However, non-scanning (stationary) framing optical cameras have major advantages which are excellent geometric fidelity and quality of the entire image with minimal distortions.

Pushbroom sensor is along-track scanner. This sensor collects one line at a time in perpendicular to the flight direction of the spacecraft; all picture elements (pixels in a line) at the focal plane are acquired simultaneously. Pushbroom imagers delete the cross-track scanning mechanism and use a linear array of detectors to cover all the pixels in the across-track dimension at the same time. Pushbroom sensors are commonly used in satellite optical cameras for the generation of 2D images of the Earth's surface. Examples of pushbroom optical sensors are SPOT, NASA EOS Terra ASTER and MISR, EROS A 1, IKONOS, QuickBird, OrbView-3 and OrbView-5, GeoEye-1, Terra MODIS, SENTINEL-2.

Whiskbroom sensor is a cross-track scanner. Unlike the pushbroom sensor, whiskbroom sensor uses rotating mechanisms with a mirror to scan a pixel array across its field-of-view covering a ground swath. The focal plane geometry of the whiskbroom and the pushbroom sensors is similar. The pixel array is usually very short and is rapidly scanned about the direction of platform motion. Thus, whiskbroom sensor records a single cross-track image line and constructs a larger image from the set of adjacent lines using the along-track motion of the sensor's collection platform. Examples of whiskbroom optical sensors are Landsat MSS, TM, and ETM+, NOAA GOES, NOAA AVHRR, NASA ATLAS, and SeaWiFS.

Differences between whiskbroom and pushbroom sensors are given in Table 5.2.

5.3.2 EARTH COORDINATE REFERENCE SYSTEM

A Coordinate Reference System (CRS) provides a standardized way of describing locations. CRS refers to spatial data or objects which are represented in the Earth's surface or near the Earth's surface referenced to a model of the Earth rather than to the Earth itself. Many different CRSs are used to describe data in terms of coordinates. Earth-centered and Earth-fixed (ECEF) coordinate system (X, Y, Z), which is used by the most satellites observation systems is shown in Figure 5.2.

TABLE 5.2
Whiskbroom vs. Pushbroom

Whiskbroom	Pushbroom
Visible/NIR/MIR/TIR	Mainly visible/NIR
Wide swath width	Narrow swath width
Complex mechanical system	Simple mechanical system
Rotating mirror	Array of sensitive sensors
Simple optical system	Complex optical system
Shorter dwell time	Longer dwell time
Pixel distortion	No pixel distortion

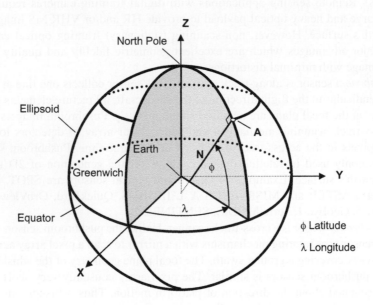

FIGURE 5.2 Earth-centered and local surface coordinate frames. (Based on NGA 2009.)

In ECEF, the X–Y plane containing the equator, **X** intersects, in the positive direction, the Greenwich Meridian, **Z** is parallel to the Earth's rotation axis with a positive direction toward the North Pole, **Y** is in the equatorial plane and perpendicular to **X** and completes a right-handed coordinate system. The cross-product of **X** and **Y** is a vector in the direction of **Z**. This system is known as a geocentric coordinate system. In this system, any point (A) on the reference surface may be described in (X, Y, Z) geocentric coordinates, or alternatively in the equivalent geodetic latitude and longitude. The point A can be referred to as either the ground or object point. Sensor observation configuration can be defined mathematically through ECEF reference frame.

5.3.3   ORBITS AND SWATHS

The ideal elliptical orbit, with the earth at one node, is described by six Keplerian elements:

α true anomaly (instantaneous angle from satellite to perigee)
ω argument of perigee (twist)
Ω longitude of the ascending node (pin)
a semi-major axis of the elliptical orbit (size)
e eccentricity of the orbital ellipse (shape)
i inclination of the orbital plane (tilt).

The Kepler parameters depicted in Figure 5.3 define an ellipse, orient that ellipse with respect to the earth, and place the satellite on the ellipse at a particular time. The size and shape of the orbital ellipse is defined by the semi-major axis of the ellipse a and the numerical eccentricity e. The orientation of the orbital plane against the equator is defined by orbital inclination i and the right ascension of the ascending node Ω. The argument of perigee ω and the true anomaly or travel angle τ define the position of the satellite on the ellipse at a particular time t. The satellite platform position is represented by the geocentric vector **R**. In terms of polar coordinates, the vector **R** is defined by the satellite geocentric latitude Ψ, satellite ground track geographic longitude λ and the geocentric radius R as functions of time.

Types of satellite orbits are listed below. Orbits of remote sensing are shown in Figure 5.4.

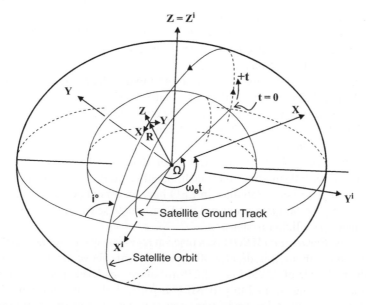

FIGURE 5.3   The Kepler orbital elements. (Based on NGA 2009.)

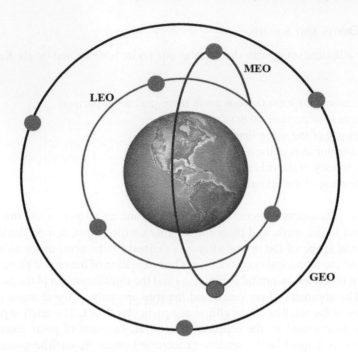

FIGURE 5.4   Types of satellite orbits.

Geostationary orbit (GEO) circles the Earth above the equator from west to east at a height of 36,000 km. Due to the Earth's rotation, which takes 23 h 56 min and 4 s, satellites in a GEO orbit appear to be "stationary" over a fixed position. Their speed is about 3 km/s. Satellites in GEO continuously cover a large portion of the Earth that makes it an ideal orbit for monitoring of weather patterns and environmental conditions. A constellation of three equally spaced satellites can provide full coverage of the Earth, except Polar Regions.

Geostationary transfer orbit (GTO) is an orbit to transfer a satellite between two circular orbits of different radii in the same plane. GTO is used to reach GEO with high-thrust chemical engines. GTO is a highly elliptical Earth orbit with an apogee of 42,164 km (26,199 mi), or 35,786 km (22,236 mi) above sea level, which corresponds to the geostationary altitude.

Low Earth orbits (LEO) are normally at an altitude of less than 1,000 km and could be as low as 160 km above the Earth. Satellites in this circular orbit travel at a speed of around 7.8 km/s. Satellites in LEO have an orbital period in the range of 90–120 min. In general, these orbits are used for remote sensing, military purposes and for human spaceflight missions.

Medium low Earth orbit (MEO) takes place at region of space around Earth above low Earth orbit with an altitude of 2,000 km (1,243 mi) above sea level and below GEO with an altitude of 35,786 km (22,236 mi) above sea level. The orbital periods of MEO satellites range from 2 to 12 h. MEO is particularly suited for constellations of satellites mainly used for navigation, communication, and space environment science. A satellite in this orbit travels at approximately 7.3 km/s.

Polar orbit (PO) passes over the Earth's polar regions from north to south. It therefore has an inclination of (or very close to) 90° to the body's equator. POs mainly take place at low altitudes of between 200 and 1,000 km. Satellites in PO provide monitoring of the Earth's entire surface and can pass over the North and South Poles several times a day. POs are used for reconnaissance and remote sensing. Satellites in POs operated at altitude of 800 km travel at a speed of approximately 7.5 km/s. Complementing the GEO satellites are the polar-orbiting satellites known as Polar Operational Environmental Satellites (POES). The POES instruments include the AVHRR/ATOVS providing visible, infrared, and microwave remotely sensed data.

Sun synchronous orbit (SSO, also called a *heliosynchronous* orbit) is a nearly PO which is synchronous with the sun. Satellites in a SSO usually operate at altitude of between 600 and 800 km. The path that a satellite has to travel to stay in a SSO is very narrow. Generally these orbits are used for remote sensing, weather forecasting, and reconnaissance. Observation is improved if the surface is always illuminated by the sun at the same angle when viewed from the satellite. The SSOs instruments include SPOT 4-5, IKONOS, WorldView, LANDSAT 4-5, TIROS, QuickBird-1 and 2.

A satellite provides a stable and, more importantly, a predictable platform. Thus one can employ constraints dictated by the Kepler's laws of motion to achieve convergence. Satellite platforms for electro-optical imaging systems are usually placed in a SSO and/or LEO. More details about satellite orbits can be found in the books (El'yasberg 1967; Montenbruck and Gill 2000; Capderou 2005).

Swath is referred to an observed area on the surface orthogonal to flight direction of a sensor. Imaging swaths for satellite sensors generally vary between tens and hundreds of kilometers wide, depending on orbit and instrument characteristics. The distance to the horizon of a planet is given by the formula $L = (2Rh)1/2$, where R is the planet's radius in km, and h is the height of the satellite above the ground in km. For example, if a satellite is in orbit at altitude of $h = 500$ km above the Earth's surface, the maximum swath width that can be viewed by satellite is $L = (2 \times 500 \times 6,378)1/2 = 2,525$ km, i.e., the swath is about equal to the maximum area of the Earth that can be viewed from horizon to horizon at the altitude of the satellite.

The *Instantaneous Field of View* (IFOV) is the angular cone of visibility of the optical sensor ϑ and determines the area on the Earth's surface which is "seen" from a given altitude at one particular moment in time (Figure 5.5). This area on the ground is called the *resolution cell* (or pixel resolution) and determines a sensor's maximum spatial resolution (FOV is called Field of View). The size of the area

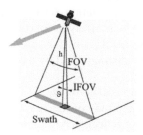

FIGURE 5.5 The Instantaneous Field of View (IFOV).

viewed is determined by multiplying the IFOV by the distance from the ground to the sensor h. At nadir, pixel size is ℓ = GIFOV = $2h\tan(\vartheta/2) \approx h\vartheta$, were $\vartheta \equiv$ IFOV is in radians. If the IFOV for all pixels at optical scanning is constant, then the ground area represented by pixels at the nadir will have a larger scale then those pixels which are off-nadir. In this case, pixel size will vary from the image center to the swath edge.

5.3.4 RESOLUTIONS

In general, the term "resolution" is defined as the smallest discernable physical unit of any signal which can be measured by a sensor. In remote sensing, resolution refers to the ability of the sensor to provide acquisition and record of fine detailed data. The following four types of resolution are known: *spatial*—relates to size in 2D space, *radiometric*—relates to electronic sensitivity and digital byte size, *spectral*—relates to a range of electromagnetic radiation, and *temporal*—relates to time and frequency of data collection. Below this will be described in more detail.

Spatial resolution (called also *geometrical resolution*) is a measure of the smallest linear dimension on Earth's ground area that can be resolved by the sensor. The angular resolution of optical lens θ is given by classical formula $\theta = 1.22\lambda/D$, where λ is the wavelength of the radiation measured, and D the diameter of the aperture (both have to be in the same units, and θ in radians). In remote sensing, spatial resolution is given in terms of the IFOV and expressed by the size of the pixel in meters. In terms of digital images, spatial resolution refers to the number of pixels utilized in construction of the image. Images having higher spatial resolution are composed with a greater number of pixels than those of lower spatial resolution. In remote sensing, several following gradations of spatial resolution are separated: less than 5 m (VHR), 5–100 m (HR), 100–1000 m (medium resolution), and 1000 m–50 km (coarse resolution or low resolution). Spatial resolution is a key parameter of optical observations significantly affected the information content of remotely sensed data and detection performance.

Spectral resolution describes the ability of a sensor to define fine wavelength intervals. Optical MS and HS systems record energy over several separate wavelength ranges at various spectral resolutions. High spectral resolution facilitates fine discrimination between different targets based on their spectral response in each of the narrow bands. Spectral band is usually defined in terms of a "central" wavelength λ_c and a "bandwidth" $\Delta\lambda$. The bandwidth is defined by lower λ_1 and upper λ_2 cutoff wavelengths. The spectral resolution $\Delta\lambda$ is given by $\Delta\lambda = \lambda_2 - \lambda_1$. Thus, the spectral resolution of an image is inversely proportional to its band width. The narrow the band, the greater the spectral resolution. Additionally, smallest separation in the wavelength that can still be distinguished by a sensor is given as fraction $\Delta\lambda/\lambda_c$ (defined by certain % criteria).

Radiometric resolution is an ability of an optical imaging system to measure the smallest changes in the magnitude of the electromagnetic energy. This is represented as the noise equivalent reflectance, radiance, or temperature change. The finer the radiometric resolution of optical sensor, the more sensitive it is to detecting small differences in reflected or emitted energy. Radiometric resolution is routinely expressed as a bit number, typically in the range of 8–16 bits. Image data are generally displayed in a range of gray tones, with black representing a digital number of 0 and white representing the maximum value 255 (in 8-bit data). Radiometric resolution depends upon the signal-to-noise ratio and determines the information content of the image.

Temporal resolution refers to the precision of a measurement with respect to time. In remote sensing, temporal resolution is defined as the amount of time needed to revisit and acquire data repeatedly for a given Earth's scene. The temporal resolution is high when the revisiting delay is low and vice versa. Temporal resolution is usually expressed in days. The actual temporal resolution depends on a variety of factors including the satellite/sensor capabilities and the orbital characteristics. For example, the images taken every 24 days have the same instrument view angle for any location, which is important so that BRDF differences do not influence the data.

In optical system design, there are always trade-offs between the four resolutions. For example, if increase the spatial resolution, i.e., reduce the IFOV, it will reduce the energy collected by the sensor, thereby producing a poor signal-to-noise ratio that leads to a poor radiometric resolution. On the other hand, keeping the constant spatial resolution, one can improve the radiometric resolution by increasing the spectral bandwidth (thereby collecting more energy), giving, however, poor spectral resolution. This last circumstance, however, is not so important at optical remote sensing of ocean hydrodynamics because spectral dependences of the water reflectivity in visible and near-infrared ranges are weak (Section 3.5). In this case, both spatial and temporal resolutions are the key parameters for the detection purposes.

5.3.5 Revisit Time

The satellite revisit time (or revisit interval or revisit period) is the time elapsed between successive observations of the same point on earth by a satellite. This time interval is also called "the repeat cycle of the satellite." It depends on the satellite's orbit, target location, and swath of the sensor. "Revisit" is related to the same ground trace—a projection on to the earth of the satellite's orbit. The revisit time can be calculated as the time between the double surveillance of the same target area with the assumption that the coverage cycle depends only on six Keplerian elements (Section 5.3.3) and the satellite position. Eventually, more frequent revisit takes place at high latitudes because the orbits coverage is near the poles. For most PO and SSO satellites, the repeat cycle ranges from twice a day to once every 16 days. Global coverage is often requires for holistic Earth observation.

5.4 IMAGE FORMATION

Formation of images is one of the most important issues in observation technology. Mathematically, an image can be defined as some function f(x,y) that represents a measure of many visual characteristics (color, brightness, contrast, etc.) of an object or selected scene. In optical remote sensing, an image is a geometrical projection of 3D surface into a 2D plane and function f(x,y) defines the light intensity at each position {x,y}.

As mentioned previously, two types of optical images—analog and digital—are divided. An analog image is characterized by continuous field of intensity variations. Such an image is produced from photosensitive chemicals in a photographic film. A digital image is characterized by discrete field of intensity; it is composed from a finite number of elements (called pixel), each of which has specified position and value. Converting an analog image into a digital image (called also digitalization) involves two operations—sampling and quantization—resulting to the creation of the array of pixels. Both analog (photographic) and digital (electro-optical) cameras can be used to obtain imaging data of different resolution; technical details and types of optical cameras are described in the books (Driggers et al. 2012; Saha 2015; Joseph 2015). In fact, differentiation of analog camera and digital camera becomes fairly common although both sensors use light-sensitive photocells which generate electrical charge, i.e., an analog signal. The brighter the light, the stronger the signal. Digital device such as the charge-coupled-device (CCD) provides only amplification and coding of analog signal into the binary system to create discrete array of brightness values which can be stored as an image file.

In general, image formation is performed using the following steps: (1) receive light energy as a continuous analog signal, (2) split signal into discrete digital level, (3) record levels in form of digital number, (4) create an array of digital numbers, (5) download and store created array in memory device, (6) specify "bits" which are 0–1 (1 bit), 0–255 (8 bits or 1 byte), 0–1023 (10 bits), 0–4095 (12 bits), (7) define a dynamic range, i.e., the difference between lowest and highest digital number, through quantization, (8) convert digital number back to measured energy through calibration, and (9) relate the energy to intrinsic property (e.g., reflectivity).

In most applications, an image formation process can be described by a linear model. In this case, output image f(x,y) is the convolution of input image g(x,y) with the impulse response h(x,y). The convolution equation is given by

$$f(x,y) = \iint g(x',y')h(x'-x,y'-y)dx'dy' = g(x,y) \otimes h(x,y), \qquad (5.1)$$

which is in discrete form becomes

$$f(i,j) = g[i,j] \otimes h(i,j) = \sum_{k=1}^{n}\sum_{l=1}^{m} g[k,l]h[i-k,j-1]. \qquad (5.2)$$

Equations (5.1) and (5.2) can be interpreted also in terms of an observation process

$$f(x,y) = ASF \cdot D\{g(x,y) \otimes h(x,y) + \eta(x,y)\}, \qquad (5.3)$$

where f(x,y) is observed image, g(x,y) is desired image, h(x,y) is blur kernel, ASF is angular sensitivity function, $\eta(x,y)$ is additive noise, and D denotes a discretization operator (see also Section 3.7.3). Thus, an optical image of an extended natural object can be represented by a convolution of the brightness distribution in the object with the diffraction pattern of a point source (impulse response) produced by the imaging system. Note that image discretization operation (5.3) involves two separate processes: discretization of the spatial domain (sampling) and discretization of the image intensity range (quantization). In more details, mathematical fundamentals and applications of the image formation theory are discussed in books (Françon 1979; Blahut 2004; Campisi and Egiazarian 2007; Milanfar 2011; Le Moigne et al. 2011).

5.5 QUANTITATIVE ASSESSMENT AND MANAGEMENT

Quantitative assessment of remotely sensed databases is an important aspect of interpretation, applications, and decision-making. In general, the goal of quantitative accuracy assessment of data and products (e.g., maps) is the identification and measurement of errors. Figure 5.6 illustrates the possible sources of errors in remotely sensed data (Congalton and Green 2009) which are multiple and compounding. Errors can be derived from the acquisition of data, their digital processing, classification, and also during uncertainties of the modeling and prediction. Accuracy assessment estimates, identifies, and characterizes the impact that arises from all of the sources of error. Books (Lyon and Lunetta 2005; Congalton and Green 2009) provides comprehensive description of the subject.

The theory and principles of quantitative accuracy assessment have been developed in the mid-1970s with the appearance of digital technologies in remote sensing, GIS, and mapping science. Nowadays, established accuracy assessment techniques include (1) generation of the error matrix, (2) the Kappa analysis, (3) discrete multivariate analysis, (4) sampling size and scheme, (5) spatial autocorrelation, (6) error budget analysis, (7) change detection accuracy assessment, and (8) fuzzy accuracy assessment.

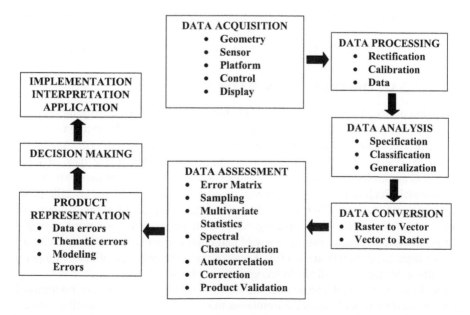

FIGURE 5.6 Sources of errors in remotely sensed data. (Updated and modified from Congalton and Green 2009.)

Quantitative accuracy assessment methods provide a very powerful scientific basis for both descriptive and analytical evaluation of the spatial remotely sensed data. In particular, optical (MS) imagery of the ocean is definitely a subject for quantitative accuracy assessment, at the first place due to the impact of a great number of environmental factors affecting the observation process and image formation. One typical example is evaluation of wave number spectra by optical or radar images using standard 1D or 2D FFT algorithms. In most practical cases, this procedure may be qualified as an assessment but not as the retrieval of oceanographic variables. At the same time, enhanced spectral processing of optical imaging data can provide remarkable results in terms of hydrodynamic detection (Chapter 7).

Work with huge archived databases (called also "big data") requires a specific management in term of data accessibility, data usability, quality of processing, and data science. In general, the big data management provides (1) discovery and characterization of data, (2) selection of relevant information, (3) selection of the area of interest, (4) specification of geodatabases and supported information, (5) initial study of data content, and (6) data integration and distribution. Most management operations include data mining and can be performed automatically using mathematical modeling methods, specially created for analysis of big data. Today, many reference publications can be found on this subject. For example, big data algorithms for satellite remote sensing are described in recent book (Swarnalatha and Sevugan 2018). Eventually, analysis and data mining of satellite optical imaging data should be done using intelligent information management system with enhanced image processing.

In recent years there has been increased focus on Research Data Management (RDM). In remote sensing, effective RDM is a multiple challenge related to generating new tools, new products and new ideas (Pryor 2012). In this connection, RDM can provide research and analysis of raw data along with processed data, selection and assessment of relevant geophysical information, data sharing, data modeling expertise, and product derivation. In fact, RDM requires certain level of scientific experience.

5.6   OCEAN OBSERVATION STRATEGY

Detection of ocean hydrodynamic features from satellites is a challenging problem which requires the development and implementation of special observation strategy. Such a strategy can be divided into two parts: short-term and long-term. According to our experience, reliable ocean data can be obtained only in the case of long-term systematic observations enable to collect statistically representative experimental material. Such data can be analyzed, evaluated, and applied for specific goals on a regular basis (e.g., detection of deep ocean processes, explosions, tsunami, or other complex events). Unfortunately, short-term fragmentary missions may not always provide adequate result in view of detection performance (although scientifically single experiments are useful). Low-volume data and lack of spatiotemporal measurements make it difficult to recognize complex situations.

Looking more broadly, basic observation strategy for the purposes of operational control and detection of ocean environments may include integrated satellite constellation system capable to provide 24/7 monitoring across the globe and cover most of the ocean's regions. Myers (2008) made estimates especially for tsunami detection

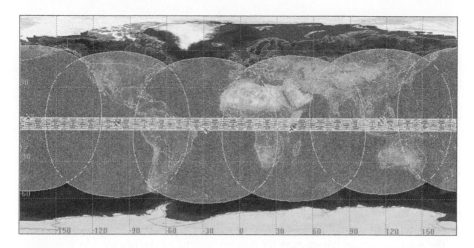

FIGURE 5.7   (See color insert.) Tsunamisat Constellation. (After Myers 2008; Myers et al. 2008.)

using a set of passive microwave radiometers. From this data we may understand that real-time global *Ocean Surveillance Detection System* (OSDS) operated with advanced remote sensors can be designed using a low-inclination and the intermediate circular-orbit (~10,000 km) satellite array. The relatively low inclination provides the large footprint of each satellite and results to fewer satellites (~5–10) in the constellation. Figure 5.7 illustrates Tsunamisat Constellation, generated using a commercial software package called Satellite Tool KitTM, STK™ (Myers 2008). Alternatively, lower orbits (~400–600 km) will increase the number of satellites in constellation up to 30–50 (depending on configuration) in order to cover entire ocean regions with suitable (~4–5 m) spatial resolution.

The strategy of OSDS may be formulated on three different levels:

- *Level 1*: Science is an approach to understand better ocean phenomena and remote sensing information. The emphasis here is on the thematic research including (1) CFD-based numerical analysis, (2) test field experiments associated with *real world* ocean hydrodynamics, (3) enhanced data processing, and (4) physics-based interpretation framework. These studies are conducted using available resources and techniques.
- *Level 2*: Using current high-resolution optical satellite capabilities (IKONOS, QuickBird, WorldView, GeoEye, and similar HR/VHR instruments) for observations of environmental and/or induced hydrodynamic phenomena/events in selected coastal or open ocean regions.
- *Level 3*: Creation of specialized OSDS, dedicated to operational control, maritime reconnaissance, and the global monitoring of ocean environments including target detection (up to depth ~200 m). This level will require the development of innovative (passive/active) multisensor observation technology, advanced infrastructure, and efficient Government management and International support. The OSDS can be designed as a low-Earth-orbit

constellation of 30–50 small fully automated satellites with flexible and reconfigurable payload architecture. OSDS provides truly global coverage with expanded capacity and high spatial/radiometric resolution, higher-speed data acquisition, real-time data processing and communication network services. Scientific support is also needed to provide *rigorous* data analysis, modeling, simulations, and prediction of oceanic scenarios, scenes, features, and signatures. The reality and future demand such kind of project anyway.

5.7 SUMMARY

In this chapter, we have described basic satellite system elements characterizing the quality and performance of optical imagery. We focused of the orbits specifications and four types of resolutions—spatial, spectral, radiometric, and temporal—that are the most important parameters for remote sensing of the oceans. As mentioned above, space-based optical detection capability in many ways depends on the choice of an appropriate satellite observation configuration. Optimization of the observation process, image formation, and quality accuracy assessment are critical factors influencing the implementation success of optical imagery. Technological aspects such as payload capacity and data acquisition also play a constructive role in satellite remote sensing of the oceans.

The current body of literature and books pertaining to satellite observation technology contains many interesting ideas and research developments that can find potential application in ocean remote sensing, in particular, in hydrodynamic detection. The overview and experiences allowed us to sketch some configuration of perspective satellite operational system (called OSDS) capable for providing the global monitoring, control and systematic targeting of ocean environments with 24/7 service. Nowadays, such a specialized *satellite constellation system*, as we know, doesn't exist due to large cost, lack of technological recourses and the absence of international scientific support. Pure military capabilities may not mean much in the solution of this complex problem.

REFERENCES

Angelo, J. A. 2006. *Satellites.* Facts on File Infobase Publishing, New York.
Baghdadi, N. and Zribi, M. 2016. *Optical Remote Sensing of Land Surface: Techniques and Methods.* ISTE Press – Elsevier, London, Oxford, UK.
Blahut, R. E. 2004. *Theory of Remote Image Formation.* Cambridge University Press, Cambridge, UK.
Campisi, P. and Egiazarian, K. (Eds.). 2007. *Blind Image Deconvolution: Theory and Applications.* CRC Press, Boca Raton, FL.
Capderou, M. 2005. *Satellites: Orbits and Missions (translated by S. Lyle).* Springer, Paris, France.
Congalton, R. G. and Green, K. 2009. *Assessing the Accuracy of Remotely Sensed Data: Principles and Practices,* 2nd edition. CRC Press, Boca Raton, FL.
Dowman, I., Jacobsen, K., Konecny, G., and Sandau, R. 2012. *High Resolution Optical Satellite Imagery.* Whittles Publishing, Dunbeath, UK.

Driggers, R. G., Friedman, M. H., and Nichols, J. 2012. *Introduction to Infrared and Electro-optical Systems*, 2nd edition. Artech House, Boston, MA.

El'yasberg, P. E. 1967. *Introduction to the Theory of Flight of Artificial Earth Satellites (translation from Russian by Z. Lerman)*. Israel Program for Scientific Translations, Jerusalem. Available on the Internet https://ia800302.us.archive.org/18/items/nasa_techdoc_19670020827/19670020827.pdf.

Françon, M. 1979. *Optical Image Formation and Processing (translated by B. M. Jaffe)*. Academic Press, New York.

Jensen, J. R. 2014. *Remote Sensing of the Environment: Pearson New International Edition: An Earth Resource Perspective*, 2nd edition. Pearson Education, Essex, UK.

Joseph, G. 2015. *Building Earth Observation Cameras*. CRC Press, Boca Raton, FL.

Kramer, H. J. 2002. *Observation of the Earth and Its Environment: Survey of Missions and Sensors*, 4th edition. Springer, Berlin, Germany.

Le Moigne, J., Netanyahu, N. S., and Eastman, R. D. (Eds.). 2011. *Image Registration for Remote Sensing*. Cambridge University Press, Cambridge, UK.

Lyon, J. G. and Lunetta, R. S. (Eds.). 2005. *Remote Sensing and GIS Accuracy Assessment (Mapping Science)*. CRC Press, Boca Raton, FL.

Maini, A. K. and Agrawal, V. 2007. *Satellite Technology: Principles and Applications*. John Wiley & Sons, Chichester, UK.

Milanfar, P. (Ed.). 2011. *Super-Resolution Imaging*. CRC Press, Boca Raton, FL.

Montenbruck, O. and Gill, E. 2000. *Satellite Orbits: Models, Methods and Applications*. Springer, Berlin, Germany.

Myers, R. G. 2008. *Potential for Tsunami Detection and Early-Warning Using Space-Based Passive Microwave Radiometry*. Master's Thesis. Massachusetts Institute of Technology, MIT, Boston, MA. Available on the Internet http://dspace.mit.edu/handle/1721.1/42913.

Myers, R. G., Draim, J. E., Cefola, P. J., and Raizer, V. Y. 2008. A new tsunami detection concept using space-based microwave radiometry. In *Proceedings of International Geoscience and Remote Sensing Symposium*, July 6–11, 2008, Boston, MA, Vol. 4, pp. IV-958–IV-961. doi:10.1109/IGARSS.2008.4779883.

NGA. 2009. *National Geospatial-Intelligence Agency NGA Standardization Document*. Pushbroom/Whiskbroom Sensor Model Metadata Profile. Supporting Precise Geopositioning, July 21, 2009. Available on the Internet http://gwg.nga.mil/documents/csmwg/PUSHBROOM_WHISKBROOM_PAPER_Version_1_0_GOLD_21JUL09.doc.

Olsen, R. C. 2007. *Remote Sensing from Air and Space*. SPIE Press, Bellingham, WA.

Pryor, G. (Ed.). 2012. *Managing Research Data*. Facet Publishing, London, UK.

Saha, S. K. (Ed.). 2015. *High Resolution Imaging: Detectors and Applications*. CRC Press, Boca Raton, FL.

Slater, P. N. 1980. *Remote Sensing: Optics and Optical Systems*. Addison-Wesley, Reading, MA.

Swarnalatha, P. and Sevugan, P. (Eds.). 2018. *Big Data Analytics for Satellite Image Processing and Remote Sensing*. IGI Global Engineering Science Reference, Hershey, PA.

Tan, S.-Y. 2014. *Meteorological Satellite Systems*. International Space University – Springer, New York.

6 Methods of Digital Analysis and Interpretation

6.1 INTRODUCTION

Analysis and interpretation of ocean optical images require an efficient digital processing. The reason is because ocean airspace pictures usually exhibit visible surface waves in form of low-contrast weakly-robust, quasi-regular geometrical patterns of different scale and orientation. To extract usable geophysical information from such *unlabeled* data, it is necessary to apply a powerful image processing algorithm that is implemented using different statistical principles and techniques. Eventually, the algorithm should provide the overall quality performance and the highest evaluation of geophysical signatures under given observation conditions. A major concern is the understanding what type of the processing is relevant in view of information content of the collected data, main objectives, and results expected.

The choice of an appropriate and ultimate processing strategy among numerous computer methods, techniques, and algorithms is always difficult task. Several textbooks (Jain 1989; Castleman 1996; Acharya and Ray 2005; Hoggar 2006; Pratt 2007; Gonzalez and Woods 2008; Bovik 2009; Jayaraman et al. 2009; Petrou and Petrou 2010; Nixon and Aguado 2012; Yaroslavsky 2012; Russ and Neal 2015) give us mathematical fundamentals of digital image processing and computer vision. Excellent books (Varshney and Arora 2004; Schowengerdt 2007; Mather and Koch 2011; Landgrebe 2003; Chen 2012; Camps-Valls et al. 2012; Richards 2013; Prost 2014; Benediktsson and Ghamisi 2015; Jensen 2015; Moser and Zerubia 2018) provide fairly broad review of image processing techniques and models giving to us comprehensive knowledge and understanding of this subject.

At the same time, we may see that specific practical recommendations and/or technical tips, how to elaborate digital processing and investigation of complex remotely sensed data are very general and sometimes, a belief is weakly justified. It is assumed, actually, that the decision should be borne in mind by the reader. Optical remote sensing of the ocean is a good example of learning. On one hand, the processing potential seems virtually to be limited simply because of apparent "monotonicity" of oceanic scenes in comparison with landscaping scenes. On the other hand, technological challenges, aims, and applications of ocean observations are quite ambitiousness. It means, in particular, that important information solutions should involve different methods of the processing and mathematical analysis of remotely sensed data. To improve or boost the reader confidence, we propose a survey of image processing techniques that are of practical interest in ocean remote sensing.

6.2 VARIOGRAM

Variogram or semivariogram is a basic component of *geostatistics* and GIS technology. The classical variogram estimator was proposed by Matheron (1963). In geostatistics, variogram is widely used for characterization and analysis of spatial structure, correlations, and data variability (Isaaks and Srivastava 1989; Goovaerts 1997; Oliver and Webster 2015). In remote sensing of the Earth's land surface, variogram is a tool for measuring image texture and subsequent image classification (de Jong and van der Meer 2006; Liu and Maso 2016).

Mathematical expression of variogram (or structure function) is

$$\gamma(h) = \frac{1}{2} E\{Z(x+h) - Z(x)\}^2, \tag{6.1}$$

where $\gamma(h)$ is call the variogram, $Z(x)$ is a function of spatial location, $Z(x+h)$ is the lagged function of spatial location, x is vector of spatial coordinates, h is lag vector representing separation between two spatial locations, and E denotes the expectation.

The experimental variogram is estimated as follows:

$$\gamma_{exp}(h) = \frac{1}{2N(h)} \sum_{i=1}^{N(h)} \{Z(x_i+h) - Z(x_i)\}^2, \tag{6.2}$$

where $\gamma_{exp}(h)$ is empirical semivariogram and $N(h)$ is the number of pairs separated by lag h (i.e., number of distant pair). By definition, the experimental variogram is half average of the squared difference between paired data values separated by vector lag h (Isaaks and Srivastava 1989; Olea 1991). Thus, the semivariogram measures the strength of statistical correlation as a function of distance. In the case of an image, variogram defines the variances of the differences between pairs of pixel values. The geostatistical analysis functions including variogram are written in the Python programming language; codes are available in PySAL open source library (Internet http://pysal.readthedocs.io/en/latest/).

A few applications of variogram to ocean satellite images have been discussed in works (Wald 1989, Legaard and Thomas 2007, and Glover et al. 2018). Additionally, variogram may have potential interest for optical/radar detection and classification of surface structural features (bizarre patterns, vortexes, eddies, wakes, slicks, etc.). Variogram can also be applied to analysis of microwave radiometric measurements of wind vector, SST, and sea surface salinity. Variogram is a simple noise-reduction method for raw experimental data that is an important issue in remote sensing.

6.3 FOURIER TRANSFORM

The Fourier transform (FT) is fundamental mathematical operation that converts a signal sampled in time or space to the same signal sampled in frequency. There are many excellent books on the FT and its mathematics (e.g., Tolstov 1976; Körner 1988; Bracewell 2000; Bloomfield 2000; Marks II 2009). Since the 1960s, Fourier analysis has become the standard technique in information technologies and applied science. The FT is widely used in digital data/image processing offering many important tools including spectral analysis, filtering, aliasing, image restoration, and compression. In particular, spectral analysis allows us to investigate spatially statistical properties of a variety of natural scenes measured by a remote sensor. Here are some practical aspects of the FT in spectral processing and interpretation.

6.3.1 BASIC DEFINITIONS

In data/image processing, the function of FT is to convert data from spatial-domain representation to the equivalent frequency-domain representation and vice versa. The definition of continuous 2D FT and the inverse transform (IFT) is given correspondingly as

$$F(u,v) = \int_{-\infty}^{\infty}\int_{-\infty}^{\infty} f(x,y)e^{-j2\pi(ux+vy)}\,dx\,dy, \tag{6.3}$$

$$f(x,y) = \int_{-\infty}^{\infty}\int_{-\infty}^{\infty} F(u,v)e^{j2\pi(ux+vy)}\,du\,dv, \tag{6.4}$$

where $F(u,v)$ is the function of spatial frequencies $\{u,v\}$ in frequency demine; $f(x,y)$ is the function of spatial coordinates $\{x,y\}$ in spatial domain. Also FT-IFT pairs are written as $f(x,y) \Leftrightarrow F(u,v)$ or $h(x,y) \Leftrightarrow H(u,v)$. Several of the essential properties of the FT are as follows (Gonzalez and Woods 2008):

- $F(u,v) = F_R(u,v) + jF_I(u,v)$ *complex* in general
- $|F(u,v)|$ *magnitude* spectrum
- $F_I(u,v)/F_R(u,v)$ *phase* angle spectrum
- $|F(u,v)|^2$ *power* spectrum
- Conjugate symmetry: $F(u,v) = F^*(-u,-v)$
$$|F(u,v)| = |F(-u,-v)|$$
- Scaling: $af(x,y) \Leftrightarrow aF(u,v),\ f(ax,by) \Leftrightarrow \dfrac{1}{|ab|}F(u/a,v/b)$
- Periodicity: $F(u,v) = F(u+M,v) = F(u,v+N) = F(u+M,v+N)$
$$f(x,y) = f(x+M,y) = f(x,y+N) = f(x+M,y+N)$$
- Separability: $f(x,y) = f(x)f(y) \Leftrightarrow F(u,v) = F(u)F(v)$
- Laplacian: $\nabla^2 f(x,y) \Leftrightarrow -(u^2+v^2)F(u,v)$
- Convolution theorem: $f(x,y) * h(x,y) \Leftrightarrow F(u,v)H(u,v)$
$$f(x,y)h(x,y) \Leftrightarrow F(u,v) * H(u,v)$$
- Correlation theorem: $f(x,y) \circ h(x,y) \Leftrightarrow F^*(u,v)H(u,v)$
$$f^*(x,y)h(x,y) \Leftrightarrow F(u,v) \circ H(u,v)$$

The discrete 2D Fourier transform (DFT) and the inverse transform (IDFT) of an image array are defined in series form as

$$F(u,v) = \sum_{x=0}^{M-1} \sum_{y=0}^{N-1} f(x,y) e^{-j2\pi\left(\frac{xu}{M} + \frac{yv}{N}\right)}, \quad u = 0,1...,N-1, \, v = 0,1...,M-1, \quad (6.5)$$

$$f(x,y) = \frac{1}{MN} \sum_{u=0}^{M-1} \sum_{v=0}^{N-1} F(u,v) e^{j2\pi\left(\frac{xu}{M} + \frac{yv}{N}\right)}, \quad x = 0,1...,N-1, \, y = 0,1...,N-1, \quad (6.6)$$

where $M \times N$ is the size of an image.

6.3.2 Fast Fourier Transform

In practice, computation of the 2D DFT is performed using the fast Fourier transform (FFT). Due to the property of separability of the 2D DFT, let's rewrite (6.5) for $N \times N$ image size as

$$F(u,v) = \sum_{x=0}^{N-1} \sum_{y=0}^{N-1} f(x,y) e^{-j2\pi(xu+yv)/N} = \sum_{x=0}^{N-1} e^{-j2\pi xu/N} \sum_{y=0}^{N-1} f(x,y) e^{-j2\pi yv/N}, \quad (6.7a)$$

$$F(u,v) = \sum_{x=0}^{N-1} F(x,v) e^{-j2\pi xu/N}. \quad (6.7b)$$

Thus, the 2D FFT can be obtained by computing the 1D DFT twice successively along each dimension: first for each column of $f(x,y)$ and then for each row on the result $F(x,v)$.

The comparison of computing resources (or time) shows that FFT versus DFT is $O\left(\frac{N}{2}\log_2 N\right)$ versus $O(N^2)$ of complex multiplications and $O(N\log_2 N)$ versus $O(N(N-1))$ of complex additions (Brigham 1988). The computational efficiency of the FFT versus the DFT becomes highly significant when the FFT point size N increases to several thousand as shown in Table 6.1. Note that the most available FFT codes (e.g., MATLAB®) use the Radix-2 algorithm, where $N = 2^M$, M is some integer, and the output power spectrum is optimized to the standard digital format, e.g., $S_{uv} = |F(u,v)|^2/N^2$.

In practice, the reconstruction of the power spectrum $S(K_x, K_y)$ of digitalized data/ image requires certain manipulations. First, an appropriate image fragment of $N \times N$ pixel size should be chosen. Second, the sampling interval Δ must be specified on the basis of the current information about geophysical size and pixel resolution of an image (there is a trade-off between resolution and accuracy of the result). The sampling interval can be either one pixel or a number of pixels. Third, the highest spatial frequency (known as the Nyquist frequency) $f_{max} = 1/2\Delta$ is computed. After that we define spectral range and the values of intermediate spatial frequencies of digital Fourier spectrum.

TABLE 6.1

Comparison of Efficiency between DFT and FFT

SN	N	DFT Multiplications, N^2	FFT Multiplications, $\frac{N}{2}\log_2 N$	FFT Efficiency
1	256	65536	1024	64:1
2	512	262144	2304	114:1
3	1024	1048576	5120	205:1
4	2048	4194304	11264	372:1
5	4096	16777216	24576	683:1

Spatial frequencies (or wavenumbers) associated with the Fourier coefficients and output spectrum are given by

$$f_m = \frac{m}{\Delta \cdot N}, \quad m = 0,1,2,\ldots,N/2, \tag{6.8a}$$

$$f_m = -\frac{(N-m+1)}{\Delta \cdot N}, \quad m = N/2+1, N/2+2, N/2+3,\ldots,N, \tag{6.8b}$$

which are knows as natural frequencies or natural wavenumbers:

$$K_m = \frac{2\pi m}{\Delta \cdot N}, \, m = 0,1,2,\ldots,N/2, \tag{6.9a}$$

$$K_m = \frac{2\pi(N-m+1)}{\Delta \cdot N}, \, m = N/2+1, N/2+2, N/2+3,\ldots,N, \tag{6.9b}$$

which are known as angular frequencies or angular wavenumbers.

For 2D FFT spectrum, plotted in wavenumber coordinates f_x, f_y or $\{K_x, K_y\}$ the value of intermediate wavenumbers are the same for both horizontal x-axis and vertical y-axes, i.e.,

$$f_m = \{f_{mx}, f_{my}\} = \frac{m}{\Delta N}, -\frac{m}{2\Delta} < f_{mx}, f_{my} < \frac{m}{2\Delta}, \tag{6.10a}$$

$$K_m = \{K_{mx}, K_{my}\} = \frac{2\pi m}{\Delta \cdot N}, -\frac{2\pi m}{2\Delta} < K_{mx}, K_{my} < \frac{2\pi m}{2\Delta}, \tag{6.10b}$$

The frequencies (wavenumbers) associated with the interval $N/2 < m < N$ are the negative frequencies and are redundant if the original function $f(x,y)$ is real. They are not redundant if the function $f(x,y)$ is complex. This is the important virtue in detection of complex-conjugation signals.

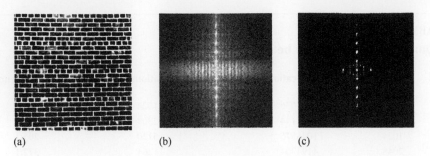

(a)   (b)   (c)

FIGURE 6.1 Example of digital 2D Fourier spectrum. (a) original image, (b) regular power spectrum, and (c) binarized log-spectrum.

Note that in further text we refer as to the *normalized power spectrum*, which is

$$S_{norm}(m) = \frac{|S(m)|^2}{MAX(|S(m)|^2)}.$$ (6.11)

The normalized power spectrum will have values between 0 and 1. However, both magnitude and phase portions of the spectrum are important for many applications. For example, in some radar (SAR) images the magnitude spectrum contains information about the shape of objects whereas the phase spectrum contains information about their actual location in the source. Example of digital 2D FFT spectrum is shown in Figure 6.1.

6.3.3 APPLICATIONS IN REMOTE SENSING

During the years, methods of spectral analysis have been extensively used in remote sensing of the ocean surface (references are in Chapter 7). Briefly, optical studies and applications have been developed in the following directions:

- measurement of surface wave parameters (periods, heights, orientation)
- estimation of surface wave (slope) spectra
- reconstruction of wave dispersion relations
- observation of the Doppler effect
- retrieval of the near-surface wind speed
- estimation of surface current velocity
- mapping seafloor bathometry
- underwater target detection.

An important application for oceanography is the obtaining geophysical *wavenumber-frequency spectra* of sea surface waves at some restricted areas. This task requires the acquisition of a large body of optical data/images or sea clutter time series. Such measurements can be provided using satellite optical instruments of high spatial resolution (~1 m), e.g., IKONOS, QuickBird, OrbView, GeoEye, and/

or the others commercial or military satellites. The retrieval of *calibrated* sea surface wave spectra from remotely sensed data/images is difficult task because of the impact of unpredictable light conditions and atmospheric effects. However, some estimates can be done using ocean reflection model. Let's consider theoretical and practical aspects of the problem.

According to existing theory, the spectrum computation is based on the FT of optical image $I(\vec{r})$ and the relationship between spectrum of image $S(\vec{k})$ and the geophysical surface slope spectrum $\Psi_{\nabla\xi}(\vec{k})$. In the simplest case, the following reflection model approach can be used:

$$g(x,y,t) = \rho(x,y)L(x,y,t) \approx \rho(\theta)\left[g_0 + \frac{\partial\xi}{\partial x}g_1 + \frac{\partial\xi}{\partial y}g_2 \right], \qquad (6.12)$$

$$I(x,y,t) = \iint g(x',y',t)h(x-x',y-y')dx'dy' \approx \Theta\left(\frac{\partial\xi}{\partial x},\frac{\partial\xi}{\partial y} \right)\Bigg|_{t=t_0}, \qquad (6.13)$$

$$S(k_x,k_y,\omega,t) = \left| \frac{1}{(2\pi)^2} \iint I(x,y,t)e^{-\imath\left[(k_x x+k_y y)\right]-\omega t}\,dx\,dy \right|^2, \qquad (6.14)$$

$$\Psi_{\nabla\xi}(k_x,k_y,t) = C\cdot M(\vec{k})\cdot H(\vec{k})\cdot S(k_x,k_y,\omega,t),\ \omega = \sqrt{gk},\ k = \left|\vec{k}\right|, \qquad (6.15)$$

$$k^2\Psi_\xi(k_x,k_y,t) = \Psi_{\nabla\xi}(k_x,k_y,t), \qquad (6.16)$$

where

$\xi(x,y)$ random field of surface elevations (a random rough sea surface)

$L(x,y,t)$ sky radiance to the surface

$\rho(\theta)$ specular reflectance of the sea surface at incidence $\theta(x,y)$

$g(x,y,t)$ geophysical optical scene (sky radiance reflected from the surface)

$h(x,y)$ *point spread function* of the optical system at offset coordinates

$I(x,y,t)$ image function (received irradiance) in the intrinsic coordinate system

$\Theta\left(\dfrac{\partial\xi}{\partial x},\dfrac{\partial\xi}{\partial y} \right)$ image function at small-slope approximation (which may not be valid)

$S(k_x,k_y,\omega,t)$ output power spectrum of an image in the spatial domain

$H(\vec{k})$ wavenumber filter (to reduce noise, artifacts, nonwave elements, etc.)

$M(\vec{k})$ retrieval (nonlinear) operator called also the transfer function

$\Psi_{\nabla\xi}(k_x,k_y,t)$ geophysical spectrum of surface elevation gradients (surface slope spectrum)

$\Psi_\xi(k_x,k_y,t)$ geophysical spectrum of surface elevations (surface wave spectrum)

$\vec{r} = \{x,y\}$ intrinsic coordinates of an image

$\vec{k} = \{k_x, k_y\}$ wave vector
ω angular frequency
t time
C unknown calibration coefficient and $\iota = \sqrt{-1}$.

The reconstruction of real world spectrum $\Psi_{\nabla\xi}(k_x, k_y, t)$ and/or $\Psi_\xi(k_x, k_y, t)$ by optical data using relationships (6.12)–(6.16) requires the knowledge of the transfer function $M(\vec{k})$ which depends on many factors and observation parameters (several references about that are available in Chapter 7). The main problem is to specify specular reflectance $\rho(\theta)$ correctly (which is not exactly the Fresnel reflection coefficient as supposed in many previous works, see Chapter 3). Moreover, to complete the reconstruction, it is necessary to invoke *in situ* calibration and/or additional oceanographic information that may not be always available at the time. Thereby, direct Fourier spectra, derived from (optical, radar) images are more adequate measure than unrigorous procedures of the wave spectrum retrieval. Direct Fourier spectra are useful for many applications (Chapter 7).

6.4 WAVELET

The word "wavelet" appeared in the early 1980s meaning "small wave." Jean Morlet (1931–2007), a French geophysicist, first introduced the term wavelet (Fr. "ondelettes") and together with the physicist Alex Grossman, conducted pioneering work in wavelet analysis. On the basis of the continuous wavelet transform (CWT), in 1987–1993 Mallat (1998) created a theory for multiresolution signal decomposition which resulted to the discrete wavelet transform (DWT) triggering further developments and applications of the wavelet theory in many fields of science. Here is a quick look at wavelets.

Wavelet transform provides time–frequency analysis of nonstationary signals at different frequencies with different resolution. Wavelet is considered as an alternative approach to the windowed or short-time FT (introduced by Dennis Gabor in 1946). Like FT, wavelets operate with orthogonal functions. Unlike FT the wavelets have special scaling properties. Every application of the FFT can be formulated using the DWT. In other words, instead a frequency spectrum, one gets a wavelet spectrum. Wavelet allows us to express any signal as a series of multiresolution approximations. According to Meyer (1992), who is among the progenitors of wavelet theory "the wavelet bases are universally applicable: 'everything that comes to hand', whether function or distribution, is the sum of a wavelet series and, contrary to what happens with Fourier series, the coefficients of the wavelet series translate the properties of the function or distribution simply, precisely and faithfully."

Wavelet analysis (called also wavelet theory or simply wavelets) is widely used in geophysics, remote sensing, data/image processing, and detection technology. Important sample applications include identifying pure frequencies, denoizing signals, detecting discontinuities and breakdown points, detecting self-similarity, subband coding. In image processing, wavelet provides image compressing, segmentation, enhancement, classification, fusion, pattern reconstruction, feature extraction, change detection. All these and other advantages of wavelet transform versus FT allow us to believe that wavelet transform is a promising flexible tool for analysis of ocean remotely sensed data simply because they exhibit explicitly non-stationary, multiscale, and stochastic nature of oceanic (hydrodynamic) processes and fields.

6.4.1 BASIC DEFINITION

Theory and applications of wavelets are discussed in numerous books and papers over the past two decades. The mathematical details are given by Daubechies (1992), Meyer (1992, 1993), Mallat (1998), and Kaiser (1994). There are many other textbooks of different reading level, devoted to fundamentals and various scientific and engineering applications of wavelets (Chui 1992; Cohen and Ryan 1995; Strang and Nguyen 1996; Erlebacher et al. 1996; Hernandez and Weiss 1997; Ogden 1997; Prasad and Iyengar 1997; Wojtaszczyk 1997; Burrus et al. 1998; Rao and Bopardikar 1998; Percival and Walden 2000; Silvermann and Vassilicos 2000; Jaffard et al. 2001; Walnut 2001; Chan and Peng 2003; Christensen 2004; Altaisky 2005; Walker 2008; Boggess and Narcowich 2009; Gao and Yan 2011; Weeks 2011; Najmi 2012; Debnath and Shah 2015; Addison 2017; Nickolas 2017; Broughton and Bryan 2018). The MATLAB codes are available in book (Misiti et al. 2007).

Since wavelets seem to be manifold and very attractive and powerful tool for geophysical and remote sensing applications including our subject as well, a brief description is given below.

As mentioned above, all wavelet transforms are divided on two parts: (1) the continuous direct and inverse wavelet transform pair (CWT) and (ICTW) and (2) the discrete wavelet transform pair (DWT) and (IDWT).

Wavelet or wavelet (basis, mother) function is defined as

$$\psi_{a,b}(t) = \frac{1}{\sqrt{a}}\psi\left(\frac{t-b}{a}\right), \tag{6.17}$$

where $a > 0$ is the scaling parameter (*scale*) and b is the shifting parameter (*shift*), which is any real number. A wavelet is a function $\psi(\cdot)$ that represents a fast-decaying oscillating wave front of finite length.

The 1D forward CWT and ICWT of a real signal $x(t)$ are given, respectively, by

$$W(a,b) = \langle x, \psi_{a,b} \rangle = \int_{-\infty}^{\infty} x(t)\psi_{a,b}^*(t)dt, \tag{6.18}$$

$$x(t) = \frac{1}{C_\psi} \int_{-\infty}^{\infty}\int_{0}^{\infty} W(a,b)\psi_{a,b}(t)\frac{1}{a^2}da\,db, \tag{6.19}$$

$$\int_{-\infty}^{\infty}|\psi(t)|^2 < \infty, \tag{6.20}$$

$$C_\psi = 2\pi \int_{-\infty}^{\infty} \frac{|\Psi(\omega)|^2}{|\omega|} d\omega < \infty, \tag{6.21}$$

where asterisk indicates a complex conjugation of $\psi(\cdot)$, Ψ is the FT of $\psi(\cdot)$, C_ψ is the *admissibility constant* (Daubechies 1992), and ω is the frequency. In (6.20) the first condition implies finite energy of the function $\psi(\cdot)$, and the second condition, the admissibility condition, implies that if $\Psi(\omega)$ is smooth then $\Psi(0) = 0$.

The Haar wavelet is the simplest kind of wavelet function with many applications (Lepik and Hein 2014). The Haar wavelet's mother function is defined as

$$\psi(t) = \begin{cases} 1 & 0 \le t \le \tfrac{1}{2}, \\ -1 & \tfrac{1}{2} \le t \le 1, \\ 0 & \text{otherwise}. \end{cases} \tag{6.22}$$

Other popular wavelet functions $\psi(\cdot)$ (wavelet family types) are: Haar (6.22), Meyer, Morlet, Mexican Hat, Daubechies, Coiflets, Biorthogonal, Symlet, Gaussian, Shannon (Addison 2017; Weeks 2011). Their choice depends on the data/image contents and the applications. The typical example of wavelet decomposition of a signal is shown in Figure 6.2 (MATLAB).

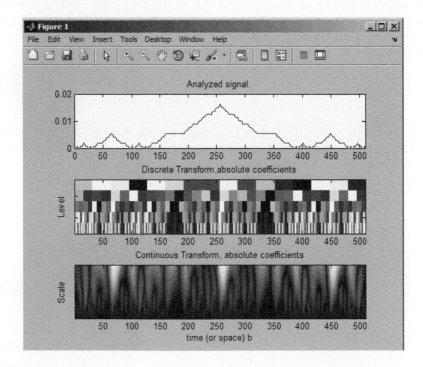

FIGURE 6.2 Example of wavelet transform. Scaleogram (MATLAB®).

The DWT has been designed to provide computer implementation of CWT (Mallat 1998). At the *dyadic grid arrangement*, parameters of the wavelet function $\psi(\cdot)$ are discretized as $a = 2^{-j}$ and $b = 2^{-j}k$, where integers j, k control the wavelet dilation and translation, respectively. In this case, the wavelet function (6.17) is written in compact form

$$\psi_{j,k}(t) = 2^{j/2}\psi\left(2^j t - k\right), \int_{-\infty}^{\infty}\psi_{0,0}(t) = 1, \tag{6.23}$$

where $\psi_{0,0}(t) = \varphi(t)$ referred to as the scaling function (or *father* wavelet).

A continuous signal $x(t)$ can be reconstructed as a sum of approximations and details:

$$x(t) = \sum_k a_x(j,k) 2^{j/2}\varphi\left(2^j t - k\right) + \sum_{j''=j}^{-\infty}\sum_k d_x(j',k) 2^{j'/2}\psi\left(2^{j'} t - k\right), \tag{6.24}$$

where $a_x(j,k)$ are approximation (scaling) coefficients and $d_x(j,k)$ are detail (wavelet) coefficients given by

$$a_x(j,k) := 2^{j/2}\int x(u)\varphi\left(2^j u - k\right) du \tag{6.25}$$

$$d_x(j,k) := 2^{j/2}\int x(u)\psi\left(2^j u - k\right) du \tag{6.26}$$

The dyadic equations for the scaling function and wavelet function are

$$\varphi(x) = \sqrt{2}\sum_k h_{k\varphi}\varphi(2x - k), \tag{6.27}$$

$$\psi(x) = \sqrt{2}\sum_k g_k\varphi(2x - k), \tag{6.28}$$

where $\{h_k\}$ and $\{g_k\}$ are a pair of low-pass (LP) and high-pass (HP) filters, respectively, that are related through $g_k = (-1)^{k-1} h_{-(k-1)}$ (Mallat 1998).

The first term in (6.24) gives the low-resolution approximation of the signal while the second term gives the detailed information at current resolutions j. The change of dilation and translation parameters in the scaling function induces a *multiresolution analysis*, MRA (Mallat 1998; Meyer 1993). The essence of MRA is the *decomposition* of a (continuous or discrete) signal or image into multitime or multiresolution scale features. A schematic example of MRA known as *wavelet image pyramid* is shown in Figure 6.3. The pyramid represents the same image at multiple scales. It is a set of layers where upper layers are lower resolution than the lower layers. Wavelet image pyramid provides information on both the spatial and frequency domains. More information about fundamentals of wavelets and MRA can be found in books

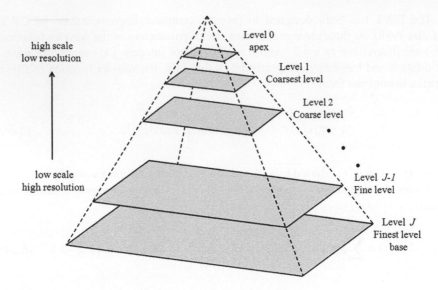

high scale
low resolution

Level 0
apex

Level 1
Coarsest level

Level 2
Coarse level

low scale
high resolution

Level *J-1*
Fine level

Level *J*
Finest level
base

FIGURE 6.3   Multiresolution analysis (MRA). Image pyramid.

(Mallat 1998; Meyer 1993; He 2000; Akansu and Haddad 2001; Rohwer 2005; Ouahabi 2012).

6.4.2 APPLICATIONS

Wavelet and MRA applications that are important for our subject cover three research areas: geophysics, hydrodynamics, and remote sensing. All these areas are connected to each other through the growth of the scientific interest in obtaining more specific and more detailed information about environmental processes. These objectives lead to developing more refined methods of data/image processing including advanced Earth science applications. Wavelet-based analysis has become a very popular frontier discipline for many researches.

Geophysical applications of wavelets refer to atmospheric and ocean (meteorological and climatic) time series, seafloor bathometry, earthquake prediction, rainfall behaviour (Meyers et al. 1993; Foufoula-Georgiou and Kumar 1994; Chandrasekhar et al. 2013; Thomson and Emery 2014), El Nino (Astaf'eva 1996; Torrence and Compo 1998) as well hydrology (Sivakumar 2017).

Hydrodynamic applications includes analysis of ocean waves (Liu 1994; Donelan et al. 1996; Gurley and Kareem 1999; Massel 2001; Haung 2004; Elsayed 2010; Nicolleau and Vassilicos 2014), wave breaking events (Liu and Babanin 2004), turbulence, intermittency (Farge 1992; Farge et al. 1996; Nicolleau and Vassilicos 1999; Ruppert-Felsot et al. 2009), fluids, vortexes, jets, wakes, and coherent structures (Everson et al. 1990; Camussi and Guj 1997; Farge et al. 1999; Li 2001; Li and Zhou 2002; Luo and Jameson 2002; Schneider and Vasilyev 2010; Khujadze et al. 2011; Indrusiak and Möller 2011; Grizzi and Camussi 2012; Watanabe et al. 2014; Fujimoto and Rinoshika 2015, 2017).

Many references and examples of wavelet analysis can also be found in books (van den Berg 2004; Debnath and Shah 2015; Addison 2017).

Remote sensing of the ocean is important but relatively limited area of wavelet applications. Several works refer to analysis of ocean (radar, optical) images (Simhadri et al. 1998; Liu et al. 1997, 2003), internal wave observations (Rodenas and Garello 1997, 1998; Arvelyna and Oshima 2004; Arvelyna et al. 2006), detection of ship wakes (Kuo and Chen 2003; Tello et al. 2005), and estimating wind field (Du et al. 2002; Fichaux and Ranchin 2002; Zecchetto and De Biasio 2008; Leite et al. 2010). Moreover, wavelet-based techniques have been used in hyperspectral target detection applications (that could be referred to remote sensing as well); however, this subject beyond the scope of our book.

6.5 FRACTAL

The term "fractal" was coined by mathematician Benoit Mandelbrot in 1975 (Mandelbrot 1983). It comes from the Latin *fractus* meaning *fractured* or *broken*. Mandelbrot gave a simple definition: a fractal is a shape made of parts similar to the whole in some way. There is a great variety of fractal objects of different geometry, structures, and origin; they all possess the following common properties:

- self-similarity
- irregularity
- non-integer dimension
- scale independence or invariance
- displacement invariance
- infinite detail or structure
- complexity
- scale symmetry
- heterogeneity, connectivity, and self-organizing.

Fractal structures are divided on two categories: deterministic and stochastic. A fractal is called deterministic when its parts are identical to original at a dilation transformation. A fractal is called stochastic when some part of the original varies randomly at rescaling. Fractal structures observed in nature belong mainly to the second category. In particular, apparent fractal properties of the radiances, registered by optical or microwave sensor can be identified in a statistical meaning only. This fact imposes certain limitations on real-time processing and interpretation of remotely sensed data.

A unique mathematical description of the fractal geometry in nature has provided a broad spectrum of the applications in Earth science (Schertzer and Lovejoy 1991; Hastings and Sugihara 1993; Barton and La Point 1995; Quattrochi and Goodchild 1997; Turcotte 1997; Dimri 2000, 2016; Lam and De Cola 2002; Franceschetti and Riccio 2007; Seuront 2010; Chandrasekhar et al. 2013; Ghanbarian and Hunt 2017). Fractal formalism has been evolved in the theory of dynamic systems and chaos theory (Schroeder 1991; Edgar 1998; Robinson 1998; Mandelbrot 2004; Feldman 2012), mathematical modeling and fractional dynamics (Barenblatt 1996; Zaslavsky

2005; Tarasov 2010), and engineering (Dekking et al. 1999). Fractal measures also stimulated a growing interest in digital processing including image modeling, image compression, image coding, segmentation, and classification of images (Peitgen and Saupe 1988; Russ 1994; Addison 1997; Turner et al. 1998; Welstead 1999; Falconer 2003; Sprott 2003; Blackledge 2005; Gao et al. 2007; Pesin and Climenhaga 2009; Barnsley 2012; Gulick 2012; Farmer 2014; Kumar et al. 2017). Today, the research on data/image processing has accomplished outstanding progress by taking advantage of the self-similarity and the fractal dimension as a powerful measuring tool.

6.5.1 SELF-SIMILARITY

Scale invariance, or self-similarity, is a fundamental well-established property of natural images. Taking into account statistical characterization of the underlying physical processes, in practice we deal with *statistical self-similarity*. Mathematical formulation of statistical self-similarity is given by the relation

$$\xi(\vec{r}) \overset{p}{=} \lambda^{-H} \xi(\lambda \vec{r}), \qquad (6.29)$$

where $\overset{p}{=}$ denotes equality in statistical sense, $\xi(\vec{r})$ is random field, $\vec{r} = \{x, y\}$ is coordinate vector, $\lambda > 0$ is scaling factor, and H is the Hurst exponent or index (Mandelbrot 1983). The Hurst exponent is a statistical measure used to classify time series; it is calculated by rescaled range analysis (R/S) (e.g.,Addison 1997; Weisstein 2003). The Hurst exponent for high-dimensional fractals can be estimated using algorithm (Carbone 2007).

An important class of statistical self-similar random processes is known as "$1/f$ process," where f is frequency. It is generally defined as a process with power spectra obeying a power law relationship

$$S(f) \propto f^{-\beta} \qquad (6.30)$$

with spectral exponent $\beta = 2H + 1$. The properties of H are summarized as the following: $0 < H < 1$; $H = 0.5$ for a Brownian motion (random walk); $H > 0.5$ for a persistent (long-term memory, correlated) process; $H < 0.5$ for an anti-persistent (short-term memory, anti-correlated) process. Furthermore, the Hurst exponent is related to a fractal dimension of a one-dimensional time series $D = 2 - H$ or $2D = 5 - \beta$ at $1 < \beta < 3$. For a self-affine surface in d-dimension space $D = d + 1 - H$ (Mandelbrot 1983).

6.5.2 THE FRACTAL DIMENSION

Fractal theory offers methods for describing the inherent irregularity (complexity) and scaling of natural objects. In fractal geometry, the dimension is characterized by a non-integer number known as the fractal (or fractional) dimension. Hausdorff and Besicovitch were the two mathematical pioneers on whose work Mandelbrot's development of fractals is based.

There are several mathematical definitions of the fractal dimension (Mandelbrot 1983). The two most commonly used are the Hausdorff dimension introduced in 1918 and capacity introduced in 1959 by Kolmogorov. Both dimensions are based on an idea of covering of the data set or an object of interest. The capacity dimension known also as *box-counting dimension* (which is more practically useful) is given by

$$D = \lim_{\varepsilon \to 0} \frac{\log N(\varepsilon)}{\log(1/\varepsilon)} \text{ or } -D = \lim_{\varepsilon \to 0} \frac{\log N(\varepsilon)}{\log(\varepsilon)}, \tag{6.31}$$

where $N(\varepsilon)$ is the smallest number of boxes with identical size ε required to completely cover fractal data set or an image object. The box-counting algorithm hence counts the number $N(\varepsilon)$ for different values of ε and plot the log of the number $N(\varepsilon)$ versus the log of the actual box size ε. The value of the box-counting dimension D is defined from the plot best fitting curve slope (see Chapter 7). In image processing, the fractal dimension can be estimated using three simple methods.

- box-counting method, $N(\varepsilon) \propto \varepsilon^{-D}$
- area-perimeter method, $P \propto \left(\sqrt{A}\right)^{D_s}$,
    D_s is fractal dimension, P is perimeter, and A is area of a single fractal object
- wavelet method (Section 6.5.4).

Although many natural systems present characteristics of scale invariance and fractal behaviour, statistical fractal measures can be considered only over certain range of parameters involved in a plot $\log N$ versus $\log \varepsilon$. Thereby, the simple box-counting method (6.31) may not give us full information about scale of the whole dynamic system. To obtain generalized scaling, *multifractal formalism* has been proposed and developed lately.

6.5.3 MULTIFRACTAL ANALYSIS

Multiracial analysis consists of determining local behaviour of complex dynamic system at different scales. In terms of box-counting measure, the multifractal scaling of an image object is characterized by

$$N(\varepsilon) \propto \varepsilon^{-f(\alpha)}, \tag{6.32}$$

where $N(\varepsilon)$ is the number of boxes of length ε required to cover an image object in a small range of singularity strength from α to $\alpha + d\alpha$, and $f(\alpha)$ is the multifractal spectrum described the statistical distribution of α (Halsey et al. 1986). The multifractal singularity spectrum is defined as

$$f(\alpha) = \lim_{\delta \to 0} \lim_{\varepsilon \to 0} \frac{\log \left[N(\varepsilon, \alpha + \delta) - N(\varepsilon, \alpha - \delta)\right]}{\log 1/\varepsilon} \tag{6.33}$$

and the generalized dimension is given by

$$D_q = \frac{1}{q-1} \lim_{\varepsilon \to 0} \frac{\log \sum_{k=1}^{N} (p_k)^q}{\log \varepsilon}, \tag{6.34}$$

where $p_k(\varepsilon) \propto \varepsilon^{\alpha_k}$ is a probability measure versus singularity strength α. The generalized dimension D_q is computed as a function of the order of the probability moment q. The singularity spectrum $f(\alpha)$ and the generalized fractal dimension D_q are connected by Legendre transform (Evertsz and Mandelbrot 1992). The multifractal formalism is a set of the following equations:

$$f(\alpha) = q\alpha(q) - (q-1)D_q, \tag{6.35}$$

$$\alpha(q) = q\frac{d}{dq}(q-1)D_q, \tag{6.36}$$

$$\tau(q) = \left[q\alpha(q) - f(\alpha(q)) \right], \tag{6.37}$$

$$\frac{d}{dq}\tau(q) = \alpha(q), \tag{6.38}$$

$$\frac{d}{d\alpha}f(\alpha) = q(\alpha). \tag{6.39}$$

According to Evertsz and Mandelbrot (1992), multifractal measure is a graph $f(\alpha)$ versus α having the shape of an inverted parabola (Figure 6.4). The singularity spectrum $f(\alpha)$ has some general properties: (1) the maximum value of $f(\alpha)$ occurs at q = 0 and $f(\alpha) = D_0$ which is the box-counting dimension or the Hausdorff

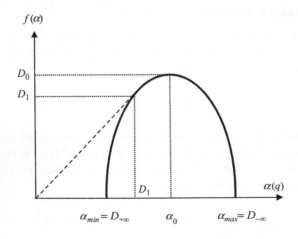

FIGURE 6.4 Multifractal analysis.

dimension, (2) $f(D_1) = D_1$, and (3) the line joining the origin to the point on the $f(\alpha)$ curve where $\alpha = D_1$ is tangent to the curve (Ott 2002). Thus, spectrum $f(\alpha)$ characterizes the original measure and its width indicates overall variability of the system. The most important application of the multifractal formalism is the study of turbulence (and turbulent cascades) occurring in dynamic systems.

Many researches use multifractal analysis for exploring experimental data (e.g., time series) showing strong multiscale spatiotemporal variability. Such data are mostly associated with nonstationary and/or nonuniformity of geophysical processes involving turbulence and energy transfer. In remote sensing, multifractal analysis actually provides segmentation and classification of multivariate data related to strongly nonlinear mixing dynamical processes. Typical example is wave breaking phenomena at high wind. Multifractal analysis allows computing a spectrum or a set of fractal dimensions to characterize and predict the behaviour of dynamic system at different timescale frames. However, many algorithms exist to evaluate this spectrum and numerical differences between the methods and results can appear.

Roughly speaking, multifractal spectrum $f(\alpha)$ characterizes the original measure and its width indicates overall variability. Detailed information about multifractal analysis and mathematical foundations can be found in books (Mandelbrot 1999; Harte 2001; Falconer 2003).

6.5.4 WAVELET-BASED FRACTAL ANALYSIS (WFA)

Fractals are complex, patterned, statistically self-similar or self-affine, scaling or scale-invariant objects with non-integer dimensions. A wavelet basis is fractal as well; therefore, wavelet can be used for fractal analysis of dynamic systems. The WFA is based on obtaining the variance of the DWT coefficients (6.26), $d_{k,j} = d_x(j,k)$, at each level of decomposition related to the scale j

$$\mathrm{Var}\{d_{\cdot,j}\} = \frac{1}{N-1}\sum_k \left(d_{k,j} - \bar{d}_{\cdot,j}\right)^2 \text{ or } \mathrm{Var}\{d_{\cdot,j}\} \approx \left(2^j\right)^{2H+1} \qquad (6.40)$$

or

$$\log_2 \mathrm{var}\{d_{\cdot,j}\} \approx (2H+1)j + \mathrm{const}, \qquad (6.41)$$

where H is the Hurst exponent related to the fractal dimension $D = d + 1 - H$ and symbol "Var" denotes variance. The plot $\log_2\{\mathrm{Var}\}$ versus j yields the estimation of H or the fractal dimension D at scale j. Thus, WFA provides a multiscale measure of dynamic systems. More advanced method known as "discrete wavelet transform-based multifractal analysis" allows us potentially deeper insights into various sophisticated problems associated with multivariate data/image processing and exploring the behaviour of complex dynamical systems.

WFA has the following benefits: (1) it performs a multiresolution decomposition suitable for scale-invariant processes analysis, (2) it performs optimally whitening and provides *Karhunen–Loeve* expansions for $1/f$-type processes, and (3) it provides good estimates of noise parameters. As a whole, WFA is a perfect estimator of scaling behaviour of nonstationary processes and fields registered by a remote sensor. For detailed information, we refer the reader to excellent books (Wornell 1995; Erlebacher et al. 1996; Welstead 1999; Abry et al. 2009; Chandrasekhar et al. 2014; Bandt et al. 2014; Massopust 2016) which cover many mathematical and practical aspects of WFA.

6.5.5 APPLICATIONS

Fractal and multifractal analysis have been applied extensively in signal, data, and image processing. Applications are grouped into the two main recurrent classes, namely segmentation and classification (characterization). Multifractal analysis is adapted to texture segmentation (Section 6.7.1.2). Its advantage is to characterize both local and global scales properties. It makes it possible to quantify the distribution of the local singularities through local morphological multifractal exponent. On the other hand, the fractal dimension can be introduced as a quantitative feature into a classification algorithm.

Scaling and complexity of dynamic environment including the ocean surface can be explored using spatiotemporal remote sensing observations and fractal-based techniques. Experimental possibilities have been demonstrated with SAR observations of marine oil spills (Benelli and Garzelli 1999; Gade and Redondo 1999; Marghany et al. 2009; Tarquis et al. 2014). The results reported are statistically significant. However, because of relatively low pixel resolution and fuzzy (noisy) texture of ocean SAR images, fractal measures may not always offer accurate estimates of topological characteristics and scaling parameters. This leads to difficulties in specification of the observed features. Additional and more reliable information can be obtained using high-resolution optical imagery.

The following ocean hydrodynamic processes and events can be investigated more adequately involving (multi)fractal analysis of optical data: (1) self-similarity and scaling of wave breaking fields (Section 7.4), (2) turbulent energy cascades at various conditions, (3) dynamics of complex surface patterns—wakes, jets, slicks, pollutions, turbulent flows, internal wave manifestations, and (4) generation and evolution of coherent structures (vortexes, eddies, streaks, wakes, jets). These developments and achievements will essentially increase remote sensing operational capabilities and overall effectiveness of ocean observations.

Multiresolution optical imagery enables detection and recognition of so-called ocean *fractal signatures* which are robust indicators of specific surface phenomena. In particular, variations of the (multi)fractal dimension derived from optical data can provide *dynamic and spatial* characterization of complex oceanic scenes to reveal distinctive self-similar hydrodynamic structures (turbulent wakes, eddies, vortexes, etc.) and explore their geometrical and statistical properties. This will offer additional detection capabilities.

As a whole, (multi)fractal analysis is a powerful method of the research with applications in various fields of geoscience (Evertsz and Mandelbrot 1992; Mandelbrot 1999; Harte 2001; Bernardara et al. 2007; Seuront 2010; Lovejoy and Schertzer 2013), medicine and biology (Rangayyan 2004; Nilsson 2007; Lopes and Betrouni

2009; Mitchell and Murray 2012) as well in forecasting of financial market volatility (Calvet and Fisher 2008).

6.6 FUSION

The term "fusion" has several definitions and means combination, composition, merging, synergy, integration and several others to express the same concept at different research areas. In information technology, fusion is a digital processing framework dealing with computer integration of multiple data sets obtaining from different sources (sensors). In particular, remote sensing image fusion provides integration, combined analysis, and joint interpretation of multiband, multispectral, hyperspectral, and/or multisensor measurements. Several excellent books (Wald 2002; Hall and McMullen 2004; Blum and Liu 2006; Stathaki 2008; Liggins et al. 2009; Chanussot et al. 2009; Raol 2009, 2016; Mitchell 2010, 2012; Chaudhuri and Kotwal 2013; Weng 2014; Alparone et al. 2015; Fourati 2016; Kedem et al. 2017; Pohl and van Genderen 2017; Chang and Bai 2018) cover comprehensively the subject-matter of multivariate and fusion researches in computer science, image processing, and remote sensing.

In this section, we will briefly overview the most important and powerful fusion techniques. There is no preference in the choice of fusion algorithm for comprehensive analysis of ocean remotely sensed data. However, some possibilities for this exist and can be outlined.

Here is some acceptable for us definition of image fusion:

The image fusion is the integration of different digital images in order to create a new image and obtain more information than can be separately derived from any of them.

There are a number of well-known fusion methods allowing us to improve the quality of multispectral and/or multisensor remote sensing imagery (e.g., Pohl and van Genderen 1998). Most of them provides enhancement of the source image, implementing *pansharpening* procedure. Others use fusion for statistical analysis and extraction of relevant information. Recent studies have shown that fusion is efficient method for spatial-spectral classification of hyperspectral images. Fusion can also be successfully applied for analysis of combined optical and microwave (radar, radiometer) observations with different information content and resolution. In Chapter 8, we consider multisensor data fusion as the most advanced processing technology capable for detection and recognition of ocean hydrodynamic features from airspace high-resolution imagery.

In general, image fusion is performed at three different processing levels according to the stage at which the fusion takes place. They are the following: (1) Pixel-level fusion, (2) Feature-level fusion, and (3) Decision level fusion. Usually, it is difficult to choose an optimal fusion algorithm which would be the most efficient and creative in analysis of multiple data. Successful fusion requires a constructive understanding of input information. Taking into account specific aspects of ocean imagery, we may refer to the following fusion methods and algorithms.

6.6.1 Average Method (AM)

AM or the *weighted average* method is a "primitive" pixel-level fusion method. The fused image is computed using arithmetic combinations:

$$I_f = (I_1 + I_2)/2 \qquad (6.42)$$

or

$$I_f = (a_1 I_1 + a_2 I_2)/(a_1 + a_2), \qquad (6.43)$$

where I_1 and I_2 are two input images of the same size and a_1 and a_2 are the weight coefficients. The weight coefficients indicate relative importance of the individual inputs on the fused image. Usually, the weight coefficients are normalized, i.e., $a_1 + a_2 = 1$. The AM works well when input images have similar brightness and contrast and collected from the same acquisition system. AM suppresses additive noise in the source imagery. The overall performance depends on the quality of input data.

6.6.2 Intensity–Hue–Saturation Method (HIS)

HIS is the standard process of image fusion. It works with the RGB (R—red, G—green, B—blue) true color space. The IHS transform separates color characteristics of the image, namely, intensity (I), hue (H), and saturation (S). The intensity refers to the total brightness of the image, hue to the dominant color, and saturation to the purity of color. The IHS substitution is given by

$$I = R + G + B, \qquad (6.44)$$

$$H = (G - B)/(1 - 3B), \qquad (6.45)$$

$$S = (I - 3B)/I. \qquad (6.46)$$

HIS is capable of quickly processing large bodies of data; this method provides spatial enhancement, i.e., preserves more spatial feature and more required functional information with no color distortion (advantage). However, HIS method involves only three (R, G, B) bands (disadvantage). More popular technique is a combination of HIS method with other fusion method (Section 6.6.9).

6.6.3 Multiplicative Technique (MT)

MT is a pixel-level fusion technique that combines two data sets by multiplying each pixel in each band of the multispectral data by the corresponding pixel of the panchromatic data. It is also grouped under the arithmetic methods—addition, subtraction, division, and multiplication. However, the only multiplication can preserve the color information. The MT fusion algorithm is based on the following relation:

$$I_f = \sqrt{PAN \times MS}, \qquad (6.47)$$

where I_f is the resulting fused image, PAN is the input panchromatic image, and MS is the each input band of multispectral image. Resampling of MS images to the spatial resolution of PAN image is an essential step in this fusion method; all data sets I_f, PAN and MS need to bring on the same pixel size. MT is simple and straightforward fusion method which works with big optical data such as IKONOS and QuickBird as well with multisensor data (advantage); however, it creates highly correlated bands that alter the spectral characteristics of original images (disadvantage).

6.6.4 PRINCIPAL COMPONENT ANALYSIS (PCA)

PCA is a statistical technique which transforms a number of correlated variables into a number of uncorrelated variables (called *principal components*). PCA was invented in 1901 by Karl Pearson. Since then, PCA has become very popular technique in data analysis and remote sensing. Books (Jolliffe 2002; Vidal et al. 2016; Naik 2018) are fairly comprehensive references giving many example applications including recent developments of PCA methods. Recent review (Jolliffe and Cadima 2016) gives the reader complete information about PCA techniques.

PCA reduces the dimensionality of multivariate datasets preserving as much of the relevant information as possible. The PCA is also known as *Karhunen–Loève* transform or the *Hotelling* transform. PCA is used extensively in image compression, image classification, and image fusion. In the fusion process, PCA generates a set of uncorrelated images $I_1, I_2, \ldots, I_n$, where n is the number of input multispectral bands. In the case of two spectral bands, the fused image represents a sum $I_f = P_1 I_1 + P_2 I_2$, where I_1, I_2 are input images and P_1, P_2 are normalized principle components $(P_1 + P_2 = 1)$ which are computed from the covariance matrix. PCA is standard pixel-level fusion algorithm available in many retail signal processing toolboxes.

6.6.5 BROVEY TRANSFORM (BT)

The Brovey Transform, also called the color normalization transform, is based on the chromaticity transform. It is a mathematical combination of the panchromatic and multispectral images, developed to avoid the disadvantages of the multiplicative method (MT). BT can be applied to fuse images with different spatial and spectral characteristics. The BT fusion involves three multispectral and panchromatic bands The BT mathematical formula is given by

$$\begin{pmatrix} R_f \\ G_f \\ B_f \end{pmatrix} = \frac{PAN}{I} \begin{pmatrix} R \\ G \\ B \end{pmatrix}, I = (R + G + B)/3, \tag{6.48}$$

where I, R, G, and B are pixel values of the original multispectral image, and PAN; R_f, G_f, B_f are the corresponding pixel values of the fused image. Before the fusion process a decorrelation process is performed to reduce the data redundancy. Preprocessing filtering can also be applied to reduce noise. The BT increases the contrast in the low and high ends of an image histogram (advantage). However, the BT method merges only three bands at a time (disadvantage). Successful application of the BT fusion may require additional information and experiences as well.

6.6.6 FREQUENCY FILTERING METHODS (FFM)

FFM includes several methods: high-pass filter additive method, high-frequency addition method, and high-frequency modulation method (e.g., Stathaki 2008). Fusion techniques in this group use filters and FFT to evaluate high-frequency components between PAN and MS images. Filtered by certain spatial domains images are fused. As a result, certain spatial features can be revealed much better in the fused image. The main problem is in optimization of high-passed filters. A subjective design of these filters may cause spectral degradations and false elements in the fused images. Therefore, this method requires additional specification of input data.

6.6.7 WAVELET-BASED FUSION (WTF)

WTF method is a pixel-feature-level fusion method which has been adopted for fusion from wavelet multiresolution (MRA) theory (Mallat 1998; Section 6.4). Wavelet MRA is the most common technique for the processing of multivariate data (He 2000; Petrosian and Meyer 2001; Barth et al. 2002; Ouahabi 2012). Wavelet yields rich scale-dependent and structural information in both spatial and frequency domains. In image processing, wavelet also provides a great enhancement, detalization, and scaling of image structural features. Moreover, DWT is used in the processing and fusion of hyperspectral and super-resolution satellite images (Chaudhuri and Kotwal 2013). Therefore, WTF seems to be very efficient and robust method for analysis of ocean multiband and/or multisensor imagery.

The information flowchart diagram of wavelet-based image fusion is shown in Figure 6.5. In this scheme, two source images I_1 and I_2 of the same size and resolution, are decomposed into the sub-images and the wavelet coefficients are computed using forward digital wavelet transform DWT. The corresponding sub-images are

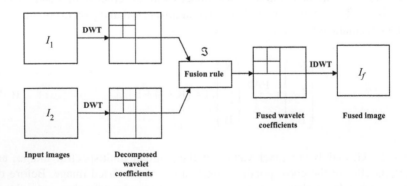

FIGURE 6.5 Flowchart diagram of wavelet-based image fusion.

fused by certain rule $\mathfrak{I}$. The result fused image I_f is obtained using the inverse digital wavelet transform IDWT. Mathematically, this process can be formulated as follows:

$$I_f = IDWT\Big[\,\mathfrak{I}\big\{DWT(I_1),DWT(I_2)\big\}\,\Big].\qquad(6.49)$$

The wavelet fusion rule can be chosen depending on complexity of source images. It can be pixel-based, window-based, average-based, PCA-based, or various statistical fusion rules (Petrosian and Meyer 2001).

6.6.8 PYRAMID

Pyramid is a pixel-feature-level fusion method based on WTF. Pyramid represents a source image as a set of multiscale images, which are fused according to pyramid structure at every scale layer (see Figure 6.3). Then the result image is reconstructed using inverse pyramid process. The implementation uses different approaches such as Gaussian, gradient, contrast, morphological, ratio-of-low-pass, and the other pyramids.

The most popular scheme is the *Laplacian pyramid* because of similarity of a Laplacian operator. Laplacian pyramid is derived from the Gaussian pyramid which is multiresolution image representation. Laplacian pyramid of an image is a set of band-pass filtered copies of an image. Each scale level of the pyramid is recursively constructed from its lower-level. For this the following steps are used: low-pass filtering, subsampling, interpolating, and differencing. The process of fusion of multiscale filtered image copies can be performed using different fusion rules; the simplest rule is AM fusion. Pyramid-based fusion provides considerable sharpening and enhancement of image texture features.

6.6.9 HYBRID FUSION

Hybrid fusion method integrates various fusion algorithms into one workflow. Hybrid method yields good results in the reconstruction of motion blurred and noisy low-resolution image into high-resolution perfect fused image. It is an option to improve information content and quality of multispectral optical data collected at dynamic environment. Most popular schemes of hybrid fusion are the following: PCA-WTF, BT-WTF, and HIS-BT. Hybrid fuzzy image processing methods are also used for selection and extraction of visually-hidden details in the original images.

6.6.10 EHLERS FUSION

Ehlers fusion is a variant of hybrid fusion. The Ehlers fusion was developed specifically for a spectral characteristics preserving image merging. It is based on the IHS transform coupled with FFT filtering (Klonus and Ehlers 2007; Ehlers et al. 2010). First, color and spatial information have to be separated. Second, the spatial information content has to be enhanced or suppressed using FFT. Then LP and HP filtering are applied correspondingly to intensity spectrum and panchromatic image spectrum. After that, panchromatic spectrum (PAN-HP) is transformed back to spatial domain using inverse IFFT and added together with intensity component (I-LP) to form new (I-LP)HS image. Finally, an inverse HIS transformation produces a fused multispectral RGB image. These steps can be repeated with successive three-band selections until all bands are fused with the panchromatic image. The Ehlers fusion shows the best spectral preservation but also the highest computation time.

6.6.11 ARTIFICIAL NEURAL NETWORK (ANN)

The ANN, also known as neural network (Haykin 1999), is a powerful and self-adaptive intelligent computer method of machine learning, pattern recognition, classification, prediction, and optimization systems. ANN methods are widely discussed in literature; a review can be found in the books (da Silva et al. 2017; Duval 2018). Since the late 1980s, these methods have been developed for various applications including geoscience and remote sensing (Benediktsson et al. 1990; Mas and Flores 2008; Haupt et al. 2009; Krasnopolsky 2013; Pohl and van Genderen 2017). In particular, the ANN has been found attractive to image fusion due to the ability to predict, analyze, and infer information from a given training data. Here is a brief description of the ANN fusion method.

ANN is feature-level fusion method. The basic idea is to define interconnected networks of simple units (called "artificial neurons") in which each connection has a weight. The networks have some inputs where the feature values are placed and they compute one or more output values. The learning takes place by adjusting the weights in the network so that the desired output is produced whenever a sample in the input data set is presented.

Figure 6.6 illustrates the ANN-based fusion diagram. The input layer has several neurons, which represent the feature factors extracted and normalized from image A and image B. The first step is the decomposition of two registered images into several blocks with size of M and N. The second step provides extraction of image features; thus, the normalized feature vector incident to neural networks is constructed.

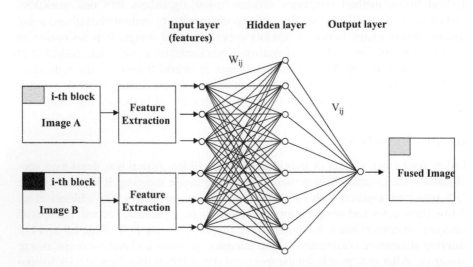

FIGURE 6.6 Artificial neural network (ANN) fusion diagram.

The third step is the selection of some vector samples to train neural networks. The ANN is a universal value function approximator that directly adapts to any non-linear function defined by a representative set of training data. Once trained, the ANN model can remember a functional relationship and be used for further calculations. For these reasons, the ANN concept has been adopted to develop strongly nonlinear models for multiple sensors data fusion. More detailed description of the ANN fusion models and algorithms can be found, e.g., in the book (Sumathi and Paneerselvam 2010).

Many researchers believe that ANN-based fusion methods have more advantages than other fusion methods, especially when input multisensor imaging data are incomplete or very noisy. The ANN is often used as an efficient decision level fusion tool for data/image classification. The ANN method provides the multiple inputs – multiple outputs capabilities framework that make it possible to fuse high dimension data such as big time-series data or hyperspectral data.

6.6.12 Applications

Fusion concept has been applied first to SAR-optical imagery conducted from the SeaSat, ALMAZ, ERS-1, and LANDSAT-4 satellite platforms (Tilley 1992; Tilley et al. 1992). Later, remote sensing data fusion is becoming of paramount importance in Earth Observations. The multivariate analysis and fusion have a long history of success in observations of the terrestrial and urban environments. Ocean remote sensing studies with date fusion are limited and include monitoring of sea surface temperature and salinity (Umbert et al. 2014, 2015), altimetry (Fok 2015), underwater topography (Pleskachevsky et al. 2011) and microwave radiometry (Raizer 2013). In particular, detection and recognition of ocean hydrodynamic features is still a challenging task because of lack of the corresponding multisensor data acquisition system. Meanwhile, modern data fusion technologies could provide capability requirements for this goal.

6.6.13 Concluding Note

Listed above fusion techniques are of primary interest for analysis of ocean remotely sensed data. Among a large number of the existing methods, IHS, PCA, and WTF are the most popular and well-respected algorithms in present. However, fast processing with HIS and PCA usually has a problem with color distortion and spectral degradation during fusion. WTF performs a better result minimizing these errors. But WTF has three serious disadvantages: shift sensitivity, poor directionality, and lack of phase information. To overcome these disadvantages, more complicated multiscale fusion algorithms such as Ridgelet, Curvelet, and Contourlet transforms have been developed and tested (e.g., Soman et al. 2010; Alparone et al. 2015; Addison 2017). Indeed, they improve performance result, but require more computational resources.

Another challenge is the information fusion processing of hyperspectral and multisensor satellite data. Multisensor data fusion has been considered in a number of books (Wald 2002; Klein 2004; Hall and McMullen 2004; Liggins et al. 2009; Blum and Liu 2006; Mitchell 2012; Fourati 2016; Chang and Bai 2018). As it follows from these and other literature sources, the ANN-based fusion seems to be one possible state-of-the-art digital technique to provide adequate joint analysis of multisensor data. The principal motivation for multisensor data fusion is to improve the quality of the information output in a process known as *synergy*. This problem is directly related to detection capability of ocean environments. More specifically, a multisensor synergy concept will be discussed in Chapter 8.

6.7   TEXTURE

Texture is an important ubiquitous cue in vision perception. Texture describes a myriad of spatial patterns. Texture has been explored by many researches in psychophysics, engineering, natural science, computer vision science, and computer graphics. Texture may be classified as being artificial or natural. Artificial textures consist of arrangements of fixed symbols (circles, squares, lines) or decorative motif placed against a neutral background. Such textures are widely used in art and design. Natural textures are images of natural scenes containing repetitive arrangements (patterns) of pixels. Examples include photographs of brick walls, terrazzo tile, sand, grass, and ocean waves. A classical album (Brodatz 1966) contains large collections of photographs of naturally occurring textures; many of them are used in texture studies and computer experiments in present.

Quantitative description and modeling of texture have started with the pioneering works published in the books (Lipkin and Rosenfeld 1970; Rosenfeld 1981). Later, more detailed investigations of texture, principles of texture analysis, computer graphics, and state-of-the-art computer vision algorithms and models have been developed for many applications. A number of books (Tomita and Tsuji 1990; Rao 1990; Gimel'farb 1999; Pietikäinen 2000; Ebert et al. 2003; Li 2009; Petrou and Sevilla 2006; Mirmehdi et al. 2008; Engler and Randle 2009; Mumford and Desolneux 2010; Haindl and Filip 2013; Heilbronner and Barrett 2014) provide valuable scientific information about texture and texture characteristics.

However, there is no universal mathematical definition for a texture. In the literature we can find different definitions of texture, formulated by different authors depending upon the particular application. One definition acceptable for remote sensing and our subject is the following: "Digital image texture is a spatial pseudo-regular fine structure of elements or pixels (called texels) having certain repetitive patterns in which elements or primitives are arranged according to a placement rule."

A major characteristic of texture is the repetition of pattern or a set of patterns over a region. Depending on the spatial distribution of texels, texture is divided on two major categories: regular (deterministic) and stochastic. In general, texture can be coarse, fine, smooth, granulated, spotted, rippled, swirl, regular, irregular, or linear. Texture can be investigated digitally using various image processing algorithms including FT, histogram thresholding, convolution filters, co-occurrence matrix, spatial autocorrelation, wavelets, fractals, etc. (Rosenfeld 1981; Petrou and Sevilla 2006; Gonzalez and Woods 2008). For example, at low and moderate wind, optical images comprise periodic-like textures. With increasing of wind, textures become more stochastic and represent complex field and patterns of variable geometry (see Figure 2.17).

6.7.1 Texture Analysis

Texture analysis is an important generic research area of machine vision. Texture analysis refers to a class of mathematical procedures and models that characterize the spatial variations within imagery as a means of extracting information. Mathematical procedures of texture analysis include statistical, geometrical, model-based methods and signal processing methods (Pratt 2007; Chen 2016). In machine vision, there are five domains of texture analysis: (1) texture classification, (2) texture segmentation, (3) shape from texture, (4) texture synthesis, and (5) compression. Let's consider briefly these subdivisions.

6.7.1.1 Texture Classification

Texture classification is a process of assigning an *unknown* texture class to a *known* set of texture class. This process involves two phases: the learning phase and the recognition phase. Correspondingly, two main classification methods known as *supervised* classification and *unsupervised* classification are distinguished. Supervised classification provides machine learning each known texture class as a training set. Unsupervised classification does not require prior knowledge and provides automatic discovering different classes from input texture. Another, more complicated method is *semi-supervised* classification with only partial prior knowledge being available.

Texture classification techniques are grouped up in five main groups in general, namely (1) structural, (2) statistical, (3) signal processing, (4) model-based stochastic, and (5) morphology-based methods. Among them, statistical and signal processing methods (2) and (3) are the most widely used because they can be directly applied onto any type of texture. The rest are not as widely used because they unable to estimate reliably parameters of complex texture features or they relatively new and require more complicated processing.

As a whole, texture classification consists of a number of state-of-the-art computer vision algorithms which enable to sort abundant random image data into more readily interpretable information. Image/texture digital classifiers are the most commonly used in interpretation and application of remotely sensed data (Tso and Mather 2009; Mather and Koch 2011). Comprehensive and major algorithms of classification and change detection expound in the book (Canty 2014).

6.7.1.2 Texture Segmentation

Texture segmentation is the most important and most usable procedure in image processing and texture analysis. The term "segmentation" means division into segments. In particular, texture segmentation provides partitioning digital image into a small set of disjoint regions with homogeneous texture properties. Segmentation is a commonly used term for identifying differences between particularly interesting and uninteresting objects as well as distinguishing foreground from background content. Results of segmentation are applied to further image processing, e.g., to automatic detection and recognition of geophysical features. Texture segmentation can also be supervised, unsupervised, or semi-supervised depending on a prior knowledge of texture properties. Supervised texture segmentation identifies and separates one or more regions that match certain texture classes. Unsupervised segmentation has to first recover different texture classes and then separating them into regions. Compared to the supervised case, the unsupervised segmentation is more flexible in remote sensing imagery but it generally uses more recourses. Segmentation of an image texture into homogeneous regions is very useful in a variety of applications of pattern recognition and machine learning (Bishop 2006).

While a number of segmentation methods and algorithms exist, they are constantly improved and developed for more efficient image processing. The most commonly used are the following: (1) Edge-based, (2) Thresholding-based, (3) Region-based, (4) Clustering-based, and (5) Graph-based. More specifically, they are summarized in book (Verma et al. 2013). Detailed review of these and other methods as well digital implementations can be found in books (Pratt 2007; Yoo 2004; Zhang 2006; Gonzalez and Woods 2008; Chen 2012).

6.7.1.3   Shape from Texture

Shape from texture is a procedure of estimating a 3D surface from 2D image. According to Gibson (1950), texture gradients are a potential source of optical information about the layout of surfaces in the environment. Since then, many different properties of optical texture have been identified that can be used to estimate various aspects of 3D surface structure. Recover the orientation of a textured 3D surface from a single image is based on the following assumptions: (1) the image projection is orthographic and texture is homogeneously distributed, (2) the 3D surface is approximately planar, (3) the 3D texels (i.e., texture elements or pixels) are small line segments, called needles and (4) the needles are distributed evenly in all orientations and evenly on the 3D surface. Shape from texture has been used in computer vision for 3D graphics, recovering true surface orientation, reconstructing surface shape, and inferring the 3D layout of objects (Lindeberg 1994). For instance, 3D objects on the sea surface can be modeled using shape from texture algorithms and real world optical imagery.

6.7.1.4   Texture Synthesis

Texture synthesis is a common computer graphics technique to create large textures from small texture samples. Texture synthesis is used for texture mapping in surface or scene rendering applications. A synthetic texture should differ from input texture samples, but should have perceptually identical texture characteristics. In computer vision, texture synthesis increases texture coherence providing superior image quality and performance. Compared to texture classification and segmentation, texture synthesis poses a challenge on texture analysis because it requires a more detailed texture description. Applications of texture synthesis include image editing, image completion, video synthesis, and computer animations (Magnenat-Thalmann and Thalmann 1987). Algorithms of texture synthesis use the following methods: (1) Pixel-based sampling, (2) Block sampling, (3) Multiresolution sampling, (4) Mosaic-based synthesis, (5) Markov-Gibbs random field, (6) Pyramided-based synthesis, (7) Cut-Primed Smart Copying. Most of synthesis approaches are based on Markov-Gibbs random field models of textures (Li 2009). A few recent algorithms designed for texture synthesis on 3D surfaces.

6.7.1.5 Compression

Compression is a way to reduce the amount of data to be received or processed. Texture compression versus image compression has important benefits. Texture compression reduces memory bandwidth usage and the storage requirements, increase texture cacheability and performance. Texture compression is an important tool for real-time texture mapping. The most known technique, called "Block truncation coding" has been used during the years to compress the resolution levels of a texture. Texture compression schemes can be applied for utilization of multiple remote sensing (optical, radar) images comprising high-complexity texture features. Detailed descriptions are beyond the scope of this book and we refer the reader to available literature sources.

6.7.2 TEXTURE MODELS

Texture models are designed to describe and measure the unique characteristics of a texture. Texture models have been proposed and studied by many authors during the past years and have ranged from simple pixel-based and/or region-based models to random field and neural network models. These developments motivated not only by dramatic innovations in computer graphics and video game industry, but also widespread scientific interest in digital image processing and computer vision. The review of recent studies in this field allows us to select several texture models suitable for analysis of ocean optical data. They are listed below.

6.7.2.1 Fourier Series

This is a simple image texture generator based on the formula

$$I(x,y) = \sum_{i=1}^{n} k_i \sin(a_i x + b_i y + \varphi_i), \qquad (6.50)$$

where $I(x,y)$ is the resulting image texture, k_i, a_i, b_i and φ_i are coefficients, and n is the number of terms in a sum. Textures are very representative visually at large n $(n > 10 \div 20)$. Using (6.50) it is also possible to generate stochastic textures, e.g., if coefficients φ_i are random. The Fourier series method works well for modeling of *non-cresting* quasi-periodic textures with different shapes, sizes, colors, and orientations. Examples are surface water waves, sand ripples, zebra skin, fence, and other quasi-periodic textures.

6.7.2.2 Markov Random Field Model

Markov Random Field (MRF) or Markov–Gibbs random field (MGRF) model, also known as an undirected graphical model, is the most prevalent tool for image and texture modeling; detailed reviews of the mathematical theory and model variants can be found in books (Brémaud 1998; Gimel'farb 1999; Winkler 2003; Rue and Held 2005; Li 2009; Blake et al. 2011). The MRF model is based on the following principle.

The image is called Markovian, if the probability distribution of gray levels g_i for each pixel X_i merely depends on the gray levels g^i of the neighboring pixels X^i. The *Hammersley–Clifford* theorem (1971) establishes the equivalence between MRF and Gibbs distributions. An MRF is characterized by its local property, whereas a Gibbs random field (GRF) is characterized by its global property. In this approach, the image is characterized by the MRF-GRF conditional probability density (Gimel'farb 1999; Li 2009)

$$P\left(X_i = g_i \middle| X^i = g^i\right) = \frac{1}{Z}\exp\left[-\frac{1}{T}U\left(g_i, g^i\right)\right], \tag{6.51}$$

$$Z = \sum_{g \in G} \exp\left[-U\left(g_i, g^i\right)\right], \tag{6.52}$$

$$U\left(g_i, g^i\right) = \sum_{c \in C} V_c\left(g_i, g^i\right), \tag{6.53}$$

where Z is normalized constant called the *partition function*, $U\left(g_i, g^i\right)$ is the *energy function* characterizing the MRF, and T is constant called the *temperature* which shall be assumed to be 1. The energy function $U\left(g_i, g^i\right)$ is a sum of *clique potential* $V_c\left(g_i, g^i\right)$ over all possible cliques C. The value $V_c\left(g_i, g^i\right)$ depends on the local clique configuration. A clique is a particular spatial configuration of pixels, in which all its members are statistically dependent of each other.

An MRGF is attractive for texture modeling because its properties are locally dependent, i.e., pixel values are only influenced by their neighbors. However, the MRF texture, viewed as only depending on its nearby pixels can be expand into a large image region. The number of pixels in a clique specifies the statistical order of a Gibbs distribution and the related texture model. Statistical dependence between pixels within a clique is also referred to pixel interaction. Texture modeling based on MGRF is mainly about to choose "appropriate potential functions for desired system behavior" (Li 2009). In order to create MGRF texture, it is necessary to compute the conditional probability (6.51). Explicit construction of such a probability density is often computationally intractable. The main problem is to estimate directly the energy function. This can be done using several options: sampler, maximum entropy principle, maximum likelihood estimation, Markov chain Monte Carlo algorithms, auto-regressive models, frame model, and others (Gimel'farb 1999; Li 2009; Haindl and Filip 2013).

MRF-based models have been successfully applied into texture analysis, texture classification and segmentation, texture synthesis, pattern reconstruction, modeling of motion textures, and microstructure modeling of heterogeneous materials as well. Hidden MRF models are also used with SAR for automatic target detection.

6.7.2.3 Fractal Models

Fractal-base texture models are employed in particular, to describe multiscale image textures having high degree of irregularity, complexity and self-similarity (Crilly et al. 1993;

Russ 1994; Kaye 1994; Haindl and Filip 2013). Therefore, a number of fractal models have been commonly used in geoscience and remote sensing to improve observability of complex natural scenes as well to provide a better feature selection and extraction.

Fractal textures are usually referred to two main categories of objects: regular and stochastic (also called random fractals). Regular fractal textures are well known and studied (Mandelbrot 1983). These textures are generated using a fixed geometric replacement rule. Typical examples of this category are the Cantor sets, fractal carpet, Sierpinski gaskets, Peano curve, Koch snowflakes, Harter-Heighway Dragon Curve, T-square, Menger sponge, and many others (see, e.g., pictures in album (Baird 2011)). Stochastic textures are modeled using different random-fractal-based methods and known statistical algorithms (Rosenfeld 1981; Russ 1994; Haindl and Filip 2013). Here we consider some of them which are of interest in remote sensing.

Fractal stochastic textures can be simulated using the following mathematical methods (in texture analysis they called *fractal descriptors*):

- Fourier fractal (FF)
- Fractional Gaussian noise (FGN)
- Fractional Brownian motion (FBm)
- Fractional Brownian motion by midpoint displacement (FBmMD)
- Weierstrass–Mandelbrot function (MW)
- Wavelet synthesis (WLS)
- Multifractal (MF).

The simplest and most commonly used model of a self-similar process is that of FBm (Mandelbrot 1983). In this case, stochastic texture can be characterized by covariance function

$$\mathrm{Cov}\left[B_H(\vec{r})B_H(\vec{r}')\right] \propto \sigma_H^2\left\{\|\vec{r}\|^{2H} + \|\vec{r}'\|^{2H} - \|\vec{r} - \vec{r}'\|^{2H}\right\}, \tag{6.54}$$

$$\sigma_H^2 = \left\langle\left[B_H(\vec{r} + \vec{\lambda}) - B_H(\vec{r})\right]^2\right\rangle \propto \|\lambda\|^{\alpha} \text{ with } \alpha = 2H, \tag{6.55}$$

where $B_H(\vec{r})$ a fractional Brownian function of zero-mean Gaussian nonstationary stochastic process, σ_H^2 is the variance, $\vec{r}$ and $\vec{r}'$ refer to random coordinate vectors, λ is the scaling factor, and H is the Hurst exponent that is directly related to the fractal dimension $D = d + 1 - H$. The Hurst exponent ranges from 0 to 1, taking the values $H = 0.5$, $H > 0.5$, and $H < 0.5$, respectively, for uncorrelated, correlated, and anti-correlated Brownian function (Section 6.5.2).

The power spectrum is given by $S_H \propto \|\omega\|^{-\beta}$ with $\beta = d + 2H$, ω is angular frequency. A number of objects $N_H(\varepsilon)$ of characteristic size ε needed to cover the fractal is $N_H(\varepsilon) \propto \varepsilon^{-D}$ with $D = d + 1 - H$ (d is the space dimension). Covariance function (6.54) describes stochastic textures depending on the value of fractal dimension D. Note that many natural stochastic textures have been modeled using fractal descriptors based on fractional Brownian function. Fractal models have also been developed for ocean microwave radiometry (Raizer 2012, 2017).

6.7.2.4 Mosaic Models

A mosaic texture model, investigated first in detail by Ahuja and Rosenfeld (1981) and Rosenfeld (1981), is a combination of geometrical texture fragments created using either structural approach or statistical approach. Correspondently, mosaic textures are divided on two types: deterministic textures and stochastic textures.

In structural mosaic model known also as cell structural model, coverage model, and bombing model, the texture is composed of primitive elements or texels, which appear in near regular repetitive spatial arrangements. The primitives (the simplest primitive is the pixel) can be chosen in the form of line segments, uniform polygons, or regular graphs in which each node is connected to its neighbors in an identical fashion. Each node corresponds to a cell in a tessellation of the plane. A structural approach "placement rule+texels" provides a modeling of homogeneous deterministic textures with distinct bordered features. Such textures are considered as "training textures" as a mean for analysis, controlling or matching of many other weakly homogeneous or inhomogeneous (borderless) textures close to stochastic configurations.

In the random mosaic models, which have more natural look, the texture region is first divided into convex polygonal cells and then every cell is independently assigned a class according to a fixed set of probabilities. There are a number of models and methods to create a random mosaic texture; most known are the following: (1) the Poisson line or Poisson polygon, (2) Voronoi Tessellation, (3) Auto-regressive models, (4) Checkboards, and (5) Delaunay triangulation.

According to Ahuja and Rosenfeld (1981), "Mosaic models…are much richer in variety, and each of these models is simpler to specify, as compared to the time series and random field model." The next generation of mosaic models is based on a structural-statistical generalization of the pure structural and statistical approaches through utilizations of Gaussian- and MRF-based models. They create a class of mixture models.

6.7.2.5 Mixture Models

Mixture texture models are the most realistic and efficient for analysis of natural scenes because they involve both deterministic and statistic texture components. In the past decade, state-of-the-art texture models have been introduced and integrated into computer vision and pattern recognition (Anwar et al. 2018; Moser and Zerubia 2018). Two classes of texture mixture models are considered: static and dynamic. The first class represents *probabilistic mixture textures models* (Haindl and Filip 2013) which are known as Gaussian-mixture, Bernoulli-distribution, and discrete-mixture texture models. The Gaussian-mixture model is the most popular; it is used for image classification and segmentation. The second class represents a number of dynamic texture models known as clustering, layered, kernel, and their combinations. In this case, Bayesian, Bayes, and other approaches are used for texture classification (e.g., Aggarwal 2015).

Dynamic texture models are complicated enough for the experimental verification but they seem promising for analysis of high-resolution ocean imagery. Eventually, the performance depends on many factors, first of all, the degree of *texturization* and spatial resolution of optical images. In any case, texture is manifested by locally variable light reflectance even if it is globally ununiformly illuminated. Therefore, texture analysis and modeling may provide a better assessment and interpretation of ocean optical imagery. Stochastic texture is also an important attribute of high-resolution passive microwave images of the ocean (Raizer 2005, 2017).

6.8 FEATURE EXTRACTION

An *image feature* is a distinguishing primitive characteristic or attribute of an image (Pratt 2007). In digital data processing and computer vision, feature extraction or texture feature extraction is of major importance in image analysis, texture analysis, segmentation, and classification. There are many methods and algorithms offering feature detection, selection, and extraction; several excellent books (Landgrebe 2003; Jähne 2004; Soille 2004; Guyon et al. 2006; Hoggar 2006; Pratt 2007; Gonzalez and Woods 2008; Maître 2008; Parker 2011; Szeliski 2011; Nixon and Aguado 2012; Goshtasby 2012; Camps-Valls et al. 2012; Umbaugh 2017) give the reader comprehensive knowledge and understanding of the subject.

In remote sensing image processing, the choice of feature extraction techniques depends on many factors, first of all, the quality, information content, and science specifics of imaging data to be analyzed. Image features can be classified into two categories: local or geometrical/structural and global or topological/statistical features. Correspondingly, feature extraction algorithms can be divided on morphological and statistical (Pratt 2007). We select and list important for us methods of feature extraction in order of their practical significance but not by generally accepted mathematical or statistical principles. Here they are.

6.8.1 BINARY THRESHOLDING

Binary thresholding (BT) is the simplest segmentation and feature extraction method. BT function (called also the *Otsu method*) divides the image data into two distinct classes. BT creates a binary image with two pixel classes: high-value class, displayed with white pixels, and a low-value class, displayed with black pixels. BT is nonlinear operation that converts a gray scale image into a binary image where the two levels are assigned to pixels those are below or above the specified threshold value. BT is used in many computer vision and image processing applications in the cases when objects to be extracted have approximately similar ranges of pixel value which differs from image background pixel value.

6.8.2 EDGE DETECTION

Edge detection (ED) is essentially the operation of detecting significant local changes in an image. Edges are local changes in the image intensity. Edges typically occur on a boundary between two image regions or between distinct objects and image background. An ED digital algorithm is based on computation of the intensity gradient in the edge neighborhood pixels. The location and orientation of the edge can be estimated with subpixel resolution using linear interpolation. Many ED approaches have been developed in the recent years. The most known edge detector are (1) Robert operator, (2) Sobel operator (3) Prewitt operator, (4) Laplacian operator, (5) Gaussian ED, (6) Canny ED, and (7) Gradient-based ED also called as masks in digital images. Object-ED methods and models are summarized, e.g., in book (Louban 2009). ED is a fundamental tool for image segmentation and local feature extraction. In ocean multispectral imagery, ED algorithms are useful for extraction of geometrical and texture features associated with strong surface currents, SST anomalies, color boundaries, sea-ice coverage, oil spills, or other distinct surface patterns.

6.8.3 FOURIER TRANSFORM

FT is standard image processing tool. In context with feature extraction, we mention two following methods: (1) Fourier descriptors, FDs, and (2) Digital Fourier-transform-based filter bank, DFTB.

- FDs are used for description of an object shape. Mathematical details and algorithms are discussed, e.g., in books (Jähne 2004; Burger and Burge 2013). This method is based on computations of coefficients of digital FT of a selected curve or single line; these coefficients are called Fourier descriptions. The object boundary line is represented as a periodic function, i.e., through the Fourier series that provides shape information. The FDs method captures both local and global features of shape; it is robust and information preserving.

- DFTB is a set of band-pass filters with either a common input or a summed output. It is multi-rate digital signal processing technique based on subband coding (Vaidyanathan 1993). The implementation includes decomposition of an input image into multiple levels and further subband representation using forward digital Fourier transform (DFT), selective HP and/or LP filtering, and reconstruction of output image using inverse digital Fourier transform (IDFT). The DFTB method can be applied for a better pansharpening of multispectral optical images (Dolecek 2018). It can also be used for selection and enhancement of relevant spectral features (e.g., spectral maxims) extracted from ocean optical imagery. DFTB is useful and powerful technique for speckle noise reduction of SAR images as well.

6.8.4 GABOR FILTER

2D Gabor filter (GF) is a quadrature filter pair. It represents a complex exponential-function modulated by a Gaussian. The 2D GF at spatial domain is written as

$$G_X(x,y) = g(x,y)\exp(\imath 2\pi u_0 x),\ G_Y(x,y) = g(x,y)\exp(\imath 2\pi v_0 y), \quad (6.56)$$

$$g(x,y) = (2\pi\sigma_x\sigma_y)^{-1}\exp\left[-\frac{1}{2}\left(\frac{x^2}{\sigma_x^2}+\frac{y^2}{\sigma_y^2}\right)\right], \quad (6.57)$$

where $G_{X,Y}(x,y)$ is the Gabor function, $g(x,y)$ is the Gaussian envelope with the spatial spread constants σ_x and σ_y at two {X,Y} directions, and u_0 and v_0 denote the corresponding radial frequencies of the Gabor function. The GF can be configured to have various shape, bandwidth, center frequency, and orientation by the adjusting of these parameters.

In remote sensing image processing, GFs play an important role in texture analyses, synthesis, segmentation and classification, feature extraction, pattern reconstruction, image coding, image representation and discrimination (Feichtinger and Strohmer 2003). 2D GFs are efficient and precise for detecting periodicity and orientation of a texture pattern. In particular, an appropriately designed GF provides decomposition and specification of multiscale surface wave patterns visible in high-resolution ocean optical images.

6.8.5 MARKOV RANDOM FIELD

MGRF is the most popular probabilistic method of image segmentation and texture modeling (Blake et al. 2011; Li 2009; see also Section 6.7.2.2). The application of MGRF to feature extraction is based on a statistical description of an image texture using the conditional probability distribution of pixels describing local spatial properties and interactions. The MGRF method evaluates the pixel level from the corresponding pixel level of its neighborhood that provides process of texture feature extraction in terms of statistical parameters in the space constrains of the image. The methodology includes several main operations: image labeling, clique matrix definition, probability measure, clustering of pixels with the same intensity, potential (MGRF) function definition (parameterization), optimization, and texture estimation.

MGRF techniques are of great interest in computer science, support vector machine, and pattern reconstruction. Some remote sensing studies (mostly with hyperspectral and SAR data) involve MGRF-based models for texture analysis, segmentation, classification, and image change detection (e.g., Varshney and Arora 2004; Tso and Mather 2009; Chen 2012). We believe also that MGRF approach is useful in extracting complex hydrodynamic information from ocean remote sensing imagery.

6.8.6 DIGITAL WAVELET TRANSFORM

Digital wavelet transform (DWT) is commonly used for feature extraction (see also Section 6.4 and references). DWT provides MRA of 2D input signal (an image) using different basis functions, most popular are Haar, Morlet, Mexican Hat, and Daubechies. The basic wavelet-based feature extraction procedure (known also as wavelet filter banks) consists of the following steps: (1) input original 2D signal, i.e., digitalized image, (2) choose wavelet function, (3) provide subband decomposition of the signal using DWT, (4) apply LP filter to get scaling functions and HP filter to get wavelet functions, (5) compute wavelet coefficients, (6) create a set of feature vectors of reduced dimension, and (7) apply classifier (and training). Some useful MATLAB codes can be found in books (Weeks 2011; Corinthios 2009; Stearns and Hush 2011).

Most advanced algorithms of feature extraction involve combinations of DWT with other statistical image processing techniques. These schemes are known as DWT-DCT (Discrete Cosine Transform), DWT-FB (Filter Bank); DWT-DFT, DWT-GF, DWT-PCA, DWT-GLSM, DWT-ANN, and integrated with SVM classifier. In particular, DWT provides the best extraction performance of relevant *spectral signatures* from fused multispectral (IKONOS) images with combinations DWT-DFT and DWT-GF (Chapter 7).

6.8.7 PRINCIPAL COMPONENT ANALYSIS

PCA is well-established feature extraction technique in data science (see also Section 6.6.4). The concept and motivation of PCA are explained very well elsewhere; nevertheless, several common recommendations concerning ocean optical imagery and PCA can be made. They are the following:

- PCA is unsupervised technique for dimension reduction; it may not capture the discrimination or selective information (while ANN and SVMs are the best-known supervised techniques).
- PCA performance strictly depends on pixel resolution of ocean imagery because of highly correlated properties of optical data.
- PCA has a better performance after preprocessing of original images (brightness and contrast adjustments).
- PCA can be organized as a multi-level processing, integrated with other known methods (PCA-DWT, PCA-SVD, PCA-MGRF, PCA-GLSM, PCA-ANN). The implementation requires statistical representativeness and the same level of sampling and quantization of input digitalized data.
- PCA has many modifications (e.g., Jolliffe 2002; Jolliffe and Cadima 2016); therefore, the choice of the optimal variant remains important in ocean remote sensing as well.
- Multiscale PCA (MSPCA) provides robustness analysis and interpretation of optical data.
- MSPCA is more efficient for detection of spatiotemporal *optical-microwave* signatures associated with ocean dynamics.

6.8.8 MOMENT-BASED FEATURE DESCRIPTORS

Moment-based feature descriptors (MBD) are derived from general moment theory introduced first by Hu in 1962. The utilization of MBD includes Moment Invariants, Rotation Moments, Orthogonal Moments, Complex Moments, and Standard Moments. Two books (Mukundan and Ramakrishnan 1998; Flusser et al. 2017) provide full review of this theory and the corresponding image analysis techniques. Here we refer just to 2D geometrical moments of order $(p+q)$ which are given by

$$m_{pq} = \iint\limits_{\Omega} x^p y^q f(x,y) \, dx \, dy, \ p,q = 0,1,2,3,\ldots, \tag{6.58}$$

where Ω is an image region or segmented object where the image intensity function $f(x,y)$ is defined. A set of moment values computed by (6.58) is used for characterization of image region/object properties such as shape, orientation, major principle axes, and area. The MBD provides their digitalized representation that is an important application in machine vision. In general, the use of MBD in image analysis is straightforward when binary or gray level image segment is considered. Moments provide characterization of an image segment (an image object) and extraction of its properties. In this context, MBD seems very useful and relatively simple computer method for global and local analysis of remote sensing images comprising complex geometrical features and/or patterns.

6.8.9 SINGULAR-VALUE DECOMPOSITION

Singular-value decomposition (SVD) is a technique came from linear algebra; it is one of the most important matrix decompositions (established in 1965). SVD is the process of factorization of a real or complex matrix (Jain 1989; Pratt 2007; Hoggar 2006; Skillicorn 2008). In data analysis, there is a very direct mathematical relationship between SVD and PCA (Jolliffe and Cadima 2016), see also Section 6.6.4.

According to SVD any arbitrary matrix $\mathbf{A}$ of dimension $n \times m$ and rank r can be transformed into three matrices:

$$\mathbf{A} = \mathbf{U}\Sigma\mathbf{V}^T, \mathbf{U}\mathbf{U}^T = \mathbf{U}^T\mathbf{U} = \mathbf{I}, \mathbf{V}\mathbf{V}^T = \mathbf{V}^T\mathbf{V} = \mathbf{I}, \tag{6.59}$$

where $\mathbf{U}$ is an $n \times r$ orthogonal matrix, $\mathbf{V}$ is an $r \times m$ orthogonal matrix, Σ is an $r \times r$ diagonal matrix, and the superscript $\mathbf{T}$ indicates matrix transpose. The columns of $\mathbf{V}$ are called the right singular vectors of $\mathbf{A}$ and the columns of $\mathbf{U}$ are called the left singular vectors of $\mathbf{A}$. The diagonal elements of matrix Σ are called the singular values of $\mathbf{A}$. The matrix $\mathbf{A}$ has the representation

$$A_{ij} = \sum_{k=1}^{r} U_{ik}\lambda_k V_{kj}, \tag{6.60}$$

where U_{ik} is the matrix characterizing the rows and V_{ik} is the matrix characterizing the columns. The r nonnegative nonincreasing entries (singular values) $\lambda_1 \geq \lambda_2 \geq \lambda_3, \ldots, \geq \lambda_r \geq 0$ appear in descending order along the main diagonal of matrix Σ. The numbers $\lambda_1^2 \geq \lambda_2^2 \geq \lambda_3^2, \ldots, \geq \lambda_r^2$, are the eigenvalues of $\mathbf{A}\mathbf{A}^T$ and $\mathbf{A}^T\mathbf{A}$. If matrix $\mathbf{A}$ represents an $n \times m$ image, then r singular values can be considered as extracted features from an image. SVD algorithms are considered in book (Canty 2014).

SVD is a powerful method of *dimension reduction of data* (Figure 6.7), image compression and enhancement of an image substructure. In remote sensing, SVD is used for increasing the brightness and contrast of multispectral satellite images (or selected image segments) and extraction of distinct features as well. In particular, the quality of low-contrast ocean optical images can be improved somehow using SVD. The result depends on illumination conditions. However, a combination of DFT and SVD provides considerable enhancements of generated 2D FFT image spectra that is an important issue in the extraction of spectral signatures (Chapter 7).

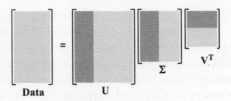

Data   **U**

FIGURE 6.7   Singular-value decomposition (SVD).

6.8.10 AUTOCORRELATION FUNCTION

Autocorrelation function (ACF) is fundamental characteristics of random processes and fields. In image processing, the ACF method provides estimating texture properties—fineness, coarseness and pattern repetitions (periodicity). The 2D ACF of an image is computed through relative shift of the position:

$$R_{ff}(x_0, y_0) \propto E\{f(x,y)f(x-x_0, y-y_0)\}, \qquad (6.61)$$

where $R_{ff}(x_0, y_0)$ is the autocorrelation function, $f(x,y)$ is the image intensity at position (x,y), x_0 and y_0 represent the distance (or lag) from the corresponding x and y position, and symbol E denotes the expected value operator.

The ACF descriptors are partially useful for detecting variations of texture parameters (size, shape, directionality) that play an important role in feature extraction process. For example, texture coarseness or the spatial size of textels can be estimated by the width of the ACF. However, ACF itself not always provides a criterion to distinguish among several texture regions because many different image ensembles can have the same ACF (Jain 1989). More adequate measure of texture features is the *moment-generating function* which is the exponential generating function of the moments of the probability distribution (Mukundan and Ramakrishnan 1998; Mirmehdi et al. 2008).

6.8.11 DECORRELATION STRETCH

Decorrelation stretch (DCS) is remote sensing image processing technique developed at Jet Propulsion Laboratory, NASA (Gillespie et al. 1986) in order to enhance (*stretch*) color contrast and reduce spectral band correlations in MS satellite image. It was used first for improving visual quality of the ASTER (the Advanced Spaceborne Thermal Emission and Reflectance Radiometer) imagery. It also has used for enhancing Mars Rover pictures. Today, DCS is a productive algorithm in many remote sensing applications (Mather and Koch 2011).

The DCS technique is based on applying the *Karhunen-Loeve* (KL) transform (formulated by Kari Karhunen, 1947 and Michel Loève, 1948) which is also known as the Hotelling transform. The KL transform is of fundamental importance in digital signal and image processing (e.g., Pratt 2007; Gonzalez and Woods 2008; Wang 2012). The KL transform is the optimal technique for input data decorrelation. The KL transform is defined as

$$U^T C U = \Lambda, \tag{6.62}$$

where U and U^T are normalized orthogonal eigenvectors and its transpose, C is covariance matrix of input data, and Λ is a diagonal matrix. The basic functions of the KL transform are orthogonal eigenvectors of the covariance matrix of input data.

The DCS produces transformation matrix that is applied to separate the color in the image. DCS supports several different color spaces. The image is converted from RGB to color space, the calculation and transformation are performed, and then the new colors are converted back to RGB image. Mathematically, the KL transform is closely related to PCA and SVD.

It is assumed, however, that color manipulations using DCS provide a better stretch performance than regular PCA or SVD methods due to the only optimal utilizing KL transform. An important application is remote sensing of variations in ocean colour scenes. The changes can be induced by many environmental factors. More specifically, it may occur under the influence of coupled thermo-hydrodynamic and biochemical (and/or *radiation*) surface effects associated with complex deep ocean events. We believe that such optical anomalies are detectable using *natural color satellite imagery* and relatively "simple" DCS processing.

6.8.12 CO-OCCURRENCE MATRIX

The gray-level co-occurrence matrix (GLCM) was introduced by Haralick et al. (1973) for texture classification. GLCM also is known as the gray-tone spatial-dependence matrix, the gray-level dependency matrix, and the concurrence matrix (Jain 1989; Pratt 2007; Gonzalez and Woods 2008). It is the most famous statistical approach for measuring spatial texture in an image. In this method texture can be seen as the spatial distribution and correlation between gray levels of an image. The 2D GLCM is defined by the join probabilities of gray levels $p(i, j | \Delta x, \Delta y)$, Δx and Δy are pixel offsets. It is the frequency of occurrence of pairs of pixels $p_{i,j}(d, \alpha)$ with gray levels i and j, separated by the pixel distanced (d) at a given direction $\alpha = 0°, 45°, 90°,$ and $135°$ (Haralick et al. 1973). The elements of the matrix, P_{ij}, are computed within the $L \times L$ neighborhood window

$$P_{ij} = \frac{p_{ij}(d, \alpha)}{\sum\limits_{i,j}^{L} p_{ij}(d, \alpha)}, \tag{6.63}$$

where the summation is over the total number of pixel pairs in the window. With N_g gray level in the image, the dimension of the matrix will be $N_g \times N_g$. To reduce computations and obtain reliable result, an input image is converted first to 8-bit gray level image $(N_g = 8)$; texture is analyzed within small fixed window of the size range from (3×3) to (9×9) pixels.

The GLCM contains a large amount of information about local texture features in an image. To characterize texture features, the following set of parameters (also known as Haralick features) is computed in a local small window: (1) angular second moment, (2) inverse difference moment, (3) contrast, (4) entropy, (5) correlation, (6) sum of squares, (7) sum of average and some others parameters (in total there is 20 Haralick texture features). Major parameters are shown in Table 6.2 (Schowengerdt 2007). It is assumed by many authors that the GLCM method has considerable advantage in the improvement of image classification accuracy and combined GLCM-PCA and GLCM-MRF approaches can provide a better feature extraction performance.

The GLCM technique is successfully used in many remote sensing applications, mostly for urban landscape classification with satellite imagery due to rich texture information of such data. Texture analysis with GLCM is often applied to SAR imagery of sea ice and oil pollutions. A few applications in ocean remote sensing can also be mentioned (e.g., Holyer and Peckinpaugh 1989; Jones et al. 2013; Harwikarya and Ramayanti 2016). We believe that GLCM as a relatively simple and powerful texture classificator enable detecting geophysical signatures from ocean optical images of high spatial resolution (IKONOS, QuickBirds, etc.). Texture variations and/or local texturization registered in the optical images can be initiated by a number of acting hydro-physical factors. Among them, we emphasize local fluctuations of wind, roughness change, turbulence, currents, wakes, fronts, sediments, and other weakly

---

TABLE 6.2

Major Gray-Level Co-occurrence Matrix (GLCM) Parameters

Name	Equation	Related Image Property
Angular second moment (ASM)	$f_1 = \sum_i \sum_j p_{ij}^2$	Homogeneity
Contrast	$f_2 = \sum_i \sum_j (i-j)^2 \, p_{ij}^2$	Semivariogram
Correlation	$f_3 = \dfrac{\sum_i \sum_j ijp_{ij} - \mu_1\mu_2}{\sigma_1\sigma_2}$	Covariance
Sun of squares	$f_4 = \sum_i \sum_j (i-\mu)^2 \, p_{ij}$	Variance
Inverse difference moment	$f_5 = \sum_i \sum_j \dfrac{p_{ij}}{1+(i-j)^2}$	
Sun average	$f_6 = \sum_{i=2}^{2^{Q+1}} ip_{x+y}$	
Entropy	$f_9 = -\sum_i \sum_j p_{ij} \log(p_{ij})$	

$p_x = \sum_j p_{ij}$ and $p_y = \sum_i p_{ij}$.

Source: After Schowengerdt (2007).

visible or "hidden" hydrodynamic events. Unfortunately, an appropriate data may not always available for scientific research.

6.8.13 Support Vector Machines

Support vector machines (SVMs) are a set of supervised statistical learning techniques which are used for classification, regression, and outlier detection. The original concept of SVMs was formulated in 1990 on the basis on statistical learning theory known also as Vapnik–Chervonenkis or VC theory (Vapnik 1995). Since that time, SVMs methods become popular because of its success a lot of classification and data mining problem. Mathematical details and recent developments are discussed in many books (e.g., Cristianini and Shawe-Taylor 2000; Schölkopf and Smola 2002; Steinwart and Christmann 2008; Alpaydin 2014).

SVMs algorithms separate high-dimensional data sets into discrete number of predefined classes using an optimal maximum margin *hyperplane* (Figure 6.8). The operation of the SVM algorithm is based on finding the hyperplane that gives the largest minimum distance, called margin, to the training data. The optimal separating hyperplane maximizes the margin of the training data. Vectors of margin lines (data points closest to margin) are called "support vectors." To construct an optimal hyperplane, SVM employs an iterative training algorithm, which is used to minimize an error function. The SVMs algorithms also use a set of mathematical functions that are defined as the kernel functions. The function of kernel is to take input multidimensional (independent) data and transform them into the required feature spaces. The process of rearranging input data is known as mapping (transformation). Different SVM algorithms use different types of kernel functions, e.g., linear, nonlinear, polynomial, radial basis function, and sigmoid. The *classification performance* of SVMs (or *kernel* machines) greatly depends on the choice of kernels and its parameters.

SVMs provide state-of-the-art accuracy in various classification tasks. Today, SVMs play increasingly important role in image segmentation, classification, feature

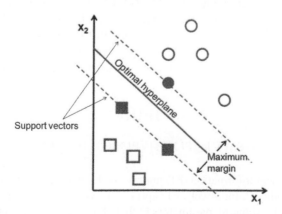

FIGURE 6.8 Support vector machine (SVM).

extraction, and pattern recognition in many scientific, engineering, and medicine, and military problems. The application of SVMs techniques in the geosciences and remote sensing area is fairly new and relatively limited (e.g., Camps-Valls and Bruzzone 2009; Tso and Mather 2009, Chen 2012). Some useful references can be found in review (Mountrakis et al. 2011). In more detail, geophysical applications of machine learning are discussed in the book (Kanevski et al. 2009).

The most important advantage of using SVMs is a generalization capability in classification and feature extraction within high-dimensional spaces. This task cannot be solved by any other methods. In comparison with ANN, the SVMs seem to be more advanced technique which may provide more accurate detection and classification of complex geophysical signatures from large bodies of satellite-based multispectral, hyperspectral, or multisensor remotely sensed data. However, the implementation of SVMs algorithms and their adaptation to current remote sensing problems (such as, e.g., environmental monitoring or target detection) require extensive software products and computational recourses.

6.9 FISHER VECTOR

The Fisher vector (FV), known also as Fisher kernel, was introduced in 1998 (Jaakkola and Haussler 1999; Cristianini and Shawe-Taylor 2000) and named after British statistician and geneticist R. Fisher (1890–1962). FV algorithms are widely used in information technology and computer vision, particularly in object recognition problems. The FV method provides state-of-the-art statistical classification and similarity analysis of *aggregative image features* (known also as Fisher encoding). For example, recently FV was successfully applied for classification of urban objects visible in SAR images. In ocean remote sensing, FV is used for ship detection by satellite optical images. Therefore, we include this novel method in our list due to potential efficiency and robustness of FV to provide detection and recognition of complex hydrodynamic features by optical data (although this application requires special consideration).

Briefly, the FV encodes the gradients of the log-likelihood of image features using a Gaussian-mixture model (GMM). GMM describes the distribution of features extracted from an image. Mathematical formulation is the following: (1) the generation process of X is modeled by a pdf $P(X|\Theta)$ with a set of parameters Θ, (2) the process X is described by the gradient of the log-likelihood function $U(X|\Theta) = \nabla \log P(X|\Theta)$ (called Fisher score), and (3) similarity between X and X′ is defined by the Fisher kernel $K(X,X')$.

$$K(X,X') = U_X^T I_\Theta^{-1} U_{X'}, \qquad (6.64)$$

$$I_\Theta := E_\Theta \left[\nabla \log P(X|\Theta) \nabla P(X|\Theta)^T \right] = E\left[UU^T \right], \qquad (6.65)$$

where I_Θ is the Fisher information (FI) matrix and E_Θ denotes expectation taken with respect to probability P. Intuitively, FI captures the variability of the gradient $\nabla \log P$ and thus allowing estimating parameters Θ that characterize image properties being explored.

As a whole, FI is used to construct hypothesis tests and confidence intervals using maximum likelihood estimators as well to define a default prior. In image processing, FI is used to measure image complexity and similarity. For example, FI provides detection and recognition of binary objects or binary patterns in the optical/radar images that is an important for us application. More about the concept of FI can be found in the book (Frieden and Gatenby 2007) and other books of one of the authors.

6.10 SUMMARY

Outlined methods of digital processing provide a wide range of possibilities for analysis and interpretation of remotely sensed data. We emphasize the importance of image enhancement, segmentation, classification, and feature extraction that is a critical issue for detection purposes. Under term "features" we understand not only local morphological or geometrical attributes visible in the image but also certain products of the processing, e.g., Fourier spectrum, fractal irregularity, wavelet map, fused pattern, or synthesized texture. In the case of multispectral optical imagery, these and other digitally measured features represent *computer vision signatures* which can be associated with certain geophysical phenomena. The choice of an appropriate algorithm and the implementation depends on the contents and quality of input data.

More advanced, synergy algorithm should involve both global and local image processing procedures including MRA and fusion as well. It is difficult to predict the performance of synergy algorithm without the corresponding multisensor databases. Some ideas are considered in the Chapter 7. Ultimately, synergy techniques may facilitate new prospective developments in automatic detection technology based on remote sensing observations.

Methods and algorithms of data/image processing are constantly improved and become more sophisticated to achieve higher level of understanding, interpretation, and application of remotely sensed data. Dramatically growing volume of the corresponding publications cover practically every topic of geosciences, remote sensing, and computer science including machine vision, detection, pattern reconstruction, and feature extraction. Eventually, among numerous techniques it is difficult to choose robust image processing algorithm without significant scientific experience and knowledge. In particular, nonacoustic detection technology requires integration of digital processing and modeling of remotely sensed data (Raizer 2017). The implementation is based on the state-of-the-art diversification methods and algorithms. Most of them were mentioned in this chapter.

REFERENCES

Abry, P., Goncalves, P., and Vehel, J. L. (Eds.). 2009. *Scaling, Fractals and Wavelets*. Wiley – ISTE, Hoboken, NJ.

Acharya, T. and Ray, A. K. 2005. *Image Processing: Principles and Applications*. John Wiley & Sons, Hoboken, NJ.

Addison, P. S. 1997. *Fractals and Chaos: An Illustrated Course*. Institute of Physics Publishing, Bristol, UK.

Addison, P. S. 2017. *The Illustrated Wavelet Transform Handbook: Introductory Theory and Applications in Science, Engineering, Medicine and Finance*, 2nd edition. CRC Press, Boca Raton, FL.

Aggarwal, C. C. (Ed.). 2015. *Data Classification: Algorithms and Applications*. CRC Press, Boca Raton, FL.

Ahuja, N. and Rosenfeld, A. 1981. Mosaic models for textures. *IEEE Transactions on Pattern Analysis and Machine Intelligence, PAMI-3*, 1:1–11.

Akansu, A. N. and Haddad, P. A. 2001. *Multiresolution Signal Decomposition: Transforms, Subbands, and Wavelets*, 2nd edition. Academic Press, San Diego, CA.

Alparone, L., Aiazzi, B., Baronti, S., and Garzelli, A. 2015. *Remote Sensing Image Fusion (Signal and Image Processing of Earth Observations)*. CRC Press, Boca Raton, FL.

Alpaydin, E. 2014. *Introduction to Machine Learning*, 3rd edition. The MIT Press, Cambridge, MA.

Altaisky, M. V. 2005. *Wavelets: Theory Applications, Implementation*. Universities Press (India) Private Limited, Hyderabad, India.

Anwar, Md. I., Khosla, A., and Kapoor, R. 2018. *Handbook of Research on Advanced Concepts in Real-Time Image and Video Processing*. IGI Global, Hershey, PA.

Arvelyna, Y. and Oshima, M. 2004. Application of wavelet transform for internal wave detection in SAR image. *International Journal of Remote Sensing and Earth Sciences*, 1(1):58–65.

Arvelyna, Y., Oshima, M., and Sumimoto, T. 2006. The application of wavelet transform for oceanic feature detection in SAR and optical image. In *Proceedings SPIE Volume 6357, Sixth International Symposium on Instrumentation and Control Technology: Signal Analysis, Measurement Theory, Photo-Electronic Technology, and Artificial Intelligence*, October 13–16, 2006, Beijing, China, Vol. 6357, pp. 63571P1–63571P4. doi:10.1117/12.716993.

Astaf'eva, N. M. 1996. Wavelet analysis: Basic theory and some applications. *Physics – Uspekhi*, 39(11):1085–1108 (translated from Russian). doi:10.1070/PU1996v039n11ABEH000177.

Baird, E. 2011. *Alt.Fractals: A Visual Guide to Fractal Geometry and Design*. Chocolate Tree Books, Brighton, UK.

Bandt, C., Barnsley, M., Devaney, R., Falconer, K. J., Kannan, V., and Vinod Kumar, P. B. (Eds.). 2014. *Fractals, Wavelets, and their Applications: Contributions from the International Conference and Workshop on Fractals and Wavelets*. Springer International Publishing, Switzerland.

Barenblatt, G. I. 1996. *Scaling, Self-Similarity, and Intermediate Asymptotics*. Cambridge University Press, Cambridge, UK.

Barnsley, M. F. 2012. *Fractals Everywhere: New Edition*. Dover Publications, Mineola, New York.

Barth, T. J., Chan, T., and Haimes, R. (Eds.). 2002. *Multiscale and Multiresolution Methods: Theory and Applications*. Springer, Berlin, Germany.

Barton, C. C. and La Point, P. R. (Eds.). 1995. *Fractals in the Earth Sciences*. Springer Science+Business Media, New York.

Benediktsson, J. A. and Ghamisi, P. 2015. *Spectral-Spatial Classification of Hyperspectral Remote Sensing Images*. Artech House, Boston, MA.

Benediktsson, J., Swain, P. H., and Ersoy, O. K. 1990. Neural network approaches versus statistical methods in classification of multisource remote sensing data. *IEEE Transactions on Geoscience and Remote Sensing*, 28(4):540–552.

Benelli, G. and Garzelli, A. 1999. Oil-spills detection in SAR images by fractal dimension estimation. In *Proceedings of International Geoscience and Remote Sensing Symposium*, Hamburg, Germany, June 28–July 2, 1999, Vol. 1, pp. 218–220. doi:10.1109/IGARSS.1999.773452.

Bernardara, P., Lang, M., Sauquet, E., Schertzer, D., and Tchiriguyskaia, I. 2007. *Multifractal Analysis in Hydrology: Application to Time Series*. Electronic Book. Éditions Quae, Versailles, Cedex.

Bishop, C. M. 2006. *Pattern Recognition and Machine Learning*, 2nd edition. Springer, New York.

Blackledge, J. M. 2005. *Digital Image Processing: Mathematical and Computational Methods*. Woodhead Publishing, Chichester, UK.

Blake, A., Kohli, P., and Rother, C. (Eds.). 2011. *Markov Random Fields for Vision and Image Processing*. The MIT Press, Cambridge, MA.

Blum, R. S. and Liu, Z. (Eds.). 2006. *Multi-Sensor Image Fusion and Its Applications*. CRC Press, Boca Raton, FL.

Bloomfield, P. 2000. *Fourier Analysis of Time Series: An Introduction*, 2nd edition. Wiley, New York.

Boggess, A. and Narcowich, F. J. 2009. *A First Course in Wavelets with Fourier Analysis*, 2nd edition. John Wiley & Sons, Hoboken, NJ.

Bovik, A. C. 2009. *The Essential Guide to Image Processing*, 2nd edition. Academic Press – Elsevier, Burlington, MA.

Bracewell, R. 2000. *The Fourier Transform & Its Applications*, 3rd edition. McGraw-Hill, New York.

Brémaud, P. 1998. *Markov Chains: Gibbs Fields, Monte Carlo Simulation, and Queues*. Springer, New York.

Brigham, E. 1988. *Fast Fourier Transform and Its Applications*. Prentice Hall, Englewood Cliffs, NJ.

Brodatz, P. 1966. *Textures: A Photographic Album for Artists and Designers*. Dover Publications, New York.

Broughton, S. A. and Bryan, K. 2018. *Discrete Fourier Analysis and Wavelets: Applications to Signal and Image Processing*, 2nd edition. John Wiley & Sons, Hoboken, NJ.

Burger, W. and Burge, M. J. 2013. *Principles of Digital Image Processing: Advanced Methods*. Springer, London, UK.

Burrus, C. S., Gopinath, R. A., and Guo, H. 1998. *Introduction to Wavelets and Wavelet Transforms. A Primer*. Prentice Hall, Upper Saddle River, NJ.

Calvet, L. E. and Fisher, A. J. 2008. *Multifractal Volatility: Theory, Forecasting, and Pricing*. Elsevier – Academic Press Advanced Finance, Burlington, MA.

Camps-Valls, G. and Bruzzone, L. (Eds.). 2009. *Kernel Methods for Remote Sensing Data Analysis*. John Wiley & Sons, Chichester, UK.

Camps-Valls, G., Tuia, D., Gómez-Chova, L., Jiménez, S., and Malo, J. 2012. *Remote Sensing Image Processing*. Morgan & Claypool Publishers, San Rafael, CA.

Camussi, R. and Guj, G. 1997. Orthonormal wavelet decomposition of turbulent flows: Intermittency and coherent structures. *Journal of Fluid Mechanics*, 348:177–199.

Canty, M. J. 2014. *Image Analysis, Classification and Change Detection in Remote Sensing: With algorithms for ENVI/IDL and Python*, 3rd edition. CRC Press, Boca Raton, FL.

Carbone, A. 2007. Algorithm to estimate the Hurst exponent of high-dimensional fractals. *Physical Review E*, 76(5) 056703. doi:10.1103/PhysRevE.76.056703.

Castleman, K. R. 1996. *Digital Image Processing*, Prentice Hall, Upper Saddle River, NJ.

Chan, A. K. and Peng, C. 2003. *Wavelets for Sensing Technologies*. Artech House, Norwood, MA.

Chandrasekhar, E., Dimri, V. P., and Gadre, V. M. (Eds.). 2014. *Wavelets and Fractals in Earth System Sciences*. CRC Press, Boca Raton, FL.

Chang, N.-B. and Bai, K. 2018. *Multisensor Data Fusion and Machine Learning for Environmental Remote Sensing*. CRC Press, Boca Raton, FL.

Chanussot, J., Collet, C., and Chehdi, K. 2009. *Multivariate Image Processing*. John Wiley & Sons – ISTE, Hoboken, NJ.

Chaudhuri, S. and Kotwal, K. 2013. *Hyperspectral Image Fusion.* Springer, New York.

Chen, C. H. (Ed.). 2012. *Signal and Image Processing for Remote Sensing,* 2nd edition. CRC Press, Boca Raton, FL.

Chen, C. H. (Ed.). 2016. *Handbook of Pattern Recognition and Computer Vision,* 5th edition. World Scientific Publishing, Singapore.

Christensen, O. 2004. *Approximation Theory, from Taylor Polynomials to Wavelets.* Birkhäuser, Boston, MA.

Chui, C. K. 1992. *An Introduction to Wavelets.* Academic Press, San Diego, CA.

Cohen, A. and Ryan, R. D. 1995. *Wavelets and Multiscale Signal Processing (Applied Mathematics and Mathematical Computation 11).* Springer, New York.

Corinthios, M. 2009. *Signals, Systems, Transforms, and Digital Signal Processing with MATLAB.* CRC Press, Boca Raton, FL.

Crilly, A. J., Earnshaw, R., and Jones, H. (Eds.). 1993. *Applications of Fractals and Chaos: The Shape of Things.* Springer, Berlin, Germany.

Cristianini, N. and Shawe-Taylor, J. 2000. *An Introduction to Support Vector Machines and Other Kernel-Based Learning Methods.* Cambridge University Press, Cambridge, UK.

da Silva, I. N., Spatti, D. H., Flauzino, R. A., Bartocci Liboni, L. H., and dos Reis Alves, S. F. 2017. *Artificial Neural Networks: A Practical Course.* Springer, Cham, Switzerland.

Daubechies, I. 1992. *Ten Lectures on Wavelets.* Society for Industrial and Applied Mathematics (SIAM), Philadelphia, PA.

de Jong, S. M. and van der Meer, F. D. (Eds.). 2006. *Remote Sensing Image Analysis: Including the Spatial Domain.* Springer, Dordrecht, the Netherlands.

Debnath, L. and Shah, F. A. 2015. *Wavelet Transformation and Their Applications,* 2nd edition. Birkhäuser, Springer, New York.

Dekking, M., Lévy-Véhel, J., Lutton, E., and Tricot, C. (Eds.). 1999. *Fractals: Theory and Applications in Engineering.* Springer, Berlin, Germany.

Dimri, V. P. (Ed.). 2000. *Application Fractals Earth Science.* CRC Press, Boca Raton, FL.

Dimri, V. P. (Ed.). 2016. *Fractal Solutions for Understanding Complex Systems in Earth Sciences.* Springer, Switzerland.

Dolecek, G. J. (Ed.). 2018. *Advances in Multirate Systems.* Springer, Switzerland.

Donelan, M. A., Drennan, W. M., and Magnusson, A. K. 1996. Nonstationary analysis of the directional properties of propagating waves. *Journal of Physical Oceanography,* 26(9):1901–1914.

Du, Y., Vachon, P. W., and Wolfe, J. 2002. Wind direction estimation from SAR images of the ocean using wavelet analysis. *Canadian Journal of Remote Sensing,* 28(3):498–509.

Duval, F. 2018. *Artificial Neural Networks: Concepts, Tools and Techniques Explained for Absolute Beginners.* CreateSpace Independent Publishing Platform.

Ebert, D. S., Musgrave, F. K., Peachey, D., Perlin, K., and Worley, S. 2003. *Texturing & Modeling: A Procedural Approach,* 3rd edition. Morgan Kaufmann Publishing, San Francisco, CA.

Edgar, G. A. 1998. *Integral, Probability, and Fractal Measures.* Springer, New York.

Ehlers, M., Klonus, S., Astrand, P. J., and Rosso, P. 2010. Multi-sensor image fusion for pansharpening in remote sensing. *International Journal of Image and Data Fusion,* 1(1):25–45.

Elsayed, M. 2010. An overview of wavelet analysis and its application to ocean wind waves. *Journal of Coastal Research,* 26(3):535–540.

Engler, O. and Randle, V. 2009. *Introduction to Texture Analysis: Macrotexture, Microtexture, and Orientation Mapping,* 2nd edition. CRC Press, Boca Raton, FL.

Erlebacher, G., Hussaini, M. Y., and Jameson, L. M. (Eds.). 1996. *Wavelets: Theory and Applications.* Oxford University Press, New York.

Everson, R., Sirovich, L., and Sreenivasan, K. R. 1990. Wavelet analysis of the turbulent jet. *Physics Letters A,* 145(6–7):314–322.

Evertsz, C. J. G. and Mandelbrot, B. B. 1992. Multifractal measures. In *Chaos and Fractals: New Frontiers of Science* (Eds. H. Peitgen, H. Jurgens, and D. Saupe). Springer, New York, pp. 921–953.

Falconer, K. 2003. *Fractal Geometry: Mathematical Foundations and Applications*, 2nd edition. John Wiley & Sons, Hoboken, NJ.

Farge, M. 1992. Wavelet transforms and their application to turbulence. *Annual Review of Fluid Mechanics*, 24(1):395–457.

Farge, M., Kevlahan, N., Perrier V., and Goirand, E. 1996. Wavelets and turbulence. *Proceedings of the IEEE*, 84(4):639–669.

Farge, M., Schneider, K., and Kevlahan, N. 1999. Non-Gaussianity and coherent vortex simulation for two-dimensional turbulence using an adaptive orthogonal wavelet basis. *Physics of Fluids*, 11(8):2187–2201.

Farmer, M. E. 2014. *Application of Chaos and Fractals to Computer Vision*. Bentham Science Publishers, Sharjah, United Arab Emirates.

Feichtinger, H. G. and Strohmer, T. (Eds.). 2003. *Advances in Gabor Analysis*. Birkhäuser, Boston, MA.

Feldman, D. P. 2012. *Chaos and Fractals: An Elementary Introduction*. Oxford University Press, Oxford, UK.

Fichaux, N. and Ranchin, T. 2002. Combined extraction of high spatial resolution wind speed and wind direction from SAR images: A new approach using wavelet transform. *Canadian Journal of Remote Sensing*, 28(3):510–516.

Flusser, J., Suk, T., and Zitov, B. 2017. *2D and 3D Image Analysis by Moments*. John Wiley & Sons, Chichester, UK.

Fok, H. S. 2015. Data fusion of multisatellite altimetry for ocean tides modelling: A spatio-temporal approach with potential oceanographic applications. *International Journal of Image and Data Fusion*, 6(3):232–248.

Foufoula-Georgiou, E. and Kumar, P. (Eds.). 1994. *Wavelets in Geophysics*. Academic Press, San Diego, CA.

Fourati, H. (Ed.). 2016. *Multisensor Data Fusion: From Algorithms and Architectural Design to Applications*. CRC Press, Boca Raton, FL.

Franceschetti, G. and Riccio, D. 2007. *Scattering, Natural Surfaces, and Fractals*. Elsevier, Academic Press, San Diego, CA.

Frieden, B. R. and Gatenby, R. A. (Eds.). 2007. *Exploratory Data Analysis Using Fisher Information*. Springer, Berlin, Germany.

Fujimoto, S. and Rinoshika, A. 2015. Wavelet multi-resolution analysis on turbulent wakes of asymmetric bluff body. *International Journal of Mechanical Sciences*, 92:121–132.

Fujimoto, S. and Rinoshika, A. 2017. Multi-scale analysis on wake structures of asymmetric cylinders with different aspect ratios. *Journal of Visualization*, 20(3):519–533.

Gade, M. and Redondo, J. M. 1999. Marine pollution in European coastal waters monitored by the ERS-2 SAR: A comprehensive statistical analysis. In *Proceedings of Geoscience and Remote Sensing Symposium*, June 28–July 2, Hamburg, Germany, Vol. 2, pp. 1375–1377. doi:10.1109/IGARSS.1999.774635.

Gao, R. X. and Yan, R. 2011. *Wavelets: Theory and Applications for Manufacturing*. Springer, New York.

Gao, J., Cao, Y., Tung, W.-W., and Hu, J. 2007. *Analysis of Complex Time Series: Integration of Chaos and Random Fractal Theory, and Beyond*. John Wiley & Sons, Hoboken, NJ.

Ghanbarian, B. and Hunt, A. G. (Eds.). 2017. *Fractals: Concepts and Applications in Geosciences*. CRC Press, Boca Raton, FL.

Gibson, J. J. 1950. *The Perception of the Visual World*. Haughton Mifflin, Boston, MA.

Gillespie, A. R., Kahle, A. B., and Walker, R. E. 1986. Color enhancement of highly correlated images. I. Decorrelation and HSI contrast stretches. *Remote Sensing of Environment*, 20(3):209–235. doi:10.1016/0034-4257(86)90044-1.

Gimel'farb, G. L. 1999. *Image Textures and Gibbs Random Fields*. Kluwer Academic Publishers, Dordrecht, the Netherlands.

Glover, D. M., Doney, S. C., Oestreich, W. K., and Tullo, A. W. 2018. Geostatistical analysis of mesoscale spatial variability and error in SeaWiFS and MODIS/Aqua global ocean color data. *Journal of Geophysical Research: Oceans*, 123(1): doi: 10.1002/2017JC013023.

Gonzalez, R. C. and Woods, R. E. 2008. *Digital Image Processing*, 3th edition. Pearson Prentice Hall, Upper Saddle River, NJ.

Goovaerts, P. 1997. *Geostatistics for Natural Resources Evaluation*. Oxford University Press, New York.

Goshtasby, A. A. 2012. *Image Registration: Principles, Tools and Methods*. Springer, New York.

Grizzi, S. and Camussi, R. 2012. Wavelet analysis of near-field pressure fluctuations generated by a subsonic jet. *Journal of Fluid Mechanics*, 698:93–124.

Gulick, D. 2012. *Encounters with Chaos and Fractals*, 2nd edition. CRC Press, Boca Raton, FL.

Gurley, K. and Kareem, A. 1999. Applications of wavelet transforms in earthquakes, wind and ocean engineering. *Engineering Structures*, 21:149–167.

Guyon, I., Gunn, S., Nikravesh, M., and Zadeh, L. A. (Eds.). 2006. *Feature Extraction: Foundations and Applications*. Springer, Berlin, Germany.

Haindl, M. and Filip, J. 2013. *Visual Texture: Accurate Material Appearance Measurement, Representation and Modeling*. Springer, London, UK.

Hall, D. L. and McMullen, S. A. H. 2004. *Mathematical Techniques in Multisensor Data Fusion*. Artech House, Norwood, MA.

Halsey, T., Jensen, M., Kadanoff, L., Procaccia, I., and Shraiman, B. I. 1986. Fractal measures and their singularities: The characterization of strange sets. *Physical Review A*, 33(2):1141–1151.

Haralick, R. M., Shanmugan, K., and Dinstein, I. 1973. Textural features for image classification. *IEEE Transaction on System, Man, and Cybernetics*, SMC-3(6):610–621. doi: 10.1109/TSMC.1973.4309314.

Harte, D. 2001. *Multifractals: Theory and Applications*. Chapman & Hall/CRC Press, Boca Raton, FL.

Harwikarya and Ramayanti, D. 2016. Study on calibration of sea wave height using synthetic aperture radar image based on grey level co-occurrence matrix: Case on multi polarization P band. *International Research Journal of Computer Science*, 3(1):1–7.

Hastings, H. M. and Sugihara, G. 1993. *Fractals: A User's Guide for the Natural Sciences*. Oxford University Press, Oxford, New York.

Haung, M. C. 2004. Wave parameters and functions in wavelet analysis. *Ocean Engineering*, 31:111–125.

Haupt, S. E., Pasini, A. P., and Marzban, C.(Eds.). 2009. *Artificial Intelligence Methods in the Environmental Sciences*. Springer, Dordrecht, the Netherlands.

Haykin, S. 1999. *Neural Networks. A Comprehensive Foundation*. Prentice Hall, Upper Saddle River, NJ.

He, T.-X. (Ed.). 2000. *Wavelet Analysis and Multiresolution Methods*. CRC Press, Boca Raton, FL.

Heilbronner, R. and Barrett, S. 2014. *Image Analysis in Earth Sciences: Microstructures and Textures of Earth*. Springer, Berlin, Germany.

Hernandez, E. and Weiss, G. A. 1997. *A First Course on Wavelets*. CRC Press, Boca Raton, FL.

Hoggar, S. G. 2006. *Mathematics of Digital Images: Creation, Compression, Restoration, Recognition*. Cambridge University Press, Cambridge, UK.

Holyer, R. J. and Peckinpaugh, S. H. 1989. Edge detection applied to satellite imagery of the oceans. *IEEE Transactions on Geoscience and Remote Sensing*, 27(1):46–56.

Indrusiak, M. A. S. and Möller, S. V. 2011. Wavelet analysis of unsteady flows: Application on the determination of the Strouhal number of the transient wake behind a single cylinder. *Experimental Thermal and Fluid Science*, 35(2):319–327.

Isaaks, E. H. and Srivastava, R. M. 1989. *An Introduction to Applied Geostatistics*. Oxford University Press, New York.

Jaakkola, T. and Haussler, D. 1999. Exploiting generative models in discriminative classifiers. In *Advances in Neural Information Processing Systems 11* (Eds. M. S. Kearns, S. A. Solla, and D. A. Cohn). The MIT Press, Cambridge, MA, pp. 487–493.

Jaffard, S., Meyer, Y., and Ryan, R. D. 2001. *Wavelets: Tools for Science & Technology*. Society for Industrial and Applied Mathematics (SIAM), Philadelphia, PA.

Jähne, B. 2004. *Practical Handbook on Image Processing for Scientific Application*, 2nd edition. CRC Press, Boca Raton, FL.

Jain, A. K. 1989. *Fundamentals of Digital Image Processing*. Prentice Hall, Upper Saddle River, NJ.

Jayaraman, S., Esakkirajan, S., and Veerakumar, T. 2009. *Digital Image Processing Paperback*. Tata McGraw-Hill Education Private Limited, New Delhi, India.

Jensen, J. R. 2015. *Introductory Digital Image Processing: A Remote Sensing Perspective*, 4th edition. Pearson Education, Glenview, IL.

Jolliffe, I. T. 2002. *Principal Component Analysis*, 2nd edition. Springer, New York.

Jolliffe, I. T. and Cadima, J. 2016. Principal component analysis: A review and recent developments. *Philosophical Transactions. Series A, Mathematical, Physical, and Engineering Sciences*, 374(2065) 20150202. doi:10.1098/rsta.2015.0202.

Jones, C. T., Sikora, T. D., Vachon, P. W., Wolfe, J., and DeTracey, B. 2013. Automated discrimination of certain brightness fronts in RADARSAT-2 images of the ocean surface. *Journal of Atmospheric and Oceanic Technology*, 30(9):2203–2215. doi:10.1175/JTECH-D-12-00190.1.

Kaiser, G. 1994. *A Friendly Guide to Wavelets*. Birkhäuser, Boston, MA.

Kanevski, M., Pozdnukhov, A., and Timonin, V. 2009. *Machine Learning for Spatial Environmental Data: Theory, Applications, and Software*. EPFL Press – CRC Press, Boca Raton, FL.

Kanjir, U., Greidanus, H., and Oštir, K. 2018. Vessel detection and classification from spaceborne optical images: A literature survey. *Remote Sensing of Environment*, 207:1–26. doi:10.1016/j.rse.2017.12.033.

Kaye, B. 1994. *A Random Walk through Fractal Dimensions*, 2nd edition. VCH Publishers, New York.

Kedem, B., De Oliveira, V., and Sverchkov, M. 2017. *Statistical Data Fusion*. World Scientific Publishing Company, Hackensack, NJ.

Khujadze, G., Nguyen van Yen, R.,Schneider, K., Oberlack, M., and Farge, M. 2011. Coherent vorticity extraction in turbulent boundary layers using orthogonal wavelets. *Journal of Physics: Conference Series*, 318(2) 022011. doi:10.1088/17426596/318/2/022011.

Klein, L. A. 2004. *Sensor and Data Fusion: A Tool for Information Assessment and Decision Making*. SPIE Press, Bellingham, WA.

Klonus, S. and Ehlers, M. 2007. Image fusion using the Ehlers spectral characteristics preserving algorithm. *GIScience & Remote Sensing*, 44(2):93–116.

Körner, T. W. 1988. *Fourier Analysis*. Cambridge University Press, Cambridge, UK.

Krasnopolsky, V. M. 2013. *The Application of Neural Networks in the Earth System Sciences: Neural Network Emulations for Complex Multidimensional Mappings*. Springer, Dordrecht, the Netherlands.

Kumar, D., Arjunan, S. P., and Aliahmad, B. 2017. *Fractals: Applications in Biological Signalling and Image Processing*. CRC Press, Boca Raton, FL.

Kuo, J. M. and Chen, K.-S. 2003. The application of wavelets correlator for ship wake detection in SAR images. *IEEE Transaction on Geoscience and Remote Sensing*, 41(6):1506–1511.

Lam, N. and De Cola, L. 2002. *Fractals in Geography*. The Blackburn Press, Caldwell, NJ.

Landgrebe, D. A. 2003. *Signal Theory Methods in Multispectral Remote Sensing*. John Wiley & Sons, Hoboken, NJ.

Legaard, K. R. and Thomas, A. C. 2007. Spatial patterns of intraseasonal variability of chlorophyll and sea surface temperature in the California Current. *Journal of Geophysical Research*, 112 C09006. doi:10.1029/2007jc004097.

Leite, G. C., Ushizima, D. M., Medeiros, F. N. S., and de Lima, G. G. 2010. Wavelet analysis for wind fields estimation. *Sensors (Basel)*, 10(6):5994–6016.

Lepik, Ü. and Hein, H. 2014. *Haar Wavelets: With Applications*. Springer, Switzerland.

Li, H. 2001. Visualization of a turbulent jet using wavelets. *Journal of Thermal Science*, 10(3):211–217.

Li, S. Z. 2009. *Markov Random Field Modeling in Image Analysis*, 3rd edition. Springer, Berlin, Germany.

Li, H. and Zhou, Y. 2002. Wavelet multi-resolution analysis of initial condition effects on near-wake turbulent structures. *Journal of Visualization*, 5(4):343–354.

Liggins, M. E., Hall, D. J., and Llinas, J. (Eds.). 2009. *Handbook of Multisensor Data Fusion: Theory and Practice*, 2nd edition. CRC Press, Boca Raton, FL.

Lindeberg, T. 1994. *Scale-Space Theory in Computer Vision*. Springer, New York.

Lipkin, B. S. and Rosenfeld, A. (Eds.). 1970. *Picture Processing and Psychopictorics*. Academic Press, New York.

Liu, P. C. 1994. Wavelet spectrum analysis and ocean wind waves. In *Wavelets in Geophysics* (Eds. E. Foufoula-Georgiou and P. Kumar). Academic Press, San Diego, CA, Vol. 4, pp. 151–166.

Liu, P. C. and Babanin, A. V. 2004. Using wavelet spectrum analysis to resolve breaking events in the wind wave time series. *Annales Geophysicae*, 22:3335–3345.

Liu, J. G. and Maso, P. J. 2016. *Image Processing and GIS for Remote Sensing: Techniques and Applications*, 2nd edition. John Wiley & Sons, Chichester, UK.

Liu, A. K., Peng, C. Y., and Chang, S. Y.-S. 1997. Wavelet analysis of SAR images for coastal watch. *IEEE Journal of Oceanic Engineering*, 22(1):9–17. doi:10.1109/48.557535.

Liu, A. K., Wu, S. Y. and Zhao, Y. 2003. Wavelet analysis of satellite images in ocean applications. In *Frontiers of Remote Sensing Information Processing* (Ed. C. H. Chen). World Scientific, Singapore, Chapter 7, pp. 141–162.

Lopes, R. and Betrouni, N. 2009. Fractal and multifractal analysis: A review. *Medical Image Analysis*, 13(4):634–649. doi:10.1016/j.media.2009.05.003.

Lovejoy, S. and Schertzer, D. 2013. *The Weather and Climate: Emergent Laws and Multifractal Cascades*. Cambridge University Press, Cambridge, UK.

Louban, R. 2009. *Image Processing of Edge and Surface Defects: Theoretical Basis of Adaptive Algorithms with Numerous Practical Applications*. Springer, Berlin, Germany.

Luo, J. and Jameson, L. 2002. A wavelet-based technique for identifying, labeling, and tracking of ocean eddies. *Journal of Atmospheric and Oceanic Technology*, 19(3):381–390.

Magnenat-Thalmann, N. and Thalmann, D. 1987. *Image Synthesis: Theory and Practice*. Springer, Tokyo, Japan.

Maître, H. (Ed.). 2008. *Image Processing*. John Wiley & Sons – ISTE, Hoboken, NJ.

Mallat, S. 1998. *A Wavelet Tour of Signal Processing: The Sparse Way*, 3rd edition. Academic Press, San Diego, CA.

Mandelbrot, B. B. 1983. *The Fractal Geometry of Nature*. W. H. Freeman, New York.

Mandelbrot, B. B. 1999. *Multifractals and 1/f Noise: Wild Self-Affinity in Physics*. Springer, New York.

Mandelbrot, B. B. 2004. *Fractals and Chaos: The Mandelbrot Set and Beyond.* Springer, New York.

Marghany, M., Cracknell, A. P., and Hashim, M. 2009. Comparison between Radarsat-1 SAR different data modes for oil spill detection by a fractal box counting algorithm. *International Journal of Digital Earth*, 2(3):237–256.

Marks, R. J. II. 2009. *Handbook of Fourier Analysis & Its Applications.* Oxford University Press, New York.

Mas, J. F. and Flores, J. J. 2008. The application of artificial neural networks to the analysis of remotely sensed data. *International Journal of Remote Sensing*, 29(3):617–663. doi: 10.1080/01431160701352154.

Massel, S. R. 2001. Wavelet analysis for processing of ocean surface wave records. *Ocean Engineering*, 28:957–987.

Massopust, P. R. 2016. *Fractal Functions, Fractal Surfaces, and Wavelets*, 2nd edition. Elsevier – Academic Press, London, UK.

Mather, P. M. and Koch, P. 2011. *Computer Processing of Remotely-Sensed Images: An Introduction*, 4th edition. John Wiley & Sons, Chichester, UK.

Matheron, G. 1963. Principles of geostatistics. *Economic Geology*, 58(8):1246–1266. doi:10.2113/gsecongeo.58.8.1246.

Meyer, Y. 1992. *Wavelets and Operators.* Cambridge University Press, Cambridge, UK.

Meyer, Y. 1993. *Wavelets: Algorithms and Applications.* Society for Industrial and Applied Mathematics (SIAM), Philadelphia, PA.

Meyers, S. D., Kelly, B. G., and O'Brien, J. J. 1993. An introduction to wavelet analysis in oceanography and meteorology: With application to the dispersion of Yanai waves. *Monthly Weather Review*, 121(10):2858–2866.

Mirmehdi, M., Xie, X., and Suri, J. 2008. *Handbook of Texture Analysis.* Imperial College Press, London, UK.

Misiti, M., Misiti, Y., Oppenheim, G., and Poggi, J.-M. 2007. *Wavelet Toolbox™ 4. User's Guide.* The MathWorks, Inc., Natick, MA. Available on the Internet www.mathworks.com/help/pdf_doc/wavelet/wavelet_ug.pdf.

Mitchell, H. B. 2010. *Image Fusion: Theories, Techniques and Applications.* Springer, Berlin, Germany.

Mitchell, H. B. 2012. *Data Fusion: Concepts and Ideas*, 2nd edition. Springer, New York.

Mitchell, E. W. and Murray, S. R. (Eds.). 2012. *Classification and Application of Fractals: New Research.* Nova Science Publishers, Hauppauge, NY.

Moser, G. and Zerubia, J. (Eds.). 2018. *Mathematical Models for Remote Sensing Image Processing: Models and Methods for the Analysis of 2D Satellite and Aerial Images.* Springer, Cham, Switzerland.

Mountrakis, G., Im, J., and Ogole, C. 2011. Support vector machines in remote sensing: A review. *ISPRS Journal of Photogrammetry and Remote Sensing*, 66(3):247–259.

Mukundan, R. and Ramakrishnan, K. P. 1998. *Moment Functions in Image Analysis: Theory and Applications.* World Scientific Publishing, Singapore.

Mumford, D. and Desolneux, A. 2010. *Pattern Theory: The Stochastic Analysis of Real-World Signals.* CRC Press, Boca Raton, FL.

Naik, G. R. (Ed.). 2018. *Advances in Principal Component Analysis: Research and Development.* Springer, Singapore.

Najmi, A.-H. 2012. *Wavelets: A Concise Guide.* Johns Hopkins University Press, Baltimore, MD.

Nickolas, P. 2017. *Wavelets: A Student Guide.* Cambridge University Press, Cambridge, UK.

Nicolleau, F. and Vassilicos, J. C. 1999. Wavelets for the study of intermittency and its topology. *Philosophical Transactions of the Royal Society A*, 357:2439–2457.

Nicolleau, F.C. G. A. and Vassilicos, J. C. 2014. Wavelet analysis of wave motions. *International Journal of Applied Mechanics*, 06(03) 1450021. doi:10.1142/S1758825114500215.

Nilsson, E. 2007. *Multifractal-Based Image Analysis with Applications in Medical Imaging.* Master's Thesis. Umeå University, Sweden.

Nixon, M. S. and Aguado, A. S. 2012. *Feature Extraction and Image Processing*, 3rd edition. Elsevier – Academic Press, Oxford, UK.

Ogden, R. 1997. *Essential Wavelets for Statistical Applications and Data Analysis.* Birkhauser, Boston, MA.

Olea, R. A. 1991. (Ed.). *Geostatistical Glossary and Multilingual Dictionary.* Oxford University Press, Oxford, NY.

Oliver, M. A. and Webster, R. 2015. *Basic Steps in Geostatistics: The Variogram and Kriging.* Springer, New York.

Ott, E. 2002. *Chaos in Dynamical Systems*, 2nd edition. Cambridge University Press, Cambridge, UK.

Ouahabi, A. (Ed.). 2012. *Signal and Image Multiresolution Analysis.* John Wiley & Sons – ISTE, Hoboken, NJ.

Parker, J. R. 2011. *Algorithms for Image Processing and Computer Vision.* John Wiley & Sons, Indianapolis, IN.

Peitgen, H. O. and Saupe, D. (Eds.). 1988. *The Science of Fractal Images.* Springer, New York.

Percival, D. B. and Walden, A. T. 2000. *Wavelet Methods for Times Series Analysis.* Cambridge University Press, Cambridge, UK.

Pesin, Y. and Climenhaga, V. 2009. *Lectures on Fractal Geometry and Dynamical Systems (Student Mathematical Library, Volume 52).* American Mathematical Society, Mathematics Advanced Study Semesters, Providence, RI.

Petrosian, A. A. and Meyer, F. G. (Eds.). 2001. *Wavelets in Signal and Image Analysis: From Theory to Practice.* Springer, Dordrecht, the Netherlands.

Petrou, M. and Petrou, C. 2010. *Image Processing: The Fundamentals*, 2nd edition. John Wiley & Sons, Chichester, UK.

Petrou, M. and Sevilla, P. G. 2006. *Image Processing: Dealing with Texture.* John Wiley & Sons, Chichester, UK.

Pietikäinen, M. (Ed.). 2000. *Texture Analysis in Machine Vision.* World Scientific Publishing, Singapore.

Pleskachevsky, A., Lehner, S., Heege, T., and Mott, C. 2011. Synergy and fusion of optical and synthetic aperture radar satellite data for underwater topography estimation in coastal areas. *Ocean Dynamics*, 61(12):2099–2120.

Pohl, C. and van Genderen, J. L. 1998. Multisensor image fusion in remote sensing: Concepts, methods and applications. *International Journal of Remote Sensing*, 19(5):823–854. doi:10.1080/014311698215748.

Pohl, C. and van Genderen, J. 2017. *Remote Sensing Image Fusion: A Practical Guide.* CRC Press, Boca Raton, FL.

Prasad, L. and Iyengar, S. S. 1997. *Wavelet Analysis with Applications to Image Processing.* CRC Press, Boca Raton, FL.

Pratt, W. K. 2007. *Digital Image Processing*, 4th edition. John Wiley & Sons, Hoboken, NJ.

Prost, G. L. 2014. *Remote Sensing for Geoscientists: Image Analysis and Integration*, 3rd edition. CRC Press, Boca Raton, FL.

Quattrochi, D. A. and Goodchild, M. F. (Eds.). 1997. *Scale in Remote Sensing and GIS.* CRC Press, Boca Raton, FL.

Raizer, V. 2005. Texture models for high-resolution ocean microwave imagery. In *Proceedings of International Geoscience and Remote Sensing Symposium*, July 25–29, 2005, Seoul, Korea, Vol. 1, pp. 268–271. doi:10.1109/IGARSS.2005.1526159.

Raizer, V. 2012. Fractal-based characterization of ocean microwave radiance. In *Proceedings of International Geoscience and Remote Sensing Symposium*, July 22–27, 2012, Munich, Germany, pp. 2794–2797. doi:10.1109/IGARSS.2012.6350852.

Raizer, V. 2013. Multisensor data fusion for advanced ocean remote sensing studies. In *Proceedings of International Geoscience and Remote Sensing Symposium*, July 21–26, 2013, Melbourne, Victoria, Australia, pp. 1622–1625. doi:10.1109/IGARSS.2013.6723102.

Raizer, V. 2017. *Advances in Passive Microwave Remote Sensing of Oceans*. CRC Press, Boca Raton, FL.

Rangayyan, R. M. 2004. *Biomedical Image Analysis*. CRC Press, Boca Raton, FL.

Rao, A. R. 1990. *Taxonomy for Texture Description and Identification*. Springer, New York.

Rao, R. M. and Bopardikar, A. S. 1998. *Wavelet Transforms: Introduction to Theory & Applications*. Addison-Wesley, Boston, MA.

Raol, J. R. 2009. *Multi-Sensor Data Fusion with MATLAB®*. CRC Press, Boca Raton, FL.

Raol, J. R. 2016. *Data Fusion Mathematics: Theory and Practice*. CRC Press, Boca Raton, FL.

Richards, J. A. 2013. *Remote Sensing Digital Image Analysis: An Introduction*, 5th edition. Springer, Berlin, Germany.

Robinson, C. 1998. *Dynamical Systems: Stability, Symbolic Dynamics, and Chaos*. CRC Press, Boca Raton, FL.

Rodenas, J. A. and Garello, R. 1997. Wavelet analysis in SAR ocean image profiles for internal wave detection and wavelength estimation. *IEEE Transaction on Geoscience and Remote Sensing*, 35(4):933–945.

Rodenas, J. A. and Garello, R. 1998. Internal wave detection and location in SAR images using wavelet transform. *IEEE Transactions on Geoscience and Remote Sensing*, 36(5):1494–1507.

Rohwer, C. 2005. *Nonlinear Smoothing and Multiresolution Analysis*. Birkhäuser Verlag, Basel, Switzerland.

Rosenfeld, A. (Ed.). 1981. *Image Modeling*. Academic Press, New York.

Rue, H. and Held, L. 2005. *Gaussian Markov Random Fields: Theory and Applications*. Chapman & Hall/CRC Press, Boca Raton, FL.

Ruppert-Felsot, J., Farge, M., and Petitjeans, P. 2009. Wavelet tools to study intermittency: Application to vortex bursting. *Journal of Fluid Mechanics*, 636:427–453.

Russ, J. C. 1994. *Fractal Surfaces*. Premium Press, New York.

Russ, J. C. and Neal, F. B. 2015. *The Image Processing Handbook*, 7th edition. CRC Press, Boca Raton, FL.

Schertzer, D. and Lovejoy, S. (Eds.). 1991. *Non-Linear Variability in Geophysics: Scaling and Fractals*. Kluwer Academic Publishers, Dordrecht, the Netherlands.

Schneider, K. and Vasilyev, O. V. 2010. Wavelet methods in computational fluid dynamics. *Annual Review of Fluid Mechanics*, 42(1):471–503 doi:10.1146/annurev-fluid-121108-145637.

Schölkopf, B. and Smola, A. J. 2002. *Learning with Kernels: Support Vector Machines, Regularization, Optimization, and Beyond*. The MIT Press, Cambridge, MA.

Schowengerdt, R. A. 2007. *Remote Sensing, Models, and Methods for Image Processing*, 3rd edition. Elsevier – Academic Press, Burlington, MA.

Schroeder, M. 1991. *Fractals, Chaos, Power Laws: Minutes from an Infinite Paradise*. W. H. Freeman and Company, New York.

Seuront, L. 2010. *Fractals and Multifractals in Ecology and Aquatic Science*. CRC Press, Boca Raton, FL.

Silvermann, B. W. and Vassilicos, J. C. (Eds.). 2000. *Wavelets: The Key to Intermittent Information*. Oxford University Press, Oxford, UK.

Simhadri, K. K., Iyengar, S. S., Holyer, R. J., Lybanon, M., and Zachary, J. M., Jr. 1998. Wavelet-based feature extraction from oceanographic images. *IEEE Transaction on Geoscience and Remote Sensing*, 36(3):767–778.

Sivakumar, B. 2017. *Chaos in Hydrology: Bridging Determinism and Stochasticity*. Springer, New York.

Skillicorn, D. 2008. *Understanding Complex Datasets: Data Mining with Matrix Decompositions.* Chapman & Hall/CRC Press, Boca Raton, FL.

Soille, P. 2004. *Morphological Image Analysis: Principles and Applications,* 2nd edition. Springer, Berlin, Germany.

Soman, K. P., Resmi, N. G., and Ramachandran, K. I. 2010. *Insight into Wavelets: From Theory to Practice,* 3rd edition. PHI Learning Private Limited, New Delhi, India.

Sprott, J. C. 2003. *Chaos and Time-Series Analysis.* Oxford University Press, Oxford, UK.

Stearns, S. D. and Hush, D. R. 2011. *Digital Signal Processing with Examples in MATLAB®,* 2nd edition. CRC Press, Boca Raton, FL.

Stathaki, T. (Ed.). 2008. *Image Fusion: Algorithms and Applications.* Elsevier, Amsterdam, the Netherlands.

Strang, G. and Nguyen, T. 1996. *Wavelets and Filter Banks.* Wellesley-Cambridge Press, Wellesley, MA.

Steinwart, I. and Christmann, A. 2008. *Support Vector Machines.* Springer, New York.

Sumathi, S. and Paneerselvam, S. 2010. *Computational Intelligence Paradigms: Theory & Applications Using MATLAB.* CRC Press, Boca Raton, FL.

Szeliski, R. 2011. *Computer Vision: Algorithms and Applications.* Springer, London, UK.

Tarasov, V. E. 2010. *Fractional Dynamics: Applications of Fractional Calculus to Dynamics of Particles, Fields and Media.* High Education Press and Springer, Beijing and New York.

Tarquis, A. M., Platonov, A., Matulka, A., Grau, J., Sekula, E., Diez, M., and Redondo, J. M. 2014. Application of multifractal analysis to the study of SAR features and oil spills on the ocean surface. *Nonlinear Processes in Geophysics,* 21:439–450.

Tello, M., López-Martínez, C., and Mallorqui, J. 2005. A novel algorithm for ship detection in SAR imagery based on the wavelet transform. *IEEE Geoscience and Remote Sensing Letters,* 2(2):201–205.

Thomson, R. E. and Emery, W. J. 2014. *Data Analysis Methods in Physical Oceanography,* 3rd edition. Elsevier, Amsterdam, the Netherlands.

Tilley, D. G. 1992. SAR-optical image fusion with Landsat, Seasat, Almaz and ERS-l satellite data. In *Proceedings of XVIIth Congress International Society for Photogrammetry and Remote Sensing Society,* August 2–14, Washington, DC, Vol. XXIX, Part B2, pp. 512–519.

Tilley, D. G., Sarma,Y. V., and Beal, R. C. 1992. Ocean data reduction and multi-sensor fusion studies of the Chesapeake region. In *Proceedings of International Geoscience and Remote Sensing Symposium,* May 26–29, 1992, Houston, TX. pp. 665–668. doi:10.1109/IGARSS.1992.576799.

Tolstov, G. P. 1976. *Fourier Series.* Dover Publications, Mineola, New York.

Tomita, F. and Tsuji, S. 1990. *Computer Analysis of Visual Textures.* Kluwer Academic Publishers – Springer, New York.

Torrence, C. and Compo, G. P. 1998. A practical guide to wavelet analysis. *Bulletin of the American Meteorological Society,* 79(1):61–78.

Tso, B. and Mather, P. M. 2009. *Classification Methods for Remotely Sensed Data,* 2nd edition. CRC Press, Boca Raton, FL.

Turcotte, D. L. 1997. *Fractals and Chaos in Geology and Geophysics,* 2nd edition. Cambridge University Press, Cambridge, UK.

Turner, M. J., Blackledge, J. M., and Andrews, P. R. 1998. *Fractal Geometry in Digital Imaging.* Academic Press, San Diego, CA.

Umbaugh, S. E. 2017. *Digital Image Processing and Analysis: Applications with MATLAB® and CVIPtools,* 3rd edition. CRC Press, Boca Raton, FL.

Umbert, M., Hoareau, N., Turiel, A., and Ballabrera-Poy, J. 2014. New blending algorithm to synergize ocean variables: The case of SMOS sea surface salinity maps. *Remote Sensing of Environment,* 146:172–187. doi:10.1016/j.rse.2013.09.018.

Umbert, M., Guimbard, S., Lagerloef, G., Thompson, L., Portabella, M., Ballabrera-Poy, J., and Turiel, A. 2015. Detecting the surface salinity signature of Gulf Stream cold-core rings in Aquarius synergistic products. *Journal of Geophysical Research: Oceans*, 120(2):859–874. doi:10.1002/2014jc010466.

Vaidyanathan, P. P. 1993. *Multi-Rate Systems and Filter Banks*. Prentice Hall, Englewood-Cliffs, NJ.

van den Berg, J. C. (Ed.). 2004. *Wavelets in Physics*, 2nd edition. Cambridge University Press, Cambridge, UK.

Vapnik, V. N. 1995. *The Nature of Statistical Learning Theory*. Springer, New York.

Varshney, P. K. and Arora, M. K. 2004. *Advanced Image Processing Techniques for Remotely Sensed Hyperspectral Data*. Springer, Berlin, Germany.

Verma, S., Khare, D., Gupta, R., and Chandelln, G. S. 2013. Analysis of image segmentation algorithms using MATLAB. In *Proceedings of the Third International Conference on Trends in Information, Telecommunication and Computing* (Ed. V. V. Das). Springer, New York, pp. 163–172.

Vidal, R., Ma, Y., and Sastry, S. S. 2016. *Generalized Principal Component Analysis*. Springer, New York.

Wang, R. 2012. *Introduction to Orthogonal Transforms: With Applications in Data Processing and Analysis*. Cambridge University Press, Cambridge, UK.

Wald, L. 1989. Some examples of the use of structure functions in the analysis of satellite images of the ocean. *Photogrammetric Engineering and Remote Sensing*, 55(10):1487–1490.

Wald, L. 2002. *Data Fusion: Definitions and Architecture: Fusion of Images of Different Spatial Resolutions*. Les Presses, Paris, France.

Walker, J. S. 2008. *A Primer on Wavelets and their Scientific Applications*, 2nd edition. Chapman & Hall/CRC Press, Boca Raton, FL.

Walnut, D. F. 2001. *An Introduction to Wavelet Analysis*. Birkhäuser, Boston, MA.

Watanabe, T., Sakai, Y., Nagata, K., Ito, Y., and Hayase, T. 2014. Wavelet analysis of coherent vorticity near the turbulent/non-turbulent interface in a turbulent planar jet. *Physics of Fluids*, 26(9) 095105.

Weeks, M. 2011. *Digital Signal Processing Using MATLAB & Wavelets*, 2nd edition. Jones and Bartlett Publishers, Sudbury, MA.

Weisstein, E. W. 2003. *CRC Concise Encyclopedia of Mathematics*, 2nd edition. Chapman & Hall/CRC Press, Boca Raton, FL.

Welstead, S. T. 1999. *Fractal and Wavelet Image Compression Techniques*. SPIE Optical Engineering Press, Bellingham, WA.

Weng, Q. (Ed.). 2014. *Scale Issues in Remote Sensing*. John Wiley & Sons, Hoboken, NJ.

Winkler, G. 2003. *Image Analysis, Random Fields and Markov Chain Monte Carlo Methods: A Mathematical Introduction*, 2nd edition. Springer, New York.

Wojtaszczyk, P. A. 1997. *Mathematical Introduction to Wavelets*. Cambridge University Press, Cambridge, UK.

Wornell, G. 1995. *Signal Processing with Fractals: A Wavelet Based Approach*. Prentice-Hall, Englewood Cliffs, NJ.

Yaroslavsky, L. P. 2012. *Theoretical Foundations of Digital Imaging Using MATLAB®*. CRC Press, Boca Raton, FL.

Yoo, T. S. (Ed.). 2004. *Insight into Images: Principles and Practice for Segmentation, Registration, and Image Analysis*. A K Peters/CRC Press, Boca Raton, FL.

Zaslavsky, G. M. 2005. *Hamiltonian Chaos and Fractional Dynamics*. Oxford University Press, New York.

Zecchetto, S. and De Biasio, F. 2008. A wavelet-based technique for sea wind extraction from SAR images. *IEEE Transaction on Geoscience and Remote Sensing*, 46(10):2983–2989.

Zhang, Y. J. 2006. *Advances in Image and Video Segmentation*. IRM Press, Hershey, PA.

7 Advanced Optical Observations

This chapter is an overview of our optical remote sensing studies conducted over the years. Among regular aerial optical observations of a wind-driven sea surface, wave breaking, and foam and whitecap fields, we present a novel experimental material obtained from space. The focus is on a digital processing, interpretation, and advanced application of multispectral satellite optical imagery of high spatial resolution (~1–4 m). The research is based on a concept of the so-called *optical spectral portrait*. It is a spatial mosaic of specific spectral features generated from selected image fragments. The portrait allows us to define *areas of the interest* in the images, which can be associated with certain ocean phenomena. By this means, spectral portrait performs the function of a robust classifier in computer vision of *weakly emergent hydrodynamic phenomena (e.g., turbulent wake)*. The concept suggested provides a methodological basis for research and development capabilities for detection of the ocean target from space.

7.1 A BRIEF HISTORICAL SURVEY

The purposeful application of optical imagery (aerial photography) for exploring sea surface parameters (called also *aerial photo hydrography*) has begun since the 1950s (Cox and Munk 1954a,b; Zdanovich 1963; Stilwell 1969; Levanon 1971; Kasevich et al. 1972; Stilwell and Pilon 1974; Kasevich 1975; Peppers and Ostrem 1978; Schau 1978). In these earlier works, collected data processed using analog methods of photogrammetry. These methods allowed the researchers to measure the main surface wave parameters—slopes, periods, heights, and directions. Many efforts have been made to create empirical formulas for calculating surface wave permanents at different wind conditions.

Later in the 1980–1990s, more detailed optical observations of the sea surface were conducted by many authors (Gotwols and Irani 1980; Lubard et al. 1980; Sheres 1981; Chapman 1981; Chapman and Irani 1981; Monaldo and Kasevich 1981a,b, 1982; Gotwols and Irani 1982; Luchinin 1982; Holthuijsen 1983; Keller and Gotwols 1983; Wald and Monget 1983; Bondur and Voliak 1984; Zuykova et al. 1985; Álvarez-Borrego and Machado-Gama 1985; Shemdin et al. 1988; Shemdin and Tran 1992; Álvarez-Borrego 1993; Borrego 1993; Walker 1994; Churnside et al. 1995; Morel et al. 1995; Dugan et al. 1996; Holland et al. 1997; Shaw 1999a). The goal of these and others studies was to explore statistical properties and reflection of the sea surface including measurements of scales, spatial spectra, dispersion relationships of surface waves as well attempts to retrieve wind speed from optical data and photographs. These works provided an important step forward in the understanding optical capabilities for remote sensing of the ocean.

Incredible material was collected in the Space Research Institute (IKI), Russian Academy of Sciences, Moscow, during the decade of 1981–1992. Complex remote sensing experiments were conducted in Pacific Ocean on the far eastern shelf of the Kamchatka Peninsula. In particular, a large body of aerial grayscale black and white aerial photo films of different type was recorded and processed. In these experiments, the six- or four-channel multispectral aerial photo camera Carl Zeiss MKF-6/MSK-4 (Chapter 1, Figure 1.3, and Table 1.1) was installed first on the Antonov An-30 aircraft (1981–1986) and then on the aircraft laboratory Tupolev Tu-134SKh (1987–1992). The Tupolev aircraft laboratory (picture available at web site https://russianplanes.net/id34257) carried several sensors: the MKF-6 camera, the Ku-band (2.25 cm wavelength) side-looking airborne radar (SLAR "Nit"), and a set of passive microwave radiometers. The plane also participated in the Joint US/Russia Internal Wave Remote Sensing Experiment (JUSREX'1992) in 1992 and made seven flights from the NASA Wallops Flight Facility, Virginia, to the New York bight test area. The most important data are published in the book (Raizer 2017). Ironically, after JUSREX'1992, as a result of the *perestroika*, this remarkable Russian aircraft laboratory was re-equipped into the passenger plane and sold to the commerce airline.

In our ocean experiments, optimal flight altitude for joint optical and radar measurements was chosen 5000 m, so that the size of each MKF-6 frame was ~3 km × 5 km with spatial resolution ~3–5 m. As a result, more than 100 films containing ~5000 frames in total have been subjected to thematic analysis. The quality of these optical data was the best at that time and the most obtained results were used for sophisticated applications. Meanwhile, several scientific papers (Grushin et al. 1986; Raizer et al. 1990; Rayzer and Novikov 1990; Kosorukova et al. 1990; Mityagina et al. 1991; Raizer 1994; Raizer et al. 1994; Etkin et al. 1994) have been published at the time when they were being worked on. Some other aerial data, collected by IKI are discussed in the book (Sharkov 2007).

The arrival of digital technologies in the 2000s has led to the development of multispectral electro-optical systems for remote sensing observations including the monitoring of ocean and atmosphere. Here we refer to papers (Gelpi et al. 2001; Dugan et al. 2001a,b; Su et al. 2002; Bondur 2004; Mount 2005; Walsh et al. 2005; Wanek and Wu 2006; Bakhanov et al. 2006; Bréon and Henriot 2006; Cureton et al. 2007; Ross and Dion 2007; Levin et al. 2008; Kay et al. 2009; Reineman et al. 2009; Borodina and Salin 2010; Romero and Melville 2010; Kosnik and Dulov 2011; Yurovskaya et al. 2013; Titov et al. 2014; Miskey 2015; Hooper et al. 2015; Melville et al. 2016; Shaw and Vollmer 2017). In these works, accurate measurements of the glint reflectance angular distribution and wave slope statistics (non-Gaussian function) were performed. Fourier spectra were calculated and investigated from the optical (and radar) images using digital processing. A problem of reconstruction of the calibrated sea surface wavenumber spectra from generated image spectra was considered as well. Airborne experiments (Stockdon and Holman 2000; Piotrowski and Dugan 2002; Dugan and Piotrowski 2003) have also demonstrated applications for retrievals of ocean bathometry and surface current.

In the past three decades, 1980–2010s, several optical ocean experiments were conducted using lidar (Bunkin and Volyak 1987; Shaw and Churnside 1997a,b; Lee et al. 2002; Churnside 2008, 2010, 2014; Churnside and Wilson 2006;

Churnside and Ostrovsky 2005; Lancaster et al. 2005; Churnside and Donaghay 2009; Lynch et al. 2011; Bunkin et al. 2011; Churnside et al. 2014; Mobley 2015) and infrared sensors (Shaw 1999b, 2001, 2002; Shaw and Marston 2000; Shaw et al. 2001; Veron et al. 2008). These observations were made from satellite, aircraft, helicopter, ship, and sea platforms. In these works, variations of lidar signals, their spectral (and fractal) characteristics, and polarimetric signatures were measured and investigated for the first time. Infrared reflectance properties of the sea surface were measured and modeled as well. However, direct applications of laser (lidar) methods for detection and recognition of induced hydrodynamic events in the upper ocean layer are not known (or not understood) so far. The exception is lidar measurements of subsurface bubbles but data are limited. Though never mentioned, there are some speculations that an airborne lidar is capable for detecting short-term disturbances produced by underwater moving vehicle. For now, lidar methods and laser spectroscopy provide previously unobtainable information on the physico-chemical properties of the sea surface microlayer (Korenowski 1997; Bunkin and Voliak 2001) including diagnostics of marine oil spills (Fingas 2017) as well.

Finally, since the 2000s opportunities to involve space-based optical imagery for ocean studies are arising. During this time a few authors (Keeler et al. 2005; Gibson et al. 2008; de Michele et al. 2012; Bondur and Murynin 2016; Bondur et al. 2016a,b; Kudryavtsev et al. 2017a,b) reported results of spectral analysis of the optical satellite images. Moreover, satellite data were used for mapping shallow water depth and surface current (Abileah 2006, 2007; Crocker et al. 2007) and detection of ship wakes (Kanjir et al. 2018).

In the past, optical data were often considered as a complementary material for a better interpretation of radar and radiometer microwave measurements. However, it has been clear for us that perfectly conducted optical observations yield more valuable information about ocean hydrodynamics than active/passive microwave sensors. Spatial resolution plays a critical role in ocean remote sensing in anyway. In most cases, the content of ocean optical images can be understood very quickly without computerized manipulations; the processing is needed only for specification of spectral features. For instance, discrete Fourier transform can be applied to extract peaks of the wavenumber spectrum. Experiences show that this operation is valid only for high-quality ocean images.

At high winds and/or stormy conditions, optical images contain two types of relevant information related to surface wave dynamics and foam/whitecap coverage. Correspondingly, image analysis can include two different parts: (1) spectral analysis of surface waves visible in the images and (2) morphological analysis of foam and whitecap objects registered in the same images. We consider both problems, focusing on advanced applications of optical data.

7.2 SPECTRAL SIGNATURES

7.2.1 INTRODUCTION. WAVE SPECTRA

Wave spectrum is an important component of the wind wave dynamic and air-sea interactions. Over the past many decades, numerous experiments with remote sensing (radar, radiometer, optics, lidar) technologies were conducted in order to investigate a variability of the sea surface wave spectrum under the influence of different factors—wind action, hurricanes, internal waves, currents, underwater acoustics, wakes, and others sources. The potential of radar and optical sensors to measure spectral characteristics of ocean waves has been convincingly demonstrated by many authors. The reconstruction of full-directional, two-dimensional (2D) wave spectrum from remotely sensed data is a formidable problem: it is required spectral analysis in spatial frequency domain covered diversified scales of surface waves—from fewer meters to hundreds of meters.

Traditionally, computation of the wave spectra from radar and/or optical images is performed using Fourier methods including the discrete FFT (Section 6.3). Although the FFT is a powerful, mathematically established and the most practically proven algorithm; however, the result of the spectral analysis depends not only on the image quality but also on FFT implementation strategy. It means, that it is necessary first to "optimize" an image and then to define the most informative and important spatial domains. In the case of multispectral (hyperspectral) optical imagery, computed 2D FFT spectrum may depend on electromagnetic band being used.

Spectral analysis of optical images provides a unique possibility to explore transformations of selected wave components (or their groups) under the influence of different phenomena or events. In this case, the spectrum itself can be used as a measure of hydrodynamic situations. In other words, spectrum computed from optical data is an integral instrument-based characteristic of ocean environment but not just a geophysical variable. The concept provides an easier way for dynamic assessment that is important for optical detection.

Selection of quality image fragment for generation of digital spectrum is an important step. Experiences show that the choice is based on the following objective criteria: (1) uniform or monotonic illumination conditions, (2) maximal contrast and maximal brightness variations, (3) low noise, (4) minimal blur effects, (5) sharpness of visible wave patterns, and (6) their statistical representatives in the images. The spectral analysis may include several preprocessing operations which provide optimization of brightness, contrast and intensity of input optical image fragment.

Figure 7.1 illustrates spectral processing of a single aerial photography divided on several fragments. This test demonstrates the performance of 2D FFT with different number of data points. Several common statements known from the experiences are the following:

- Generated spectrum is sensitive to lighting conditions
- Spectrum varies considerably depending on sampling interval (number of pixels)
- Spectrum anisotropy causes by azimuthal distributions of large-scale surface waves
- Low-frequency features (spectral peaks) are perfectly evaluated using enhanced visualization
- High-frequency features are masked due to *penumbral blurring.*

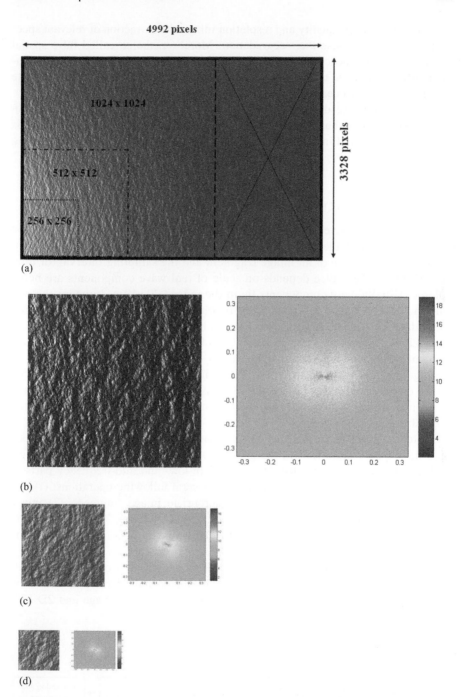

FIGURE 7.1 (See color insert.) Variants of 2D FFT digital processing of sea aerial photography. (a) Basic image, (b) fragment 1024 × 1024, (c) fragment 512 × 512, (d) fragment 256 × 256.

Losses of the image quality and resolution inhibit the extraction of relevant spectral features and/or cause their degradation. Moreover, the so-called false spatial isotropy in high-frequency interval of the spectrum occurs. These and other degradation effects should be taken into account at accurate spectral analysis of satellite optical images of the ocean.

In this section, we develop a framework for advanced spectral analysis in context with hydrodynamic detection. A novel idea of *spectral signatures* is introduced, illustrated, and elaborated. The performance is demonstrated using IKONOS data.

7.2.2 ENHANCED SPECTRA

As we mentioned above, the goal of 2D FFT is to provide the best extraction and visualization of the spectral features. To make this possible, it is necessary to specify the size of an image fragment and the sampling interval (or total pixel number). The choice depends on scale of real wave components are being explored. Usually these components lie in the range ~20–40 m. It means that the sampling interval sometimes may not be reconcilable with actual pixel resolution of raw data. If so, the sampling interval can be matched up through resampling (or resizing). Another option is to execute 2D FFT with non-equal to input image fragment pixel number. Both manipulations are equivalent to low-pass spatial filtering; they also provide reduction of blurring and noise effects in the original images (if they are). To optimize the processing and obtain reliable result, the compromise between the resolution and brightness/contrast characteristics of the image should be found.

Figure 7.2 shows block diagram of refinement spectral analysis. This processing is developed especially with the goal to extract multiple spectral components (peaks) and investigate their distributions within low- and mid-frequency parts of the entire spectrum. The processing includes several following operations: (1) display of original image, (2) selection an appropriate image fragment, (3) image optimization—adjusting brightness and contrast, (4) computing 2D FFT spectrum, (5) selection of relevant spectral features, (6) their enhancement, and (7) extraction and measurement.

Unlike analog methods of spectral analysis which is based on laser-beam diffraction on film transparencies, digital discrete FFT formally has no limitation by pixel number. Indeed, we can vary pixel numbers of an input image and 2D FFT independently.

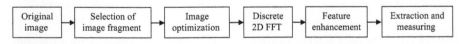

FIGURE 7.2 Block diagram of refinement spectral analysis.

7.2.3 SIGNATURE CLASSIFICATION

To investigate the content of generated image spectra, standard image processing is applied. The following procedures such as histogram equalization, speckle-noise reduction (median filtering), color filtering, zoom, and bi-level thresholding, provide an enhancement of the complete spectrum as a unit image. Figure 7.3 demonstrates common procedure of digital evaluation of spectral features—peaks, their groups, and angular distributions.

For the processing, original IKONOS image fragment (a) of size 2048×2048 pixels is chosen specifically to display several different wave systems. In generated 2D FFT spectrum of size 512×512 pixels (b) two types of features exist: (1) mid-frequency components (red-yellow bundles) with estimated wavelengths $\Lambda = 512 \cdot 4/(80 \div 120) \sim 25-17$ m and azimuth $\varphi \sim 145°$; they correspond to the group structure of the wind waves and (2) low-frequency components with estimated wavelength $\Lambda = 512 \cdot 4/(40 \div 50) \sim 50-40$ m and azimuth $\varphi \sim 30°$ that probably corresponds to the swell. Relevant spectral features are enhanced using an image processing (c). This last procedure is shown in Figure 7.4 in more detail. Note that the revealing of spectral features depends on the size of original image fragment and the number of data points (sampling interval) which is used for computing 2D FFT.

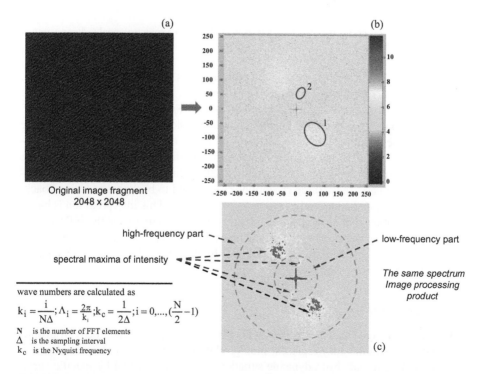

wave numbers are calculated as

$$k_i = \frac{i}{N\Delta}; \Lambda_i = \frac{2\pi}{k_i}; k_c = \frac{1}{2\Delta}; i = 0,...,\left(\frac{N}{2}-1\right)$$

N   is the number of FFT elements
Δ is the sampling interval
k_c is the Nyquist frequency

FIGURE 7.3 (See color insert.) Enhancement and extraction of spectral features from 2D FFT. (a) Original IKONOS image, (b) 2D FFT spectrum 512×512, (c) enhanced spectrum and extraction of spectral maxima.

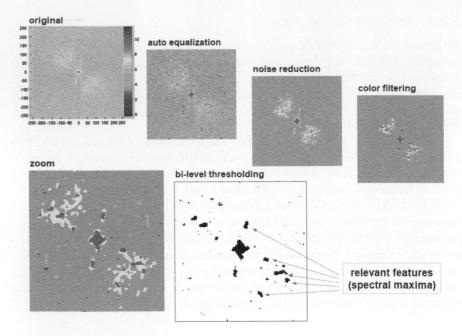

FIGURE 7.4 **(See color insert.)** Extraction and display of spectral peaks during image processing.

Signature classification is based on configurations of extracted spectral peaks. It is considered certain types of their group structure. According to the level of the research, several categories (gradations) of the *optical spectral signatures* (Figure 7.5 and Table 7.1) can be specified. This is a phenomenological classification (because some categories require statistical verification), based on so-called multimode characterization of 2D FFT spectra. The suggested classification makes physical sense because we explore *fine*-scale transformations (splitting) of the surface wave components. In our case, these three gradations correspond to background, mixed background/complex, and complex hydrodynamic situations. This interpretation is based on the assumption that such spectral transformations occur under the influence of certain hydrodynamics processes. Among them, wave interactions, excitations, modulations, bifurcations, and/or instabilities are the most probable causes (Chapter 2). All these factors may result to spectrum split.

7.2.4 Summary

Optical spectral signatures are defined by 2D FFT spectra generated directly from optical images. Signatures may represent multigradation configuration of spectral peaks. Detection of "unusual" hydrodynamic situations can be performed by specific variations of spectral signatures. No any manipulations with wave-spectrum retrieval are necessary to perform. Digital 2D FFT spectra of the images are perfect indicators itself. The problem is how to distinguish and evaluate different hydrodynamic situations by optical spectral signatures? In the next section, we will demonstrate some options.

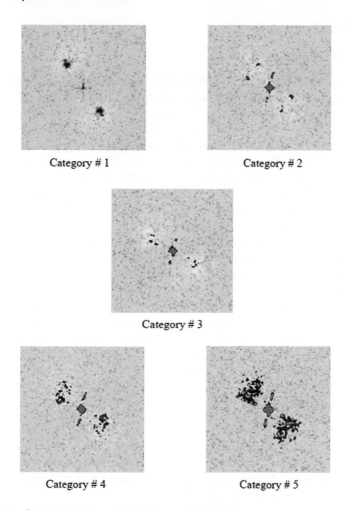

Category # 1   Category # 2

Category # 3

Category # 4   Category # 5

FIGURE 7.5   Categories of spectral signatures (Table 7.1).

TABLE 7.1
Categories of the Optical Spectral Signatures

# Category	Description	Background	Complex
1. One-mode spectrum	Single distinct maximum	✓	
2. Two-four-mode spectrum	Distinct maxima; (may satisfy 4-wave interaction scheme)	✓	
3. A split spectrum (intermediate)	Sets of several closely localized maxima	✓	✓
4. Distinct multi-mode spectrum	Composition system of discrete maxima; (may satisfy multi-mode interaction schemes)		✓
5. Spread, multi-split (multi-mode) spectrum	Distributed system (a set) of many different maxima		✓

7.3 INTERACTIONS

7.3.1 INTRODUCTORY REMARK

As mentioned in the Section 2.5, wave interactions play an important role in ocean dynamics. Interaction mechanism causes energy transfer in the space-time wave spectrum, and effects its broadening at wind-generated wave conditions. The first description of the energy transfer between four different wave components was given by Phillips (Section 2.5.1). Books (Phillips 1980; Yuen and Lake 1982; Craik 1985; Komen et al. 1996; Zakharov 1998; Janssen 2004) provide comprehensive review. Laboratory experiments (e.g., Liu et al. 2015; Waseda et al. 2015; Bonnefoy et al. 2016) demonstrated the existence of resonant wave-wave interactions as well. SeaSat SAR observations (Beal et al. 1983) were the first evidence that wave-wave interactions exist in the real ocean. Resonant interactions between surface gravity waves were also detected using an airborne side-looking radar (Volyak et al. 1987) and aerial photo camera (Grushin et al. 1986). These experimental data revealed nonlinear wave synchronism predicted by the four-wave interaction theory.

Many years ago Phillips (1981) noted that "Perhaps the simple ideas developed in the past twenty years by many people have reached their natural limit, at least in this direction, and further progress will be dependent on new mathematics, new physics and new intuition." His words have a good chance to come true. In present, high-resolution optical observations from space give us a unique opportunity to explore hydrodynamic interactions in the ocean.

7.3.2 DIAGRAM TECHNIQUE

In this section, we consider a framework for derivation of wave interaction diagrams from optical images. In general, these empirical diagrams reflect spatial transformations of the wavenumber spectrum through energy distribution of primary and resonant wave components. Several examples of wave vector diagrams will be presented and their practical sense will be explained.

To create a wave vector diagram using optical images, it is necessary to apply an improve data processing (Figure 7.6) and complete the following operations:

- Select an appropriate image fragment with clear visualization of wave systems
- Optimize brightness, contrast, and intensity of an image fragment
- Compute digital 2D FFT spectra from the selected image fragment
- Enhance computed spectra using image processing and extract relevant spectral peaks
- Define positions of each spectral peak in wavenumber coordinates $\{K_x, K_y\}$
- Place the corresponding wave vectors and estimate their magnitude and orientation
- Combine wave vectors into wave vector diagrams
- Test diagrams on available wave synchronisms.

Figure 7.7 illustrates schematic design of classical four-wave interaction diagram, delivered from digital 2D FFT spectrum (only symmetric part of the spectrum

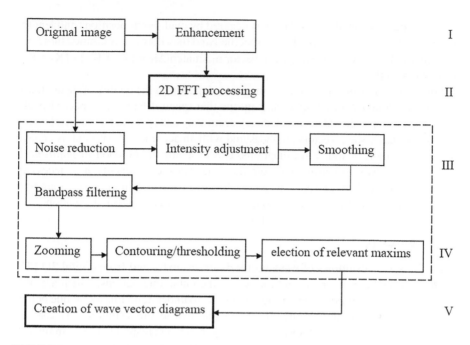

FIGURE 7.6 Block diagram of improved data processing. I, preprocessing; II, 2-D FFT; III, spectrum enhancement; IV, extraction of spectral maxims; V, release. Dashed line block denotes additional operations.

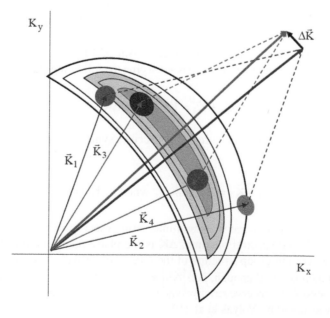

FIGURE 7.7 Schematic illustration of wave vector diagram design. Relevant spectral maxims are shown by ovals.

is shown). Each selected spectral peak is denoted by oval. First, positions and lengths of four wave vectors $\vec{K}_1, \vec{K}_2, \vec{K}_3, \vec{K}_4$ are specified and then they are combined into four-wave vector diagram. A possible wave vector mismatch $\Delta\vec{K} = \left(\vec{K}_3 + \vec{K}_4\right) - \left(\vec{K}_1 + \vec{K}_2\right)$ is shown as well.

Note that the performance of wave diagrams depends on the quality of input optical data and generated 2D FFT spectra. It means that spectral features—spectral peaks and/or their groups, should be revealed very accurately and specified in the pixel equivalent. In our case, wave vectors are measured with an accuracy of ~5% that is enough for the realization of four-wave diagrams. However, for the creation of *multi-mode or multi-wave diagrams* involving more than four wave components, it is necessary to invoke additional processing. Moreover, the development of multi-mode diagrams requires a physics-motivated approach in accordance with the detected situation.

7.3.3 Examples

Figure 7.8 shows examples of wave vector diagrams extracted from digital 2D FFT spectra. These diagrams represent four-wave and cubic interactions, which are well known in hydrodynamics and nonlinear physics. To understand this technique better, we have to consider enhanced digital spectra and their features (Figure 7.5). Usually these spectra have one large spread "fat" peak which corresponds to the main surface wave system having certain direction. We assume that this "fat" peak may split into a group of individual spectral subpeaks. In order to specify positions and magnitudes of subpeaks, we employ an improved image processing (Figure 7.6) providing additional enhancement, contouring, and color filtering of the low-frequency part of 2D FFT spectra. As a result, relevant wave components (subpeaks) are defined very precisely. Then extracted wave vectors are combined into a diagram and tested on available wave synchronisms.

Diagrams in Figure 7.8a describe synchronism of 10–30 m wavelength surface waves. It is four-wave interaction process satisfying the following classical conditions:

$$\vec{K}_1 + \vec{K}_2 = \vec{K}_3 + \vec{K}_4 \pm \Delta\vec{K}. \tag{7.1}$$

The second type of diagram (Figure 7.8b) describes synchronism of coincident wave components:

$$2\vec{K}_1 - \vec{K}_2 = \vec{K}_3 \pm \Delta\vec{K} \quad \text{when} \quad \vec{K}_1 = \vec{K}_4, \tag{7.2}$$

where $\vec{K}_1, \vec{K}_2, \vec{K}_3, \vec{K}_4$ are wave vectors of interacting wave systems which are defined through the selected spectral maxima; $\Delta\vec{K}$ is the phase mismatch (or phase differential) characterizing group structure of the interacting waves. It is assumed that the phase mismatch is small enough, i.e., $\left|\Delta\vec{K}\right| \ll \left|\vec{K}_{1,2,3,4}\right|$. In a practical sense, it means that it is impossible to observe strictly-synchronized surface gravity waves in the real world that was noted by Volyak et al. (1987).

In our consideration, four or three main interacting wave systems are revealed perfectly; they have different spatial scales and propagation directions. As it follows

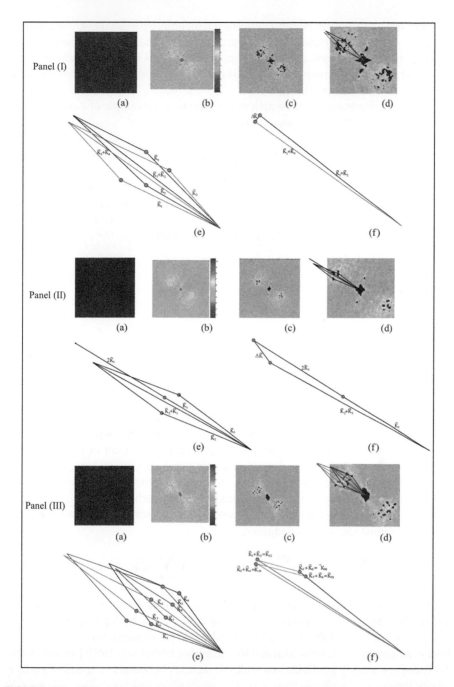

FIGURE 7.8 (See color insert.) Examples of wave interaction diagrams extracted from digital 2D FFT spectra. Panel (I): four-wave interaction, Equation 7.1. Panel (II): four-wave interaction with two coincident waves, Equation 7.2. Panel (III): multiple-wave interaction, Equation 7.3. (a) original image (IKONOS-PAN), (b) 2D FFT spectrum, (c) enhanced spectrum, (d) spectral maxima with zoom, (e) wave vector diagram, (f) wave synchronism.

from spectral analysis, we, perhaps, observe interactions between *weakly-modulated nonlinear surface wave trains*. Such interpretation does make sense for environmental (background) situations, e.g., in the case of interactions between wind-generated waves and swell. Both group synchronisms and spatial modulations therewith occur in a narrow spectral interval.

Figure 7.8c illustrates more complicated experimental example of a multiple diagram which satisfy the resonance conditions:

$$\left(\vec{K}_1 + \vec{K}_2\right) - \left(\vec{K}_3 + \vec{K}_4\right) = \left(\vec{K}_5 + \vec{K}_6\right) - \left(\vec{K}_7 + \vec{K}_8\right) \pm \Delta\vec{K}. \qquad (7.3)$$

Theoretical interpretation may be made on the basis of the Zakharov's equation for surface gravity waves with resonance conditions (7.1) if suppose that

$$\left(\vec{K}_{12} + \vec{K}_{34}\right) = \left(\vec{K}_{56} + \vec{K}_{78}\right) \pm \Delta\vec{K}, \qquad (7.4)$$

$$2\vec{K}_{12} = \vec{K}_1 + \vec{K}_2, \qquad (7.4a)$$

$$2\vec{K}_{34} = \vec{K}_3 + \vec{K}_4, \qquad (7.4b)$$

$$2\vec{K}_{56} = \vec{K}_5 + \vec{K}_6, \qquad (7.4c)$$

$$2\vec{K}_{78} = \vec{K}_7 + \vec{K}_8, \qquad (7.4d)$$

where the phase mismatch $\left|\Delta\vec{K}\right| \ll \left|\vec{K}_{12,34,56,78}\right|$. This particular diagram involves eight surface waves of ~5–10 m wavelength. We call this configuration *dual four-wave interaction diagram*. The reasonable question here, is what kind of hydrodynamic event can cause such an unusual *multi-mode spectral signature*? Actually, if environment unknown, the answer may vary. But in this particular case, we, probably, deal with hydrodynamic interactions induced by strong turbulence.

In the ideal case of resonance multiple interactions, synchronism between n-wave components must satisfy the conditions:

$$\sum \left|\vec{K}_n\right| = 0 \text{ or } \sum \left(\left|\vec{K}_n\right| \pm \left|\Delta\vec{K}_n\right|\right) = 0, \quad \sum \omega\left(\left|\vec{K}_n\right|\right) = 0, \qquad (7.5)$$

where $\left|\Delta\vec{K}_n\right|$ is the wave number mismatch for the corresponding interacting spectral components. However, it seems *unlikely* that a theory of multiple interactions (if any) will be suitable for the description of real world phenomena. So far, the analytic theory (Krasitsky 1994; Krasitskii and Kozhelupova 1995; Lvov 1997; Lin and Perrie 1997) defines weakly nonlinear resonant interactions between five trains of surface gravity waves.

We believe that nonlinear quasi-resonant multiple interactions in the ocean can be described statistically only. For example, the relevant wavevector components extracted from optical data should satisfy to certain rules associated with a number of the physics-based interaction diagrams. Their statistical characterization can be made using multiplicative (or multifractal) cascade models which may fit experimental diagrams. If it is possible at all, the main type of the interaction process could be specified and investigated. This research, however, requires the use of advanced remote sensing imaging technology and perfect experiment. In any case, it is necessary to measure relevant spectral peaks and their spatial distributions with highest accuracy. This is not always possible due to image degradation at high-frequency domain. The situation can be improved thought a multispectral data *fusion* (Section 7.5).

The implementation of diagram techniques can be better understood if consider test remotely sensed data related to well-known environmental phenomena. One interesting example is optical observation of surface manifestations of oceanic internal waves (Mityagina et al. 1991; Etkin et al. 1994). This is a purely nonlinear interaction event. In the field of gravity internal waves, a multiple split of the main spectral peak into a number of subpicks is observed. This process can be associated with strong surface-internal wave interactions.

7.3.4 SUMMARY

High-resolution satellite optical imagery such as IKONOS provides a unique opportunity for exploring nonlinear hydrodynamic processes in the ocean. For this goal, it is necessary to collect dynamic sets of optical images and apply an enhanced processing. We demonstrated that under certain conditions, it is possible to extract low-frequency spectral components of ~10–30 m wavelength and develop wave vector diagrams describing resonant wave-wave interactions.

Empirical wave vector diagrams represent more sophisticated type of "optical spectral signatures." In this context, certain multiple-wave diagrams may reflect complex situations causing by specific (induced) hydrodynamic phenomena or events. These principles endow with detection capabilities. Indeed, it will be real ocean hydrodynamics from space.

7.4 WAVE BREAKING ENVIRONMENT

7.4.1 INTRODUCTION

It is well known that wave breaking, foam, and whitecap are important factors contributing to the remote sensing measurements at high winds. Microwave properties of foam and whitecap have been studied over the past three decades by many authors (e.g., Cherny and Raizer 1998; Raizer 2017). On the other hand, spatial and statistical characteristics of foam and whitecap coverage can be investigated using optical remote sensing methods. For example, there is empirical relationship between foam/whitecap area fraction W and wind speed V, given by formula $W \propto aV^b$ (a, b are constants). A number of dependencies W(V) was obtained using optical data. This particular problem has been discussed in historical works (Monahan 1971; Ross and Cardone 1974; Bondur and Sharkov 1982; Bortkovskii 1987; Wu 1988). The $\Lambda(c)$-rate distributions of breaking-wave crests were also measured using video data (Melville and Matusov 2002; Kleiss and Melville 2011; Melville et al. 2016).

Oceanic foam and whitecap structures are clearly registered in aerial photographs (and digital optical images) in the form of distinct white objects of variable geometry and structure (see Figure 2.17). At high winds and stormy conditions, whitecap and foam area fractions vary significantly according to the Beaufort scale that can be recorded using aerial photography as well.

Geometrical parameters and statistics of foam and whitecap coverage were investigated in the 1980s using aerial photography. For this goal, original automated algorithm was created and applied (Novikov and Raizer 1988). In this section, we describe a technique which we have experienced in the past (1980s). However, some original ideas can be reproduced using available today software codes. With updated algorithm, space-based observations may offer new possibilities for exploring wave breaking and foam/whitecap dynamics. Satellite data are also important for development and improvement of detection technology especially at high winds.

7.4.2 Metrics

To characterize foam/whitecap objects visible in the optical images, the following metrics are introduced:

$$
\begin{array}{lll}
\text{number of objects} & N & \text{in a single image} \\[4pt]
\text{area} & A & \\
\text{perimeter} & P & \\
\text{sphericity} = (\text{circularity})^2 & P^2/A & \left.\right\} \text{metrics for a single object} \\
\text{aspect ration} & L_{min}/L_{max} & \\
(\text{roundness factor})^{-1} & A/(L_{max}L_{min}) & \\[4pt]
\text{distributions} & \multicolumn{2}{l}{F(A),\ F(P),\ F(P^2/A),\ F(L_{min}/L_{max}),} \\
& \multicolumn{2}{l}{F\big[A/(L_{min}L_{max})\big]}
\end{array}
\tag{7.6}
$$

(L_{max} and L_{min} are maximum and minimum linear size of the object).

These computer vision metrics provide an extensive analysis and description of geometrical properties of the images with a detailed *taxonomy* (i.e., classification) for local, regional, and global image features. In other words, chosen metrics work as a detector of foam/whitecap geometrical objects registered in the images. In present, the morphological processing can be performed using machine vision algorithms, described, e.g., in the book (Krig 2014).

7.4.3 Morphological and Statistical Processing

Figure 7.9 shows basic algorithm for combined morphological and statistical analyses of geometrical objects visible in the images. It consists of three main parts.

Part (I) is preprocessing which is applied to improve the quality of original optical images. It is performed using standard image processing operations such as histogram equalization, sharpening, contrast enhancement, noise reduction, and the other adjustments.

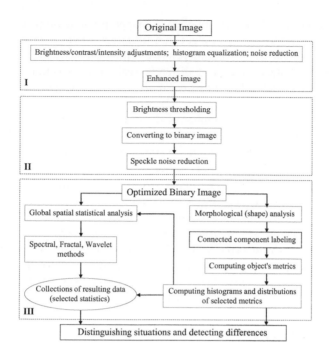

FIGURE 7.9 Flow chart of combined morphological and statistical image processing.

Part (II) provides *binarization* of optical images, i.e., converting grayscale original (or improved) image into two-level black and white image. This can be done using brightness thresholding. However, this operation may lead to the change of number and geometry of selected objects in a resulting binary image. The reason of that is because visible whitecap/foam objects usually have variable size, brightness, and contrast; therefore, the threshold level should be optimized in order to keep their original properties. To improve a binary image and erase artifacts (noise and non-relevant pixels), spatial filtering is applied.

Part (III) is the main executive algorithm providing morphological and statistical analyses of a binary image containing objects. Morphological analysis is based on the fundamental operation in image processing known as *connected component labeling in a binary image*. Set of pixels is grouping into components using connectivity of neighborhood pixels by certain rules. Algorithm performs automatic identification, numbering, and computation of geometrical parameters for all objects existing in a binary image.

The implementation of morphological processing was tested and optimized for different types of aerial photography. Several important cases are as follows:

- Analysis of the images containing of multiple-contrast blurred objects connected to each other (i.e., foam+whitecap unit objects). *Separationability* of whitecap and foam objects depends on brightness gradient and texture gradient which are defined automatically.

- Analysis of optical images obtained at inappropriate lighting and weather conditions. In this situation, algorithm illuminates light effects and foam/whitecap objects using selecting priority-setting geometrical criteria.
- Analysis of optical images of different scale and resolution. This is an important issue related to *statistical representativity* of optical data.

An important parameter is the number of the objects per unit (called surface density), $N/(L_x L_y)$, where L_x and L_y are linear sizes of an image, which characterizes wave breaking intensity dependent on sea state. This number, actually, is a quantitative measure of the visual Beaufort scale. Statistical analysis was performed for $N \geq 50 \div 100$ photo frames with the linear size $L_x \times L_y = 3 \text{ km} \times 5 \text{ km}$; in this case, classification accuracy is enough for decision making.

It is important to note that the suggested algorithm can work in the cases of fuzzy and blurred optical images. This usually takes place at very high winds and/or strong gales. Although foam/whitecap objects in the blurred images are recognizable; however, their geometry cannot be uniquely determined due to blurring. This leads to the errors in calculating foam/whitecap area fractions that is common problem in the wind vector retrieval by microwave remote sensing data.

7.4.4 Selected Results

Here we reproduce *schematically* selected results of our optical experiments conducted during the decade of 1981–1992 in the Pacific. The obtained results were used as for environmental studies as well for detection purposes. The following processing-based characteristics have been defined and explored.

7.4.4.1 Histograms and Statistical Distributions

Histograms of metrics $F(A)$, $F(P)$, $F(P^2/A)$, $F(L_{min}/L_{max})$, and $F[A/(L_{min}L_{max})]$ are computed for certain experiments, environmental conditions, and/or situations by a large volume of digitalilized image databases. These histograms define completely spatially statistical properties of foam/whitecap objects and their fields registered in the images. Note that the most experimental histograms can be approximated by the gamma distribution (Kosorukova et al. 1990; Raizer et al. 1994). As a whole, these metrics and statistics reflect *large-scale hydrodynamic wave breaking surface environment* at high winds and/or other exceptional situations. Variations of the descriptive statistics of the metrics provide quantitative criteria for optical detection of different situations. Thus, the only way to explore global dynamic wave breaking processes in the ocean is panoramic optical remote sensing imagery.

7.4.4.2 Fractal Geometry

The application of fractal analysis for exploring wave breaking fields by optical and microwave remotely sensed data is still a challenging task. Fractal geometry describes spatial self-similarity and scaling of dynamic systems and natural objects (Mandelbrot 1983). In image processing, fractal algorithms offer considerable flexibility in optimization of the pixel resolution that is important for statistical characterization of multiple featured images. Although the realization of fractal algorithms requires large databases, in many practical cases, (multi)fractal mathematics become a universal tool for analysis of variable geophysical processes and fields (Section 6.5). Specifically, fractal geometry of sea surface and breaking waves have been discussed in earlier works (Glazman 1988; Glazman and Weichman 1989; Rayzer and Novikov 1990; Raizer et al. 1994; Kerman and Bernier 1994; Sharkov 2007). Here we update fractal-based observation concept and present selected results.

To compute the fractal dimension of foam/whitecap fields visible in optical images, two following methods are used: (1) the box-counting method, based on the formula $D_0 = \lim_{r \to 0}\left[\log N(r)/\log(1/r)\right]$, where $N(r)$ is the smallest number of squares with the side r required to completely cover the set of image objects; and (2) the "area-perimeter relation" method, based on the formula $P = \left(\sqrt{A}\right)^{D_S}$. Here D_0 and D_S are the corresponding fractal dimensions. The first method provides global characterization of entire image or an image region containing multiple features, which in our case, are foam and whitecap objects. The second method is used for exploring complexity and self-similarity of geometrical objects itself. Figure 7.10 illustrates box-counting method and numerical examples.

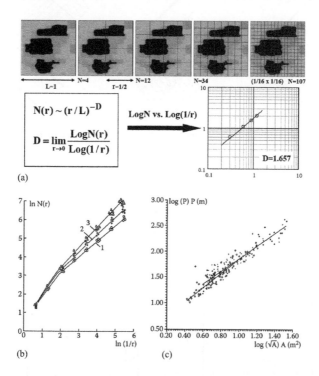

FIGURE 7.10 Fractal dimension of foam/whitecap fields computed by optical images. (a) box-counting method. Computations: (b) ln $N(r)$ vs. ln $(1/r)$ for total foam and whitecap coverage. The value of fractal dimension corresponds approximately to three gradations of the Beaufort wind force: $1 - 3 \div 4$ ($D_0 = 1.05$); $2 - 4 \div 5$ ($D_0 = 1.15$); and $3 - 5 \div 6$ ($D_0 = 1.25$) and (c) log-log plots of perimeter (P) as a function of square root from area (A) for foam streaks (*) and whitecaps (◊). Sold lines are linear least-square fits in different ranges of area A. Beaufort wind force is 4. (After Cherny and Raizer 1998.)

7.4.4.3   Distinguishing Situations and Detecting Differences

In Figure 7.11, we show two realistic scenarios and statistical diagrams describing the differences between situations (all plots are shown schematically). Their specification is the following. Situation # 1 which we call "background" represents regular stochastic wave breaking (foam/whitecap) field having one type of the statistics (e.g., stationary distribution). Situations # 2 (along track) and # 3 (across track)which we called "complex" includes specific hydrodynamic "wave breaking patterns" having another type of the statistics (e.g., nonstationary probability density). The differences between situations are defined through the variations of the corresponding optical parameters: metrics and statistics (7.6) as well the fractal dimension D_0 and D_S. Mixed statistics of the metrics may also be employed in some cases. To explain the differences shown in Figure 7.11, several options can be considered.

The first option is trivial. The wave breaking intensity, i.e., the number of acts per unit area is changed by some reasons. For simplicity, wave breaking field in

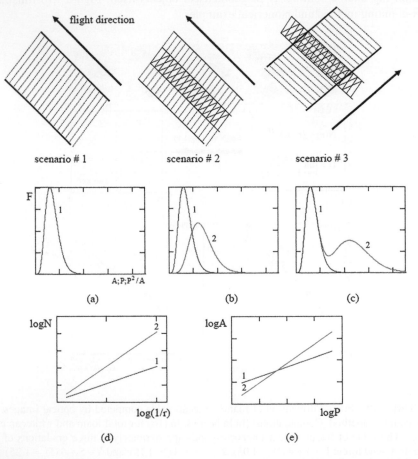

FIGURE 7.11   Possible scenarios and diagrams. Scenario #1 – background, scenarios # 2 and 3 – complex along and across. Plot lines: (1) background and (2) complex. Distributions: one-mode (a), (b); two-mode (c). Box-counting fractal diagram (d). Area-perimeter fractal diagram (e).

the images is represented by a discrete set of points. Variation of the number of points is recorded using regular optical sensor, e.g., high-resolution video camera. The differences are defined in terms of the number of points and point statistics. The corresponding point-type semi-empirical models exist (Snyder and Kennedy 1983; Snyder et al. 1983) but they seem reasonable at moderate winds (<10–12 m/s) only. As mentioned above, at very high wind and strong gale, foam/whitecap coverage represents stochastic fields of multiple foam/whitecap geometrical objects that may not be described by point-type models and point statistics.

The second option is the most common scenario. It is based on the suggestion that geometrical and statistical properties of wave breaking fields vary by space and time. These variations may occur under the influence of many environmental factors, e.g., due to wind fetch, surface wave-current interactions, or internal wave motions. Such observations have been made using our aerial photography. In this case, geometrical and statistical parameters of whitecap/foam objects registered in the images are measured very accurately because we consider a field of geometrical objects but not field of points. Situations can be distinguished by variations of the metric's statistics.

And third, more sophisticated but realistic option is the generation of large-scale wave breaking structures, evaluated in the optical images as a fractal. It means that under certain conditions, there occur *statistical self-similarity* and *scaling* of ocean wave breaking and foam/whitecap fields. Although fractal properties can be revealed in both situations, however, there are the difference between fractal dimensions computed by optical data for situations # 1 and # 2 (Figure 7.11).

From the physics point of view, wave breaking fields represent extreme type of *large-scale nonlinear dynamic system* with stochastic properties (not to be confused with a random wind-generated surface). Variations of the fractal dimension and/ or metrics (7.6) computed by optical data, reflect specific hydrodynamic situations, e.g., "complex" situations. Ocean experiments show that the most probable signatures of the complex situations include *self-similar quasi-regular wave breaking patterns* which looks like to those shown in Figure 7.12. However, it is difficult or even

FIGURE 7.12   Aerial photograph of wave breaking pattern. (After Romero et al. 2017.)

impossible to predict the appearance of such a localized pattern at variable surface conditions. The only way to survey on the situation is to provide operational monitoring and real-time data processing.

7.4.5 SUMMARY

Optical manifestations of ocean wave breaking environment represent geometrical sets of foam and whitecap objects visible in the images. Because of a great variability and *unpredictability* of natural wave breaking phenomena, optical observations are only a tool that can provide reliable manifestations and detection of sea state by statistics of foam/whitecap metrics. No any other remote sensing technique enables to complete this particular task. More specifically, the main results are formulated as the following:

- A general vision metrics taxonomy for feature description is developed for analysis of wave breaking and foam/whitecap fields visible in optical images
- Suggested morphological shape metrics (7.6) characterize adequately geometrical properties of foam/whitecap objects visible in the images
- Variations of descriptive statistics of the metrics provide robustness criteria for detection of complex hydrodynamic situations involved wave breaking environment
- Optical data reveal intrinsically *stochastic nature* of wave breaking fields in open ocean
- Experimental results support our hypothesis about self-similarity and scaling of wave breaking fields visible in optical images.

Stochastic dynamics of wave breaking fields and foam/whitecap coverage can be investigated more efficiently and effectively using combined processing including spectral, fractal, and wavelet methods. In particularly, our experiences have demonstrated that the differences between "background" and "complex" situations can be defined much better using fractal-based techniques than descriptive statistical analysis.

7.5 SPECTRAL PORTRAIT

7.5.1 INTRODUCTION

Remote sensing of ocean hydrodynamics requires the formulation of a methodological concept. In the case of active (radar) microwave observations, detection, and identification of large-scale features—surface waves, roughness patterns, fronts, internal waves, strong currents, slicks, eddies, rainstorms, ship wakes, oil pollutions, and atmospheric boundary layer phenomena, are based on the Bragg-scattering theory and oceanographic expertise. In the case of passive (radiometer) microwave observations, extraction of hydro-physical (hydrodynamic) information is more complicated process because of an integral impact of ocean-atmosphere environment and multiparametric description of the ocean surface emissivity itself.

High-resolution optical imagery provides visualization of ocean surface features directly. For example, the IKONOS multispectral (MS) images of 4-m resolution and panchromatic (PAN) image of 1-m resolution both contain the similar visual attributes involving manifestations of surface waves of different scale and propagation direction. Sometimes, atmospheric turbulence, variable and unpredictable lighting conditions, and natural blur effects reduce the quality of multispectral optical data *unevenly* that makes it difficult to complete information processing. Perhaps, this is a reason why multispectral high-resolution imagery of the vast open ocean spaces at clear air is not so exciting product for scientific research unlike, e.g., land cover imagery or coastal imagery. Meanwhile, ocean IKONOS data contain a bunch of hidden hydrodynamic signatures simply because of the best optical resolution available today and the evidence of the ocean environmental impacts. The problem is how to extract and use this information properly.

These and other technical circumstances demand new idea about utilization of airspace optical data for ocean hydrodynamic studies. It would be also necessary to have a scientifically applied concept, aimed to improve system analysis, interpretation, and application of optical data. The goal of this section is to demonstrate possibilities for exploring ocean IKONOS data containing specific hydrodynamic information (about IKONOS imagery and products see primary source paper (Dial et al. 2003)). For this, a concept of the so-called optical spectral portrait is proposed. Here is a description.

7.5.2 Basic Concept

The concept is based on spectral classification of optical data involving computer vision. Optical spectral portrait is a final product which is a geolocated mosaic of spectral features extracted from optical data. The features can be recognized using computer vision. The key point here is the manifestation of the differences in feature disposition and their frequency-domain characteristics. This way provides possibilities for detection and recognition of the areas of interest which may be associated with certain hydrodynamic events (situations) occurring in the ocean. Spectral features can also characterize a spatial variability of the ocean surface including transformations and/or excitations of the wavenumber spectrum.

The argumentation of this novel, state-of-the-art optical remote sensing concept is based on our common knowledge what types of spectral features might be related to one or another hydrodynamic phenomenon including the target event occurrence. In this context, terminologies "the area of interest" and "event" are not synonymous; environmental processes can also produce analogical spectral features as the event at some cases. Technically, the strategy of spectral processing and interpretation remains the same independently on situation. However, because *a priori* information usually is not available, we consider an expanded methodology in terms of *open research*. To realize this concept practically, the following technique is applied.

7.5.3 Algorithmic Framework

Figure 7.13 shows block diagram of the framework. It consists of several parts and operates with standard software products.

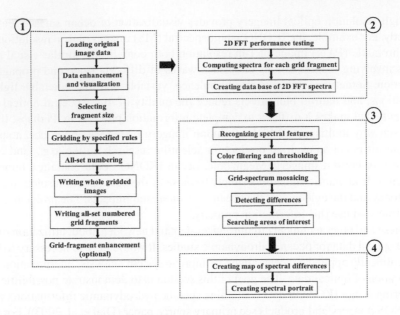

FIGURE 7.13 Algorithmic framework for creating optical spectral portrait.

The first part (1) provides preprocessing of original images including enhancement, gridding, fragmentation, numbering, and storage. An original image is divided into a large number of image fragments of the same size in order to perform then 2D FFT for each image fragment separately. The second part (2) provides computing Fourier spectra for the selected image fragments. Detalization of spectra, selection, and measurement of relevant spectral components (peaks) are performed using enhanced image processing. The third part (3), called "multiplication" provides reconstruction of the spectrum mosaic and searching of the differences between features by certain criteria. The final part (4) creates a spectral portrait which is a map of the features manifested in the spectrum mosaic. In common case, it is a multigradation portrait which is displayed as a continuous color image (picture).

The implementation of the technique is shown in Figure 7.14. To obtain reliable results, it is necessary to complete the following operations:

- Create test area image (Figure 7.14a)
- Create a set of image fragments, gridding (Figure 7.14b)
- Generate the corresponding set of Fourier spectra
- Investigate in detail each spectrum
- Provide classification of spectra
- Generate mosaic from 2D FFT spectra (Figure 7.14c)
- Perform searching for the differences
- Generate mosaic of spectral features
- Create portrait by selected spectral features (Figure 7.14d).

Spectral classification of optical data is based on structural characterization of the generated 2D Fourier spectra. The classification yields some gradations of spectral

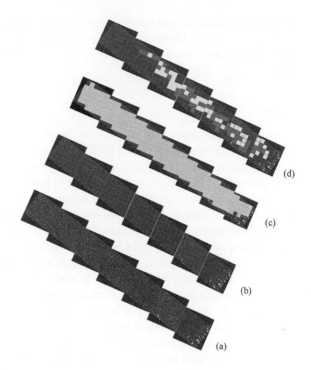

(d)

(c)

(b)

(a)

FIGURE 7.14 **(See color insert.)** Optical spectral portrait implementation (see text).

features which are defined by a configuration of relevant spectral components (peaks) and their disposition (i.e., the distribution of the corresponding wave vectors). Hydrodynamic situations are distinguished by variations of spectral features or their gradations in the portrait.

Figure 7.14d shows the realization of optical spectral portrait designed using three robust spectral gradations chosen from set of the spectra (Figure 7.5). We specify them as "background" (black squares), "background/complex (yellow squares), and "complex" (red squares). Considering these particular data, we would assume that this particular portrait is an optical manifestation of complex hydrodynamic situation involving multi-mode transformations of the surface wave spectrum with the generation of side wave components. Such a situation may occur under the influence of, e.g., wave-wave or wave-turbulence interactions as well as a result of spatiotemporal periodic modulations (Section 7.3). From this viewpoint, the applied technique is unique for detection purposes.

The suggested optical-based framework is organized as interactive, semi-automated, and multi-step routine processing. It is a general scientific approach which can be used for many geophysical applications. Spectral portrait is a result of detailed analysis and utilization of optical data. Ultimately, the portrait is a tool which can be used for detection and recognition of *optico-hydrodynamic signatures* related to certain hydrodynamic situations (phenomena/events).

In principle, satellite optical data of high resolution allow us to develop a multigradation portrait in order to explore possible situations in more detail. It depends on the level of research. An advanced algorithm can be created using *image fusion* and computer vision. However, this option requires some experimental effort, multispectral data acquisition, and additional resources. Examples are considered below.

7.5.4 IMAGE FUSION

As shown above, certain hydrodynamic applications can be developed using panchromatic IKONOS ocean imagery and digital 2D FFT. In this section, we consider more advanced approach based on fusion of multispectral IKONOS data. Here we emphasize that *multispectral data fusion provides sometimes sophisticated hydrodynamic information which may not be obtained using panchromatic data.*

In general, any ocean optical data of high resolution are much better than data with low resolution. As it follows from theoretical and empirical studies, spectral dependences of the reflection coefficient of light (spectral reflectance) for free pure water surface in the range of 0.45–0.90 μm is <5% at nadir (Chapter 3), i.e., the difference within IKONOS multispectral bands is small. It is also known from aerial experiments that sea surface reflectance is color neutral and modulations of the light by wave slopes are very weak. It means, in particular, that spectral processing is more efficient for PAN data with 1-m resolution than for any MS band data with 4-m resolution because of PAN data have a better quality and highest pixels intensity. If necessary, PAN data can be integrated into low resolution data and in this case, they will be also better than multispectral data. However, fusion between individual multispectral bands (Blue, Green, Red, NIR) and PAN band makes sense to investigate in terms of spectral characterization of ocean optical data. This operation known as "pansharpening" is often used in multispectral imagery for land cover classification (Alparone et al. 2015; Pohl and van Genderen 2017). Although fusion methods do not necessarily improve visual quality of ocean PAN images, in our case, *pansharpening* provides much better *detalization* and specification of generated digital 2D FFT spectra (Figure 7.15). *Fused spectra are more informative than non-fused spectra.*

Figure 7.16 demonstrates realizations of fused multispectral portrait created using IKONOS data. PCA-based pixel-level image fusion algorithm is applied for this particular processing (e.g., Raol 2009).

A central idea here is to accomplish robust *spectral tracking* of features in a multispectral portrait. This option gives us sophisticated hydrodynamic information concerning *optico-hydro-physical* characterization of ocean environment (Section 7.6).

Briefly, the conclusion is the following (for ocean IKONOS data only):

- Fusion provides an enhancement of digital 2D FFT spectra depending on the combination of bands
- Fusion between Blue, Green, Red, NIR bands is not effective because of low resolution
- Fusion between NIR (4-m resolution) and PAN (1-m resolution) is the most efficient; this improves visual quality of 2D FFT spectra significantly
- Multispectral portrait based on fusion provides more information than panchromatic portrait

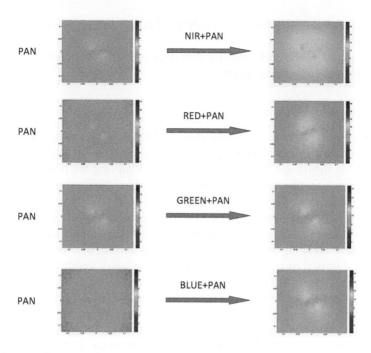

FIGURE 7.15 **(See color insert.)** Examples of fused spatial spectra generated from multi-spectral optical images.

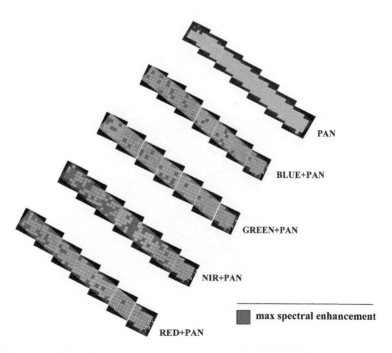

FIGURE 7.16 **(See color insert.)** Realizations of fused spectral portrait.

- Fusion optimization facilitates analysis of output data, rather than characterize input data
- The performance and interpretation may depend on fusion technique. Follow-up studies are required.

7.5.5 Fusion as Detection Tool

It is well known that fusion is an effective and efficient technique for analysis of MS and HS remotely sensed imaging data. The main and novel contribution here is accomplishing hybrid *spectral-spatial fusion* and achieving the best overall performance of the processing that is important for detection of hydrodynamic signatures. Some specific aspects of our analysis are the following.

Utilization of single-band optical data can be applied in cases of background situations at variable and high wind ocean environment when multi-scale surface wave patterns, wave breaking and foam/whitecap fields are registered very well on all multispectral optical images independently. Hydrodynamic conditions can be specified through spectral and morphological image processing.

In the case of complex situations, e.g., in the presence of so-called *weakly emergent hydrodynamic phenomena*—multiple hydrodynamic patterns, localized turbulence, wakes, or other induced surface disturbances, the processing and interpretation require more advanced approach. The solution can be found using fusion methods that provide enhanced computer vision process and more reliable detection and recognition of optico-spectral signatures.

State-of-the-art optical technique for detection purposes can be developed using a *multispectral statistical network* shown in Figure 7.17. In this diagram, the input layer represents multispectral images, the hidden layer includes different metrics and representations (working as "black box"), and the output layer represents resulting data generated using a multivariable statistical algorithm. Since multispectral ocean

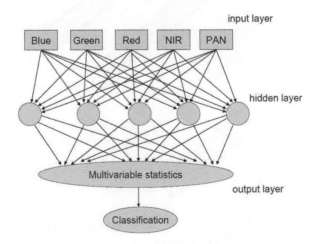

FIGURE 7.17 Multispectral statistical neural network for utilization of high-resolution optical imagery.

images have different pixel-by-pixel characteristics, this network may provide analysis of individual or combined multispectral optical data using regression, correlation, texture synthesis, or other available algorithms and techniques.

The most promising option for detection is *multisensor fusion* involving big data technologies. Although different sensors (optical and microwave) have different capabilities, acquisition geometry and resolutions, fusion product benefits are obvious due to incredibly increasing the volume of remotely sensed data that provides more flexibility and recourses for analysis, interpretation, and application.

7.5.6 SUMMARY

A concept of optical geolocated (multi)spectral portrait is developed with the goal to characterize and explore hydrodynamic situations. A spectral portrait is a mosaic of spectral features, generated using 2D FFT and enhanced image processing. Portrait is a product of computer vision but not human vision. Fusion techniques improve the quality of 2D FFT spectra and provide their detalization much better. The portrait allows us to select the areas of interest and reveal *implicit* effects in optical images. Ultimately, (multi)spectral portrait is a robust classifier for computer vision of *weakly emergent hydrodynamic phenomena (events)*. The technique is also intended for event/target detection using advanced optical observations.

7.6 APPLICATIONS FOR ADVANCED STUDIES

Several applications of HR/VHR optical imagery will be briefly described in this section. We focus on three realistic scenarios: (1) turbulent wake, (2) surface currents, and (3) internal waves. These examples may give the reader the understanding of optical capabilities for hydrodynamic detection. Note that in our description, no any real world experimental data and/or materials are involved.

7.6.1 TURBULENT WAKE

One of the most sophisticated applications of HR/VHR optical imagery is detection of turbulent wake generated by the moving submerged vehicle or other internal sources. Unlike V-type ship wake, turbulent wake (at the moment) may not be visible or recognizable directly in the optical images. The reason of that is because optical signatures of oceanic turbulence usually represent randomized superfine texture features. Such features can be revealed and explores in terms of "spectral portraits."

As we know, at certain environmental conditions (specifically, depending on thermocline depth and stability of the atmospheric boundary layer), turbulent wake may exist in the form of a narrow striped pattern with spatiotemporal scales ~1 hour and several kilometers in length. In this case, large-scale interactions between turbulence and surface waves (also known as wave-driven turbulence) may occur. Generally speaking, wave-driven turbulence may create a mixed energy-active zone in the upper ocean that may trigger the development of Longmire circulation and/or thermohaline convection that can lead to increasing the level of overall turbulence at the ocean-atmosphere interface. Wave-driven turbulence is also a source of transformations of surface wave spectrum in a wide interval of spatial frequencies. These mechanisms as well turbulence-induced hydrodynamic instabilities provide an opportunity for detection of the wake using a concept of spectral portrait. In the portrait of the wake we may see the differences in the generated 2D FFT spectra which are characterized by multi-mode spectral variations. In other words, the split-type spectrum located at some places could be an indicator of turbulent-wake signatures (see Figures 7.14 and 7.16).

7.6.2 Induced Surface Current

The fact that surface current can be an indicator of deep ocean wave processes is well known. Strong wave-current interactions may cause two effects: transformation of surface wave spectrum and amplification of wave breaking (whitecapping) activities. Both processes are detectable by optical sensor. The problem here is to measure "correlation dynamics" of these two hydrodynamic processes in terms of optical signatures and their statistics. A concept of spectral portrait provides a solution.

For this, both spectral and morphological analysis of optical data is performed for the same geolocated set of optical images (i.e., at the same locations). Correlation dynamics are defined by spectral and morphological metrics computed from the images. Some specific spectral and morphological features (if they are), found in the portrait can be indicators of surface currents. In this case, distributions of the metrics can provide additional criteria for decision making (detection) and specification of the areas of the interest.

7.6.3 Internal Waves

Study of internal waves in the oceans is a classical subject of physical oceanography. Most airspace observations of internal waves have been conducted using radar (SAR) instruments. Optical remote sensing techniques are also capable to register internal (solitary) gravity waves by their surface manifestations. In oceanography, these manifestations are known as spatial modulations of surface roughness. The corresponding variations of sun glint (or specular) reflectance cause periodic-like features in the images that may be perceived as optical signatures of internal waves.

Spectral portrait of oceanic internal waves can be created using a spatial mosaic from digital 2D FFT spectra dependent on the content of optical images. An important operation here is the correct choice of image size and sampling interval that should match the scales of internal wave patterns being investigated. A portrait provides specification of internal wave signatures representing periodic-like variations of brightness and contrast in the images. It is quite possible to detect a "grating" of spectral features associated with internal wave manifestations. Additionally, transformations (modulations) of the surface wave spectrum in the field of internal waves can be investigated in great detail. Finally, a spectral portrait can be used as robust optical classifier of nonlinear internal solitary waves.

7.7 SUMMARY

In this chapter, we described a novel remote sensing concept related to optical observations of ocean environment. The concept is based on the creation of digital spectral portrait, generated from high-resolution optical images. The portrait is a product of computer vision but not a human vision. Indeed, it is a quintessential approach which is capable to provide detection and recognition of so-called *weakly emergent hydrodynamic phenomena*. In particular, the most important and exclusive part of our research is the demonstration of space-based optical capabilities to distinguish critical hydrodynamic situations (called "background" and "complex") using a concept of spectral portrait. The concept and methodology reported in this chapter can be adapted on a number of other types of advanced sensors (e.g., sophisticated radar or lidar) which potentially enable to provide hydrodynamic detection. This concept can also be useful for machine analysis of multisensor remotely sensed data. Our preliminary results show that satellite quality optics with ~1 m resolution is *efficient* and *productive* tool for hydrodynamic detection. Further efforts and developments are required to elaborate operational technology.

REFERENCES

Abileah, R. 2006. Mapping shallow water depth from satellite. In *Proceedings of the American Society for Photogrammetry and Remote Sensing (ASPRS) Annual Conference*, May 1–5, Reno, Nevada.

Abileah, R. 2007. Mapping ocean currents with IKONOS. In *Proceedings of IEEE OCEANS 2007 – Europe Conference*, June, 18–21, 2007, Aberdeen, Scotland, paper ID 061206-002, pp. 1–5. doi:10.1109/OCEANSE.2007.4302203.

Alparone, L., Aiazzi, B., Baronti, S., and Garzelli, A. 2015. *Remote Sensing Image Fusion (Signal and Image Processing of Earth Observations)*. CRC Press, Boca Raton, FL.

Álvarez-Borrego, J. 1993. Wave height spectrum from sun glint patterns: An inverse problem. *Journal of Geophysical Research*, 98(C6):10245–10258.

Álvarez-Borrego, J. and Machado-Gama, M. A. 1985. Optical analysis of a simulated image of the sea surface. *Applied Optics*, 24(7):1064–1072.

Bakhanov, V. V., Zuikova, E. M., Kemarskaya, O. N., and Titov, V. I. 2006. Determining the sea roughness spectra from an optical image of the sea surface. *Radiophysics and Quantum Electronics*, 49(1):53–63 (translated from Russian).

Beal, R. C., Tilley, D. C., and Monaldo, F. M. 1983. Large- and small-scale evolution of digitally processed ocean wave spectra from SEASAT synthetic aperture radar. *Journal of Geophysical Research*, 88(C3):1761–1778.

Bondur, V. G. 2004. Aerospace methods in modern oceanology. In *New Ideas in Oceanology. P.P. Shirshov Institute of Oceanology, Vol. 1: Physics. Chemistry. Biology* (Eds. M. E. Vinogradov and S. S. Lappo). Nauka, Moscow, pp. 55–117 (in Russian). Available on the Internet http://www.aerocosmos.info/pdf/2006/Bon_aero_2004.pdf.

Bondur, V. G. and Murynin, A. B. 2016. Methods for retrieval of sea wave spectra from aerospace image spectra. *Izvestiya, Atmospheric and Oceanic Physics*, 52(9):877–887 (translated from Russian).

Bondur, V. G. and Sharkov, E. A. 1982. Statistical properties of whitecaps on a rough sea. *Oceanology*, 22(3):274–279 (translated from Russian).

Bondur, V. G. and Voliak, K. I. 1984. Optical spatial spectral analysis of sea surface images. In book *Voliak K I. Selected Papers. Nonlinear Waves in the Ocean. 2002*. Nauka, Moscow, pp. 188–213 (in Russian).

Bondur, V. G., Dulov, V. A., Murynin, A. V., and Ignatiev, V. Y. 2016a. Retrieving sea-wave spectra using satellite-imagery spectra in a wide range of frequencies. *Izvestiya, Atmospheric and Oceanic Physics*, 52(6):637–648 (translated from Russian).

Bondur, V. G., Dulov, V. A., Murynin, A. V., Yurovsky, Yu. Yu. 2016b. A study of sea-wave spectra in a wide wavelength range from satellite and in-situ data. *Izvestiya, Atmospheric and Oceanic Physics*, 52(9):888–903 (translated from Russian).

Borodina, E. L. and Salin, M. B. 2010. Estimation of space-time characteristics of surface roughness based on video images. *Izvestiya, Atmospheric and Oceanic Physics*, 46(2):239–248 (translated from Russian).

Borrego, J. A. 1993. Wave height spectrum from sun glint patterns: An inverse problem. *Journal of Geophysical Research*, 98(C6):10245–10258.

Bortkovskii, R. S. 1987. *Air-Sea Exchange of Heat and Moisture During Storms*. D. Reidel, Dordrecht, the Netherlands.

Bonnefoy, F., Haudin, F., Michel, G., Semin, B., Humbert, T., Aumaître, S., Berhanu, M., and Falcon, E. 2016. Observation of resonant interactions among surface gravity waves. *Journal of Fluid Mechanics*, 805:R3-1–R3-12. doi:10.1017/jfm.2016.576.

Bréon, F. and Henriot, N. 2006. Spaceborne observations of ocean glint reflectance and modeling of wave slope distributions. *Journal of Geophysical Research*, 111(C6) C06005.

Bunkin, F. V. and Volyak, K. I. (Eds.). 1987. *Oceanic Remote Sensing*. Nova Science Publishers, Commack, NY (translated from Russian).

Bunkin, A. F. and Voliak (Volyak), K. I. 2001. *Laser Remote Sensing of the Ocean: Methods and Applications*. John Wiley & Sons, New York.

Bunkin, A. F., Klinkov, V. K., Lukyanchenko, V. A., and Pershin, S. M. 2011. Ship wake detection by Raman lidar. *Applied Optics*, 50(4):A86–A89. doi:10.1364/AO.50.000A86.

Chapman, R. 1981. Visibility of RMS slope variations on the sea surface. *Applied Optics*, 20(11):1959–1966.

Chapman, R. D. and Irani, G. B. 1981. Errors in estimating slope spectra from wave images. *Applied Optics*, 20(20):3645–3652.

Cherny, I. V. and Raizer, V. Y. 1998. *Passive Microwave Remote Sensing of Oceans*. John Wiley & Sons, Chichester, UK.

Churnside, J. H. 2008. Polarization effects on oceanographic lidar. *Optics Express*, 16(2):1196–1207.

Churnside, J. H. 2010. Lidar signature from bubbles in the sea. *Optics Express*, 18(8):8294–8299.

Churnside, J. H. 2014. Review of profiling oceanographic lidar. *Optical Engineering*, 53(5) 051405. doi:10.1117/1.oe.53.5.051405.

Churnside, J. H. and Donaghay, P. L. 2009. Thin scattering layers observed by airborne lidar. *ICES Journal of Marine Science*, 66(4):778–789. doi:10.1093/icesjms/fsp029.

Churnside, J. H. and Ostrovsky, L. A. 2005. Lidar observation of a strongly nonlinear internal wave train in the Gulf of Alaska. *International Journal of Remote Sensing*, 26(1):167–177.

Churnside, J. H. and Wilson, J. J. 2004. Airborne lidar imaging of salmon. *Applied Optics*, 43(6):1416–1424.

Churnside, J. H. and Wilson, J. W. 2006. Power spectrum and fractal dimension of laser backscattering from the ocean. *Journal of the Optical Society of America A*, 23(11):2829–2833.

Churnside, J. H., Hanson, S. G., and Wilson, J. J. 1995. Determination of ocean wave spectra from images of backscattered incoherent light. *Applied Optics*, 34(6):962–968.

Churnside, J. H., Sullivan, J. M., and Twardowski, M. S. 2014. Lidar extinction-to-backscatter ratio of the ocean. *Optics Express*, 22(15):18698–18706.

Cox, C. and Munk, W. 1954a. Measurement of the roughness of the sea surface from photographs of the sun's glitter. *Journal of the Optical Society of America*, 44(11):838–850.

Cox C. and Munk, W. 1954b. Statistics of the sea surface derived from sun glitter. *Journal of Marine Research*, 13(2):198–227.

Craik, A. D. D. 1985. *Wave Interactions and Fluid Flows*. Cambridge University Press, Cambridge, UK.

Crocker, R. I., Matthews, D. K., Emery, W. J., and Baldwin, D. G. 2007. Computing coastal ocean surface currents from infrared and ocean color satellite imagery. *IEEE Transactions on Geoscience and Remote Sensing*, 45(2):435–447.

Cureton, G. P., Anderson, S. J., Lynch, M. J., and McGann, B. T. 2007. Retrieval of wind wave elevation spectra from sunglint data. *IEEE Transactions on Geoscience and Remote Sensing*, 45(9):2829–2836.

de Michele, M., Leprince, S., Thiébot, J., Raucoules, D., and Binet, R. 2012. Direct measurement of ocean waves velocity field from a single SPOT-5 dataset. *Remote Sensing of Environment*, 119:266–271.

Dial, G., Bowen, H., Gerlach, F., Grodecki, J., and Oleszczuk, R. 2003. IKONOS satellite, imagery, and products. *Remote Sensing of Environment*, 88(1–2):23–36. doi:10.1016/j.rse.2003.08.014.

Dugan, J. P. and Piotrowski, C. C. 2003. Surface current measurements using airborne visible image time series. *Remote Sensing of Environment*, 84(2):309–319.

Dugan, J. P., Piotrowski, C. C., and Williams, J. Z. 2001a. Water depth and surface current retrievals from airborne optical measurements of surface gravity wave dispersion. *Journal Geophysical Research*, 106(C8):16903–16915.

Dugan, J., Suzukawa, H., Forsyth, C., and Farber, M. 1996. Ocean wave dispersion surface measured with airborne IR imaging system. *IEEE Transactions on Geoscience and Remote Sensing*, 34(5):1282–1284.

Dugan, J. P., Fetzer, G. J., Bowden, J., Farruggia, G. J., Williams, J. Z., Piotrowski, C. C., Vierra, K., Campion, D., and Sitter, D. N. 2001b. Airborne optical system for remote sensing of ocean waves. *Journal of Atmospheric and Oceanic Technology*, 18(7):1267–1276.

Etkin, V., Raizer, V., Stulov, A., and Zhuravlev, K. 1994. Optical measurements of wind-wave spectral perturbations induced by ocean internal waves. In *Proceedings of Combined Optical-Microwave Earth and Atmosphere Sensing*, April 3–6, 1995, Atlanta, GA, pp. 81–83. doi:10.1109/COMEAS.1995.472333.

Fingas, M. 2017. *Oil Spill Science and Technology*, 2nd edition. Gulf Professional Publishing, Elsevier, Oxford, UK.

Gelpi, C. G., Schuraytz, B. C., and Husman, M. E. 2001. Ocean wave height spectra computed from high-altitude, optical, infrared images. *Journal of Geophysical Research*, 106(C11):31403–31413.

Gibson, C. H., Bondur, V. G., Keeler, R. N., and Leung, P. T. 2008. Energetics of the beamed zombie turbulence maser action mechanism for remote detection of submerged oceanic turbulence. *Journal of Applied Fluid Mechanics*, 1(1):11–42.

Glazman, R. 1988. Fractal properties of the sea surface manifested in microwave remote-sensing signatures. In *Proceedings of International Geoscience and Remote Sensing Symposium*, September 13–16, 1988, Edinburgh, Scotland, Vol. 3, pp. 1623–1624. doi:10.1109/IGARSS.1988.569545.

Glazman, R. E. and Weichman, P. B. 1989. Statistical geometry of a small surface patch in a developed sea. *Journal of Geophysical Research*, 94(C4):4998–5010.

Gotwols, B. L. and Irani, G. B. 1980. Optical determination of the phase velocity of short gravity waves. *Journal of Geophysical Research*, 85(C7):3964–3970.

Gotwols, B. L. and Irani, G. B. 1982. Charge-coupled device camera system for remotely measuring the dynamics of ocean waves. *Applied Optics*, 21(5):851–860.

Grushin, V. A., Raizer, V. Y., Smirnov, A. V., and Etkin, V. S. 1986. Observation of the non-linear interaction of gravity waves by optical and radiolocation methods. *Doklady Akademii Nauk SSSR (The Proceedings of the USSR Academy of Sciences)*, 290(2):458–462 (in Russian).

Holland, K. T., Holman, R. A., and Lippmann, T. C. 1997. Practical use of video imagery in nearshore oceanographic field studies. *IEEE Journal of Oceanic Engineering*, 22(1):81–92.

Holthuijsen, L. H. 1983. Stereophotography of ocean waves. *Applied Ocean Research*, 5(4):204–209.

Hooper, B. A., Van Pelt, B., Williams, J. Z., Dugan, J. P., Yi, M., Piotrowski, C. C., and Miskey, C. 2015. Airborne spectral polarimeter for ocean wave research. *Journal of Atmospheric and Oceanic Technology*, 32(4):805–815. doi:10.1175/jtech-d-14-00190.1.

Janssen, P. A. E. M. 2004. *The Interaction of Ocean Waves and Wind*. Cambridge University Press, Cambridge, UK.

Kanjir, U., Greidanus, H., and Oštir, K. 2018. Vessel detection and classification from space-borne optical images: A literature survey. *Remote Sensing of Environment*, 207:1–26. doi:10.1016/j.rse.2017.12.033.

Kasevich, R. S. 1975. Directional wave-spectra from day-light scattering. *Journal of Geophysical Research*, 80(C33):4533–4541.

Kasevich, R. S., Tang, C.-H., and Henriksen, S. W. 1972. Analysis and optical processing of sea photographs for energy spectra. *IEEE Transactions on Geoscience and Electronics*, GE-10(1):51–58.

Kay, S., Hedley, J. D., and Lavender, S. 2009. Sun glint correction of high and low spatial resolution images of aquatic scenes: A review of methods for visible and near-infrared wavelengths. *Remote Sensing*, 1(4):697–730.

Keller, W. C. and Gotwols, B. L. 1983. Two-dimensional optical measurement of wave slope. *Applied Optics*, 22(22):3476–3478.

Keeler, R. N., Bondur, V. G., and Gibson, C. H. 2005. Optical satellite imagery detection of internal wave effects from a submerged turbulent outfall in the stratified ocean. *Geophysical Research Letters*, 32(12) L12610. doi:10.1029/2005GL022390.

Kerman, B. R. and Bernier, L. 1994. Multifractal representation of breaking waves on the ocean surface. *Journal of Geophysical Research*, 99(C8):16179–16196.

Kleiss, J. M. and Melville, W. K. 2011. The analysis of sea surface imagery for whitecap kine-matics. *Journal of Atmospheric and Oceanic Technology*, 28(2):219–243.

Komen, G. J., Cavaleri, L., Donelan, M., Hasselmann, K., Hasselmann, S., and Janssen, P. A. E. M. 1996. *Dynamics and Modelling of Ocean Waves*. Cambridge University Press, Cambridge, UK.

Korenowski, G. M. 1997. Applications of laser technology and laser spectroscopy in studies of the ocean microlayer. In *The Sea Surface and Global Change* (Eds. P. S. Liss and R. A. Duce), Cambridge University Press, Cambridge, UK, pp. 445–470.

Kosnik, M. V. and Dulov, V. A. 2011. Extraction of short wind wave spectra from ste-reo images of the sea surface. *Measurement Science and Technology*, 22(1) 015504. doi:10.1088/0957-0233/22/1/015504.

Kosorukova, A. I., Novikov, V. M., and Raizer, V. Yu. 1990. Digital recovery of spatial characteristics of foam formations from sea surface video imagery. *Soviet Journal of Remote Sensing*, 6(6):1033–1040 (translated from Russian).

Krasitsky (Krasitskii), V. P. 1994. Five-wave kinetic equation for surface gravity waves. *Physical Oceanography*, 5(6):413–421 (translated from Russian).

Krasitskii, V. P. and Kozhelupova, N. G. 1995. On conditions for five wave resonant interactions of surface gravity waves. *Oceanology*, 34(4):435–439 (translated from Russian).

Krig, S. 2014. *Computer Vision Metrics: Survey, Taxonomy, and Analysis*. Apress OPEN, New York.

Kudryavtsev, V., Yurovskaya, M., Chapron, B., Collard, F., and Donlon, C. 2017a. Sun glitter imagery of ocean surface waves. Part 1: Directional spectrum retrieval and validation. *Journal of Geophysical Research. Oceans*, 122(2):1369–1383.

Kudryavtsev, V., Yurovskaya, M., Chapron, B., Collard, F., and Donlon, C. 2017b. Sun glitter imagery of surface waves. Part 2: Waves transformation on ocean currents. *Journal of Geophysical Research. Oceans*, 122(2):1384–1399.

Lancaster, R. S., Spinhirne, J. D., and Palm, S. P. 2005. Laser pulse reflectance of the ocean surface from the GLAS satellite lidar. *Geophysical Research Letters*, 32(2) L22S10.

Lee, K. J., Park, Y., Bunkin, A., Nunes, R., Pershin, S., and Voliak, K. 2002. Helicopter-based lidar system for monitoring the upper ocean and terrain surface. *Applied Optics*, 41(3):401–406. doi:10.1364/AO.41.000401.

Levanon, N. 1971. Determination of the sea surface slope distribution and wind velocity using sun glitter viewed from a synchronous satellite. *Journal of Physical Oceanography*, 1(3):214–220.

Levin, I., Savchenko, V., and Osadchy, V. 2008. Correction of an image distorted by a wavy water surface: Laboratory experiment. *Applied Optics*, 47(35):6650–6655.

Lin, R. Q. and Perrie, W. 1997. A new coastal wave model. Part V: Five-wave interactions. *Journal of Physical Oceanography*, 27(10):2169–2186.

Liu, Z., Xu, D. L., Li, J., Peng, T., Alsaedi, A., and Liao, S. J. 2015. On the existence of steady-state resonant waves in experiments. *Journal of Fluid Mechanics*, 763:1–23.

Lubard, S. C., Krimmel, J. E., Thebaud, L. R., Evans, D. D., and Shemdin, O. H. 1980. Optical image and laser slope meter intercomparisons of high-frequency waves. *Journal of Geophysical Research*, 85(C9):4996–5002.

Luchinin, A. G. 1982. The brightness fluctuation spectrum of the natural light field escaping from under a wavy sea surface. *Izvestiya, Atmospheric and Oceanic Physics*, 18(5):431–434 (translated from Russian).

Lvov, Y. V. 1997. Effective five-wave Hamiltonian for surface water waves. *Physics Letters A*, 230(1–2):38–44.

Lynch, D. K., Dearborn, D. S. P., and Lock, J. A. 2011. Glitter and glints on water. *Applied Optics*, 50(28):F39–F49.

Mandelbrot, B. B. 1983. *The Fractal Geometry of Nature*. W. H. Freeman, New York.

Melville, W. K. and Matusov, P. 2002. Distribution of breaking waves at the ocean surface. *Nature*, 417:58–63.

Melville, W. K., Lenain, L., Cayan, D. R., Kahru, M., Kleissl, J. P., Linden, P. F., and Statom, N. M. 2016. The modular aerial sensing system. *Journal of Atmospheric and Oceanic Technology*, 33(6):1169–1184.

Miskey, C. 2015. Airborne spectral polarimeter for ocean wave research. *Journal of Atmospheric and Oceanic Technology*, 32(4):805–815.

Mityagina, M. I., Pungin, V. G., Smirnov, A. V., and Etkin, V. S. 1991. Changes of the energy-bearing region of the sea surface wave spectrum in an internal wave field based on remote observation data. *Izvestiya, Atmospheric and Oceanic Physics*, 27(11):925–929 (translated from Russian).

Mobley, C. D. 2015. Polarized reflectance and transmittance of properties of windblown sea surfaces. *Applied Optics*, 54(15):4828–4849.

Monahan, E. C. 1971. Oceanic whitecaps. *Journal of Physical Oceanography*, 1:139–144.

Monaldo, F. M. and Kasevich, R. S. 1981a. Daylight imagery of ocean surface waves for wave spectra. *Journal of Physical Oceanography*, 11(2):271–283.

Monaldo, F. M. and Kasevich, R. S. 1981b. Measurement of short-wave modulation using fine time-series optical spectra. *Journal of Physical Oceanography*, 11(7):1034–1037.

Monaldo, F. M. and Kasevich, R. S. 1982. Optical determination of shortwave modulation by long ocean gravity waves. *IEEE Transactions on Geoscience and Remote Sensing*, GE-20(3):254–259.

Morel, A., Voss, K. J., and Gentili, B. 1995. Bidirectional reflectance of oceanic waters: A comparison of modeled and measured upward radiance fields. *Journal of Geophysical Research*, 100(C7):13143–13150.

Mount, R. 2005. Acquisition of through-water aerial survey images: Surface effects and the prediction of sun glitter and subsurface illumination. *Photogrammetric Engineering and Remote Sensing*, 71(12):1407–1415.

Novikov, V. M. and Raizer, V. Yu. 1988. *Digital Analysis of the Physical Objects in Images*. Algorithm complex. Preprint no. 1444. Space Research Institute, Moscow (in Russian).

Peppers, N. and Ostrem, J. 1978. Determination of wave slopes from photographs of the ocean surface: A new approach. *Applied Optics*, 17(21):3450–3458.

Phillips, O. M. 1980. *The Dynamics of the Upper Ocean*, 2nd edition. Cambridge University Press, Cambridge, UK.

Phillips, O. M. 1981. Wave interactions - the evolution of an idea. *Journal of Fluid Mechanics*, 106:215–227.

Piotrowski, C. C. and Dugan, J. P. 2002. Accuracy of bathymetry and current retrievals from airborne optical time series imaging of shoaling waves. *IEEE Transaction on Geoscience and Remote Sensing*, 40(12):2606–2618.

Pohl, C. and van Genderen, J. 2017. *Remote Sensing Image Fusion: A Practical Guide*. CRC Press, Boca Raton, FL.

Raizer, V. Yu. 1994. Wave spectrum and foam dynamics via remote sensing. In *Satellite Remote Sensing of the Ocean Environment* (Eds. I. S. F. Jones, Y. Sugimori, and R. W. Stewart). Seibutsu Kenkyusha, Japan, pp. 301–304.

Raizer, V. 2017. *Advances in Passive Microwave Remote Sensing of Oceans*. CRC Press, Boca Raton, FL.

Rayzer (Raizer), V. Yu. and Novikov, V. M. 1990. Fractal structure of breaking zones for surface waves in the ocean. *Izvestiya, Atmospheric and Oceanic Physics*, 26(6):491–494 (translated from Russian).

Raizer, V. Y., Novikov, B. M., and Bocharova, T. Y. 1994. The geometrical and fractal properties of visible radiances associated with breaking waves in the ocean. *Annales Geophysicae*, 12:1229–1233.

Raizer, V. Yu., Smirnov, A. V., and Etkin, V. S. 1990. Dynamics of the large-scale structure of the disturbed surface of the ocean from analysis of optical images. *Izvestiya, Atmospheric and Oceanic Physics*, 26(3):199–205 (translated from Russian).

Raol, J. R. 2009. *Multi-Sensor Data Fusion with MATLAB®*. CRC Press, Boca Raton, FL.

Reineman, B. D., Lenain, L., Castel, D., and Melville, W. K. 2009. A portable airborne scanning lidar system for ocean and coastal applications. *Journal of Atmospheric and Oceanic Technology*, 26(12):2626–2641.

Romero, L. and Melville, W. K. 2010. Airborne observations of fetch-limited waves in the Gulf of Tehuantepec. *Journal of Physical Oceanography*, 40(3):441–465.

Romero, L., Lenain, L., and Melville, W. K. 2017. Observations of surface wave-current interaction. *Journal of Physical Oceanography*, 47(3):615–632. doi:10.1175/jpo-d-16-0108.1.

Ross, D. B. and Cardone, V. 1974. Observation of oceanic whitecaps and their relation to remote measurements of surface wind speed. *Journal of Geophysical Research*, 79(3):444–452. doi:10.1029/jc079i003p00444.

Ross, V. and Dion, D. 2007. Sea surface slope statistics derived from sun glint radiance measurements and their apparent dependence on sensor elevation. *Journal of Geophysical Research*, 112(C9) C09015.

Schau, H. C. 1978. Measurement of capillary wave slopes on the ocean. *Applied Optics*, 17(1):15–17.

Sharkov, E. A. 2007. *Breaking Ocean Waves: Geometry, Structure and Remote Sensing*. Praxis Publishing, Chichester, UK.

Shaw, J. A. 1999a. Glittering light on water. *Optics and Photonics News*, 10(3):43–45.

Shaw, J. A. 1999b. Degree of linear polarization in spectral radiances from water-viewing infrared radiometers. *Applied Optics*, 38(15):3157–3165.

Shaw, J. A. 2001. Polarimetric measurements of long-wave infrared spectral radiance from water. *Applied Optics*, 40(33):5985–5990.

Shaw, J. A. 2002. The effect of instrument polarization sensitivity on sea surface remote sensing with infrared spectroradiometers. *Journal of Atmospheric and Oceanic Technology*, 19(5):820–827.

Shaw, J. A. and Churnside, J. H. 1997a. Fractal laser glints from the ocean surface. *Journal of the Optical Society of America*, 14(5):1144–1150.

Shaw, J. A. and Churnside, J. H. 1997b. Scanning-laser glint measurements of sea-surface slope statistics. *Applied Optics*, 36(18):4202–4213.

Shaw, J. A. and Marston, C. 2000. Polarized infrared emissivity for a rough water surface. *Optics Express*, 7(11):375–380.

Shaw, J. A. and Vollmer, M. 2017. Blue sun glints on water viewed through a polarizer. *Applied Optics*, 56(19):G36–G41.

Shaw, J. A., Cimini, D., Westwater, E. R., Han, Y., Zorn, H. M., and Churnside, J. H. 2001. Scanning infrared radiometer for measuring the air–sea temperature difference. *Applied Optics*, 40(27):4807–4815.

Shemdin, O. H. and Tran, H. M. 1992. Measuring short surface waves with stereophotography. *Photogrammetric Engineering and Remote Sensing*, 58(3):311–316.

Shemdin, O., Tran, H., and Wu, S. 1988. Directional measurement of short ocean waves with stereophotography. *Journal of Geophysical Research*, 93(C11):13891–13901.

Sheres, D. 1981. Remote and synoptic water-wave measurements by aerial photography: A model, experimental results, and an application. *IEEE Journal of Oceanic Engineering*, OE-6(2):63–69.

Snyder, R. L. and Kennedy, R. M. 1983. On the formation of whitecaps by a threshold mechanism. Part I: Basic formalism. *Journal of Physical Oceanography*, 13(8):1482–1492.

Snyder, R. L., Smith, L., and Kennedy, R. M. 1983. On the formation of whitecaps by a threshold mechanism. Part III: Field experiment and comparison with theory. *Journal of Physical Oceanography*, 13(8):1505–1518.

Stilwell, D. Jr. 1969. Directional energy spectra of the sea from photographs. *Journal of Geophysical Research*, 74(8):1974–1986.

Stilwell, D. Jr. and Pilon, R. O. 1974. Directional spectra of surface waves from photographs. *Journal of Geophysical Research*, 79(9):1277–1284.

Stockdon, H. F. and Holman, R. A. 2000. Estimation of wave phase speed and nearshore bathymetry from video imagery. *Journal of Geophysical Research*, 105(C9):22015–22033.

Su, W., Charlock, T. P., and Rutledge, K. 2002. Observations of reflectance distribution around sunglint from a coastal ocean platform. *Applied Optics*, 41(35):7369–7383.

Titov, V. I., Bakhanov, V. V., Ermakov, S. A., Luchinin, A. G., Repina, I. A., and Sergievskaya, I. A. 2014. Remote sensing technique for near-surface wind by optical images of rough water surface. *International Journal of Remote Sensing*, 35(15):5946–5957.

Veron, F., Melville, W. K., and Lenain, L. 2008. Infrared techniques for measuring ocean surface processes. *Journal of Atmospheric and Oceanic Technology*, 25(2):307–326.

Volyak, K. I., Lyakhov, G. A., and Shugan, I. V. 1987. Surface wave interaction. Theory and capability of oceanic remote sensing. In *Oceanic Remote Sensing* (Eds. F. V. Bunkin and K. I. Volyak). Nova Science Publishers, Commack, NY, pp. 107–145 (translated from Russian).

Wald, L. and Monget, J.-M. 1983. Sea surface winds from sun glitter observations. *Journal of Geophysical Research. Oceans*, 88(C4):2547–2555.

Walker, R. E. 1994. *Marine Light Field Statistics.* John Wiley & Sons, New York.

Walsh, E. J., Banner, M. L., Churnside, J. H., Shaw, J. A., Vandemark, D. C., Wright, C. W., Jensen, J. B., and Lee, S. 2005. Visual demonstration of three-scale sea-surface roughness under light wind conditions. *IEEE Transaction on Geoscience and Remote Sensing*, 43(8):1751–1762.

Wanek, J. M. and Wu, C. H. 2006. Automated trinocular stereo imaging system for three-dimensional surface wave measurements. *Ocean Engineering*, 33(5–6):723–747.

Waseda, T., Kinoshita, T., Cavaleri, L., and Toffoli, A. 2015. Third-order resonant wave interactions under the influence of background current fields. *Journal of Fluid Mechanics*, 784:51–73.

Wu, J. 1988. Variations of whitecap coverage with wind stress and water temperature. *Journal of Physical Oceanography*, 18(10):1448–1453.

Yuen, H. C. and Lake, B. M. 1982. Nonlinear dynamics of deep-water gravity waves. In *Advances in Applied Mechanics* (Ed. Chia-Shun Yih). Academic Press, New York, Vol. 22, pp. 67–229.

Yurovskaya, M. V., Dulov, V. A., Chapron, B., and Kudryavtsev, V. N. 2013. Directional short wind wave spectra derived from the sea surface photography. *Journal of Geophysical Research*, 118(9):4380–4394.

Zakharov, V. E. (Ed.). 1998. *Nonlinear Waves and Weak Turbulence (Advances in the Mathematical Sciences, Series 2, Volume 182)*. American Mathematical Society, Providence, Rhode Island.

Zdanovich, V. G. (Ed.). 1963. *Primenenie aérometodov dlia issledovaniia morei̇̆ (Application of Aerial Methods for the Study of the Sea)*. Academy of Sciences, U.S.S.R, Moscow-Leningrad (in Russian). Library of Congress, LC classification (full) GC57.R8 (Internet address https://lccn.loc.gov/63043871).

Zuykova, E. M., Luchinin, A. G., and Titov, V. I. 1985. Determination of the characteristics of the space-time spectra of waves from an optical image of the sea surface. *Izvestiya, Atmospheric and Oceanic Physics*, 21(10):833–837 (translated from Russian).

8 Multisensor Concept and Detection

8.1 INTRODUCTION

In this chapter, we consider the integration (synergy) concept in context with multisensor remote sensing system (MSS) observations of ocean environments. The application strategy concerns distinguishing so-called background and complex situations (Chapter 7). The first situation is a regular wind-driven ocean surface without any induced disturbances; the second situation is characterized by localized surface disturbances—wakes, wave patterns, precursors, burst-type disturbances, turbulent flows, or other weakly emergent natural or induced events. Remote sensing of such oceanic phenomena or *hydrodynamic detection* is connected with nonacoustic detection by standard terminology.

For instance, payload consisting of three different instruments—optical sensor, radar, and microwave radiometer provides data acquisition at the same ocean area simultaneously. If suppose that such MSS databases exist, a problem arises what type of the processing should be applied to provide detection? Actually, this is a common question for many remote sensing missions. Perhaps, an ultimate method or "recipe" does not exist, and the choice depends on many objective and subjective factors—instrument performance, acquisition capability, the content of the collected data, product quality, applications, and most importantly, on the researcher's experience. An understanding some specific issues allows us to approach a solution of this problem.

We refer to the term "synergy," which means that all participating sensors, acquisition systems, and databases are configured (integrated) into a unit remote sensing technology system. We name this "Synergy System Approach" (SSA). To provide target detection using SSA, it is necessary first to optimize parameters of the synergy system. We also assume that the performance of the entire synergy remote sensing system is much higher than performances of single-band sensors even if some of them are very sophisticated. Unlike land cover remote sensing, in ocean remote sensing, this is absolutely correct: the greater number of different sensors are involved into the experiment, the better detection performance (simply due to the presence of large number of dynamic variables that must be considered). This statement has been confirmed experimentally many times.

Standard data processing of remotely sensed data usually includes three main algorithms (Chapter 6): (1) cross-correlation analyses of MSS data, (2) PCA and its variations, and (3) spectral analysis of selected realizations. These well-known methods of the processing demonstrate effectiveness in many remote sensing applications; however, they may not provide reliable result in terms of detection performance vs. probability of false alarm. In other words, uncertainties in thematic analysis and interpretation of ocean MSS data remain and this influences on detection ability. MSS data fusion can reduce the risk and improve analysis significantly providing the ultimate utilization and classification of MSS data. Below we consider two methodologies: non-fusion and fusion.

8.2 NON-FUSION STRATEGY

Non-fusion MSS observations have been tested in the past experiments (1980s); nowadays, with new remote sensing technology, this concept can be efficient as well. Figure 8.1 demonstrates schematically methodology of parallel signal processing of N-sensor data representing 1D time series (called also records of profiles). Usually, large bodies of MSS records (optical/radar/radiometer) are analyzed using conventional statistical, spectral, and correlation methods. For detection purposes, parallel processing of MSS records can be organized by the following schemes:

1. Specify background ("B") and complex ("C") situations
2. Select the corresponding N-sensor digital input records/profiles
3. Calibrate, normalize to maximum of intensity, and scale all input profiles $I_1(x),...,I_N(x)$, x is the distance
4. For each profile compute mean values $M_1,...,M_N$ and variances (or standard deviations) $var_1,...,var_N$
5. For each profile compute power spectral density $S_1(f),...,S_N(f)$
6. For each profile compute autocorrelation functions $corr_{11}(x),...,corr_{NN}(x)$
7. For N-sensor input records/profiles compute correlation functions (and/or matrix of correlation coefficients) $corr_{ij}(x)$, $i=1,...N, j=1,...N, i \neq j$
8. Provide analysis and interpretation of the result.

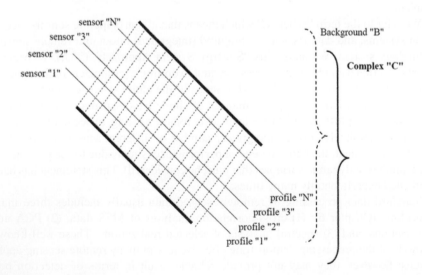

FIGURE 8.1 Parallel methodology.

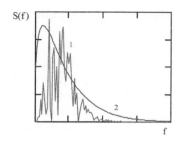

FIGURE 8.2   Generated spectra of signals.

All statistical characteristics are computed separately for each particular experiment which we define as "background + complex" (with indexes "B" + "C"). The most valuable differences can be found in terms of the power spectral density $S_N^C(f)$ and $S_N^B(f)$ and correlation function $corr_{ij}^C(x)$ and $corr_{ij}^B(x)$, delivered from the parallel processing. Note that these statistical measures are functions of many geophysical parameters characterizing sea state, atmospheric conditions, etc. Figure 8.2 illustrates some result schematically. The differences between two power spectra, generated for a single-type sensor (not specified) signal, are as follows: (1) complex, $S_N^C(f)$ and (2) background, $S_N^B(f)$. The MSS differences provide selective detection of situations.

8.3   MULTISENSOR DATA FUSION

Let's remember that data fusion is a process of combination of multiple data sources into a unit source with the goal to obtain improved information. In recent years, data fusion systems and algorithms are extensively used for military and nonmilitary purposes including various remote sensing applications as well. A comprehensive review of MSS data fusion can be found in books (Waltz and Llinas 1990; Hall and McMullen 2004; Blum and Liu 2006; Mitchell 2007; Liggins et al. 2009; Raol 2009; Appriou 2014; Fourati 2016; Chang and Bai 2018). From the current literature, it follows that image fusion is a robust and efficient tool for detection and recognition of geophysical features in multiband, multispectral, hyperspectral, and multisensory images.

The development of fusion algorithms for hydrodynamic detection is a challenging task; first of all, because needed MSS data and materials are sophisticated and not available in the meantime; second, MSS imaging data have *a priori* different properties and contents; and third, ocean (optical/radar/radiometer) data have relatively low signal-to-noise ratio. These objective circumstances limit the possibilities of hydrodynamic detection; thereby, advanced digital methods of data processing and interpretation are required. Ultimately, this means that we have to deal with computer vision products and their thematic analysis. Geophysical interpretation of computer vision products is always difficult task due to indirect visualization of the relevant information. This may lead to incorrect interpretation and even to inadequate understanding of the problem as a whole. An example is space-based detection of submerged wakes in the ocean. So far, no one knows for certain is it possible or not...

As mentioned above (Section 6.6), there are three levels of fusion—pixel/data fusion, feature level fusion, and decision level fusion. Let's consider these categories in more detail in context with detection problem.

The first (pixel) level is applied for statistical or morphological processing of imaging data collected by a single-type sensor, e.g., passive microwave radiometer-spectrometer, multiband scatterometer or radar, or multispectral optical imager. The processing includes correlation, spectral, or PCA in order to reveal relevant information and define dominant components. The pixel level fusion can also be applied for analysis of MSS imagery in terms of *binary fusion*. For this, following steps can be applied: (1) MSS images are reformatted and/or sampled to match the same geolocation coordinates and pixel resolutions. This operation is called *sensor value normalization* which is defined as the conversion of all sensor output values to a common scale, (2) MSS color images are converted into binary (black and white) images using brightness threshold, (3) binary images are fused, (4) binary image features are selected and extracted from the resulting fused binary image. This operation corresponds to the pixel level of MSS binary image processing, (5) binary features are used for detection purposes, i.e., identification of the binary signatures of the interest, and (6) specification and/or classification of binary signatures is performed.

The second (feature) and third (decision making) levels are advanced fusion methods which are more complicated than pixel level fusion. Feature level fusion provides first the extraction of visible image features and then performs their fusion. Typical examples of remotely sensed image features are lines, blobs, edges, circles, mosaic textures, or geometrical sets of distinct objects or spots. The fused image features are analyzed using classification and/or statistical algorithms. The decision level fusion defines relevant information and provides final decision making. The detection is performed on the basis of statistical treatment of fused image features. This may require the use of multivariable analysis.

As a whole, advanced fusion strategy may include a number of state-of-the-art methods and algorithms such as probabilistic fusion, fuzzy fusion, hybrid fusion, adaptive fusion and learning, and some others. The choice of an appropriate fusion technique depends on many factors; among them the most important are the number of sensors, the type, quality, and information content of the collected data and the purposes of applications for which the fused information is provided. In our case, the result directly depends on the fusion strategy being developed. The successful implementation and application of fusion requires certain scientific research, creation of a specialized fusion software platform, and interactive programming as well.

8.4   A SYNERGY FLOWCHART

Figure 8.3 shows a common MSS fusion flowchart in more detail. Fusion process consists of three levels: (I) pixel/data level fusion, (II) feature level fusion, and (III) decision level fusion. The pixel level fusion is performed to generate a fused image with improved characteristics from input MSS images. The feature level fusion provides an extraction of similar MSS image features and their fusion. The decision

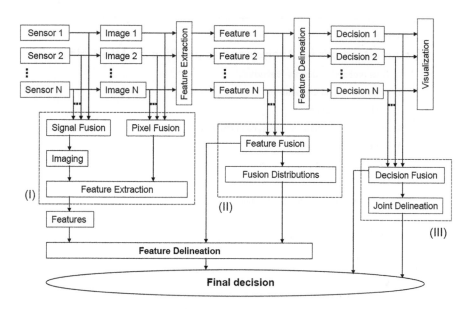

FIGURE 8.3 MSS fusion flowchart.

level fusion provides highest insight level of information which is used for decision making. Detection tool operates with certain statistical criteria and/or rules.

In our case, the goal is to specify and investigate target-related signatures associated with hydrodynamic phenomena/events. In ocean images, the signatures of the interest may be invisible, hidden, or unclear; therefore, they cannot be extracted directly from the images using conventional image processing methods. To provide detection and extraction of the signatures, first, second, and third levels of fusion are applied consecutively. This operation can be accompanied by reduction of instrument and geophysical noises and illumination of an image background as well.

To investigate statistical properties of the extracted (fused) signatures, spectral, fractal, or wavelet methods can be applied (Chapter 6). The statistical estimates can be performed in terms of probability distribution of *fused data*. These distributions and their variations give us valuable information for decision making and also the quantitative criteria needed for signature classification and specification. As a whole, the three-level fusion is the sate-of-the-art technique for target detection and recognition by MSS data. Table 8.1 presents a summary of available fusion methods and signature specifications. Possible application of MSS for wake detection is demonstrated schematically in Figure 8.4.

8.5 SUMMARY

In this chapter, we described a conceptual methodology of MSS for detection purposes. The application is based on so-called SSA which means (either non-fusion or fusion) the integration of all instrumentations, data acquisition, and processing

TABLE 8.1

Fusion Methods for Analysis and Applications of MSS Data

1. Total probability density function
2. Principal component analysis
3. Laplacian pyramid
4. Filter-subtract-decimate hierarchical pyramid
5. Ratio pyramid
6. Gradient pyramid
7. Discrete wavelet transform
8. Shift invariant discrete wavelet transform
9. Contrast pyramid
10. Morphological pyramid
11. Bio-inspired

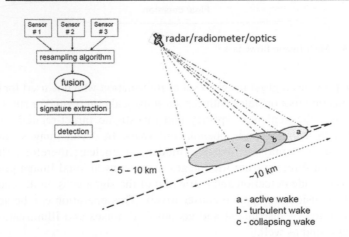

FIGURE 8.4 MSS detection concept.

techniques into a unit synergy technological (smart) remote sensing system. A physics-based motivation of the SSA is simple:

1. Target-related complex events represent *localized nonstationary synergetic hydrodynamic phenomena* consisting of many geophysical effects and variables
2. The signatures of the interest appear only in certain space-time-frequency domains
3. Detection is the entire roaming process
4. Because of stochastic, unpredictable, and multivariable environment, the number of participating sensors to be placed as much as possible
5. Detection performance is much higher for MSS than for selected single-band sensor.

Actually, these principles are well known; they follow from the applied information theory and practice. We just reformulate them for a better understanding of a whole picture. Synergy between physical phenomena, detection technology, and data analyses may require more detailed scientific research including field and laboratory experiments. Nowadays, purposeful studies can be organized and conducted using advanced MSS and numerical (CFD) analysis. We believe that the MSS–SSA may provide an innovative way for the solution of the problem.

REFERENCES

Appriou, A. 2014. *Uncertainty Theories and Multisensor Data Fusion.* ISTE – John Wiley & Sons, London, UK.

Blum, R. S. and Liu, Z. (Eds.). 2006. *Multi-Sensor Image Fusion and Its Applications.* CRC Press, Boca Raton, FL.

Chang, N.-B. and Bai, K. 2018. *Multisensor Data Fusion and Machine Learning for Environmental Remote Sensing.* CRC Press, Boca Raton, FL.

Fourati, H. (Ed.). 2016. *Multisensor Data Fusion: From Algorithms and Architectural Design to Applications.* CRC Press, Boca Raton, FL.

Hall, D. L. and McMullen, S. A. H. 2004. *Mathematical Techniques in Multisensor Data Fusion.* Artech House, Norwood, MA.

Liggins, M. E., Hall, D. J., and Llinas, J. (Eds.). 2009. *Handbook of Multisensor Data Fusion: Theory and Practice,* 2nd edition. CRC Press, Boca Raton, FL.

Mitchell, H. B. 2007. *Multi-Sensor Data Fusion: An Introduction.* Springer, Berlin, Germany.

Raol, J. R. 2009. *Multi-Sensor Data Fusion with MATLAB®.* CRC Press, Boca Raton, FL.

Waltz, E. L. and Llinas, J. 1990. *Multisensor Data Fusion.* Artech House, Norwood, MA.

Terminology

Digital imagery: representation of the phenomenon through electro-optical technique

Background: the state of environment without induced events

Complex: the state of environment with induced events

Feature: geophysical natural manifestation of the event

Image feature: object of the interest in the image

Digital signature: computer vision product

Signature of the interest: potential signature related to certain hydrodynamic event

Spectral portrait: mosaic of spectral features, generated from digital 2D FFT spectra

Image fusion: integration of multiple data sources

Detection: process of target/object identification and visualization

Recognition: process of target/object specification and classification

Index

Printed and bound by CPI Group (UK) Ltd, Croydon, CR0 4YY

24/10/2024

01778308-0011